CHEMISTRY

Hans van Kessel
Bellerose Composite High School

Frank Jenkins
Ross Sheppard High School

Michael V. Falk
Harry Ainlay High School

Oliver Lantz
Harry Ainlay High School

Dick Tompkins
Old Scona Academic High School

Contributing Authors

Michael Dzwiniel
Harry Ainlay High School

George H. Klimiuk
McNally Composite High School
(retired)

Solutions Manual

Nelson Canada

© Nelson Canada,
A Division of Thomson Canada Limited, 1993

Published in 1993 by
Nelson Canada,
A Division of Thomson Canada Limited
1120 Birchmount Road
Scarborough, Ontario M1K 5G4

All rights reserved. No part of this publication may be reproduced, stored in a retrieval system, or transmitted in any forms or by any means, electronic, mechanical, photocopying, recording, or otherwise, without the prior written permission of the publisher.

ISBN 0-17-603976-7

Canadian Cataloguing in Publication Data

Main entry under title:
Nelson chemistry.   Solutions manual

Supplement to: Jenkins, Frank, 1944 – . Nelson chemistry
ISBN   0-17-603976-7

1. Chemistry – Problems, exercises, etc.
I. Van Kessel, Hans.   II. Jenkins, Frank, 1944 – .
Nelson chemistry.

QD33.J453 1993   540   C93-093229-3

Printed and Bound in Canada
1 2 3 4 5 6 7 8 9 WC 9 8 7 6 5 4 3

This book is printed on acid-free paper. The choice of paper reflects Nelson Canada's goal of using, within the publishing process, the available resources, technology, and suppliers that are as environment friendly as possible.

COVER:
Daryl Benson/Masterfile
Dr. E. Keller/Kristallographisches Institut der Universität Freiburg, Germany (inset photo)

CREDITS

Executive Editor: **Lynn Fisher**
Project Co-ordinator: **Jennifer Dewey**
Project Editor: **Winnie Siu**

Supervising Production Editor: **Cecilia Chan**
Art Director: **Bruce Bond**
Illustrations: **VISU*TronX***
              Suzanne Peden

THE *NELSON CHEMISTRY* PROGRAM
Nelson Chemistry, Student's Edition
Nelson Chemistry, Teacher's Edition
Nelson Chemistry, Solutions Manual
Nelson Chemistry, Blackline Masters Package
Nelson Chemistry, Computer Test Bank

Table of Contents

INTRODUCTION TO THE TEACHER	7

UNIT I   MATTER AND ITS DIVERSITY

Chapter 1   Chemistry, Technology, and Society

Investigation 1.1 Demonstration of Combustion of Magnesium	8
Exercise (1 – 28)	8
Investigation 1.2 Classifying Pure Substances	9
Lab Exercise 1A Decomposition Using Electricity	10
Exercise (29 – 31)	11
Overview	11

Chapter 2   Elements

Lab Exercise 2A Testing the Periodic Law	14
Exercise (1 – 16)	14
Overview	16
Lab Exercise 2B Testing the Theory of Ions	18

UNIT II   CHANGE AND STRUCTURE

Chapter 3   Compounds

Lab Exercise 3A Empirical Definitions of Compounds	19
Lab Exercise 3B Testing the Classification of Compounds	19
Exercise (1 – 7)	20
Overview	21
Lab Exercise 3C A Chemical Analysis	22

Chapter 4   Chemical Reactions

Exercise (1 – 6)	23
Lab Exercise 4A The Combustion of Butane	23
Exercise (7)	23
Investigation 4.1 Testing Replacement Reaction Generalizations	24
Exercise (8)	26
Overview	26
Lab Exercise 4B Testing Reaction Generalizations	28

UNIT III   MATTER AND SYSTEMS

Chapter 5   Solutions

Lab Exercise 5A Identification of Solutions	29
Investigation 5.1 Chemical Analysis	29
Exercise (1 – 6)	30
Lab Exercise 5B Qualitative Analysis	31
Investigation 5.2 The Iodine Clock Reaction	31
Investigation 5.3 Sequential Qualitative Analysis	32
Exercise (7 – 11)	32
Lab Exercise 5C Qualitative Analysis by Color	33
Exercise (12 – 20)	33
Investigation 5.4 Solution Preparation	34
Exercise (21 – 28)	34
Lab Exercise 5D Qualitative Analysis	35
Investigation 5.5 The Solubility of Sodium Chloride in Water	36
Exercise (29 – 32)	37
Lab Exercise 5E Solubility and Temperature	37
Exercise (33 – 35)	38
Overview	38
Lab Exercise 5F Cation Analysis	41

Chapter 6   Gases

Lab Exercise 6A Pressure and Volume of a Gas	43
Lab Exercise 6B Temperature and Volume of a Gas	43
Exercise (1 – 30)	44
Investigation 6.1 Determining the Molar Mass of a Gas	47
Overview	48
Lab Exercise 6C Analyzing Gas Samples	53

Chapter 7   Chemical Reaction Calculations

Lab Exercise 7A Chemical Analysis Using a Graph	54
Investigation 7.1 Analysis of Sodium Carbonate Solution	54
Investigation 7.2 Decomposing Malachite	56
Exercise (1 – 14)	58
Lab Exercise 7B Testing Gravimetric Stoichiometry	60
Lab Exercise 7C Designing and Testing Stoichiometry	61
Investigation 7.3 Testing the Stoichiometric Method	62
Exercise (15 – 17)	64
Lab Exercise 7D Testing a Chemical Process	65
Exercise (18 – 22)	66
Investigation 7.4 The Universal Gas Constant	67
Exercise (23 – 25)	68
Lab Exercise 7E Testing Solution Stoichiometry	69
Lab Exercise 7F Determining a Solution Concentration	70
Lab Exercise 7G Solution and Gas Stoichiometry	71
Exercise (26 – 43)	71
Lab Exercise 7H Titration Analysis of Vinegar	74
Investigation 7.7 Titration Analysis of Hydrochloric Acid	75
Lab Exercise 7I Filtration Analysis	76
Lab Exercise 7J Gas Volume Analysis	76
Lab Exercise 7K Titration Analysis	77
Exercise (44 – 45)	77
Overview	78
Lab Exercise 7L Chemical Analysis	83

UNIT IV   CHEMICAL DIVERSITY AND SYSTEMS OF BONDING

Chapter 8   Chemical Bonding

Investigation 8.1 Activity Series	85
Lab Exercise 8A Metal and Nonmetal Oxides	86
Exercise (1 – 32)	86
Lab Exercise 8B Chemical Analysis	89
Exercise (33 – 37)	90
Investigation 8.2 Molecular Models	91
Investigation 8.3 Evidence of Double Bonds	92
Exercise (38 – 40)	92
Lab Exercise 8C Predicting Molecular Formulas	92
Investigation 8.4 Evidence for Polar Molecules	93
Lab Exercise 8D London Forces	93
Investigation 8.5 Hydrogen Bond Formation	95
Overview	96
Lab Exercise 8E Bonding Theory	99

Chapter 9   Organic Chemistry

Exercise (1 – 4)	100
Investigation 9.1 Models of Organic Compounds	100
Investigation 9.2 Classifying Organic Compounds	101
Exercise (5 – 7)	103
Investigation 9.3 Structures and Properties of Isomers	104
Exercise (8 – 13)	105
Lab Exercise 9A Molecular Structure of Unknown Liquid	106
Exercise (14 – 19)	106
Investigation 9.4 Synthesizing an Ester	108
Exercise (20 – 23)	108
Lab Exercise 9B Chemical Analysis of an Organic Compound	111
Exercise (24 – 30)	111
Investigation 9.5 Some Properties and Reactions of Organic Compounds	112
Overview	113
Lab Exercise 9C Determining Structure	117
Lab Exercise 9D Determining Percent Yield	117

UNIT V ENERGY AND CHANGE

Chapter 10 Energy Changes

Exercise (1 – 13)	118
Investigation 10.1 Designing and Evaluating a Water Heater	119
Exercise (14 – 26)	121
Investigation 10.2 Molar Enthalpy of Solution	124
Exercise (27 – 31)	125
Lab Exercise 10A Designing a Calorimetry Lab	126
Lab Exercise 10B Molar Enthalpy of a Phase Change	127
Investigation 10.3 Molar Enthalpy of Reaction	127
Exercise (32)	129
Investigation 10.4 Designing a Calorimeter for Combustion Reactions	129
Exercise (33 – 35)	131
Lab Exercise 10C Calibrating a Bomb Calorimeter	132
Lab Exercise 10D Energy Content of Foods	132
Overview	132
Lab Exercise 10E Solidification of Wax	138

Chpater 11 Reaction Enthalpies

Exercise (1 – 10)	139
Lab Exercise 11A Analysis Using Hess's Law	141
Lab Exercise 11B Testing Hess's Law	141
Investigation 11.1 Applying Hess's Law	142
Exercise (11 – 14)	144
Lab Exercise 11C Testing $\Delta H_r°$ from Formation Data	145
Lab Exercise 11D Determining Standard Molar Enthalpies of Formation	146
Lab Exercise 11E Calorimetry of Nuclear Reaction	147
Exercise (15 – 24)	147
Lab Exercise 11F Molar Enthalpy of Formation	150
Overview	150
Lab Exercise 11G H_f for Calcium Oxide	156
Lab Exercise 11H H_c by Four Methods	157

UNIT VI CHANGE AND SYSTEMS

Chapter 12 Electrochemistry

Exercise (1 – 5)	159
Investigation 12.1 Single Replacement Reactions	159
Exercise (6 – 13)	160
Investigation 12.2 Spontaneity of Redox Reactions	162
Exercise (14)	163
Lab Exercise 12A Spontaneity of Reactions	163
Lab Exercise 12B Redox Tables — Design 1	164
Exercise (15 – 18)	164
Lab Exercise 12C Redox Tables — Design 2	165
Exercise (19 – 32)	165
Lab Exercise 12D Testing Redox Concepts	169
Investigation 12.3 Demonstration with Sodium Metal	169
Lab Exercise 12E Standardizing Potassium Permanganate	170
Lab Exercise 12F Analyzing for Tin	171
Lab Exercise 12G Analysis of Chromium in Steel	171
Lab Exercise 12H Analyzing for Iron	172
Exercise (33 – 38)	173
Lab Exercise 12I Analyzing for Tin(II) Chloride	174
Investigation 12.4 Analysis of Hydrogen Peroxide Solution	175
Exercise (39 – 47)	177
Lab Exercise 12J Analyzing Blood for Alcohol Content	180
Exercise (48)	181
Overview	181
Lab Exercise 12K Analyzing Antifreeze	187
Lab Exercise 12L Redox Indicators	189

Chapter 13 Voltaic and Electrolytic Cells

Investigation 13.1 Demonstration of a Simple Electric Cell	191
Exercise (1 – 5)	191
Investigation 13.2 Designing an Electric Cell	192
Exercise (6 – 13)	193
Lab Exercise 13A Evaluating Batteries	194
Investigation 13.3 Demonstration of a Voltaic Cell	194
Exercise (14 – 27)	195
Lab Exercise 13B Creating a Table of Redox Half-Reactions	198
Lab Exercise 13C Series Cells (Enrichment)	198
Investigation 13.4 Testing Voltaic Cells	199
Investigation 13.5 A Potassium Iodide Electrolytic Cell	200
Exercise (28 – 29)	201
Investigation 13.6 Demonstration of Electrolysis	201
Exercise (30 – 33)	203
Investigation 13.7 Copper Plating	203
Exercise (34 – 37)	204
Lab Exercise 13D Quantitative Electrolysis	206
Overview	206
Lab Exercise 13E Testing a Voltaic Cell	211
Lab Exercise 13F Cell Competition	212

UNIT VII CHEMICAL SYSTEMS AND EQUILIBRIUM

Chapter 14 Equilibrium in Chemical Systems

Investigation 14.1 Extent of a Chemical Reaction	213
Exercise (1 – 8)	214
Lab Exercise 14A The Synthesis of an Equilibrium Law	215
Exercise (9 – 11)	215
Lab Exercise 14B Determining an Equilibrium Constant	216
Investigation 14.2 Demonstration of Equilibrium Shifts	216
Investigation 14.3 Testing Le Châtelier's Principle	217
Exercise (12 – 22)	218
Lab Exercise 14C The Chromate-Dichromate Equilibrium	220
Exercise (23 – 25)	220
Investigation 14.4 pH of Common Substances	221
Lab Exercise 14D Strengths of Acids	221
Lab Exercise 14E Qualitative Analysis	221
Exercise (26 – 33)	222
Overview	223
Investigation 14.5 Studying a Chemical Equilibrium System	228

Chapter 15 Acid-Base Chemistry

Investigation 15.1 Testing Arrhenius's Acid-Base Definitions	229
Exercise (1 – 15)	230
Lab Exercise 15A Using Indicators to Determine pH	232
Lab Exercise 15B Designing an Indicator Experiment	233
Lab Exercise 15C Position of Acid-Base Equilibria	233
Exercise (16 – 23)	233
Lab Exercise 15D Testing the Five-Step Method	234
Investigation 15.2 Testing Brønsted-Lowry Predictions	235
Lab Exercise 15E An Acid-Base Table	238
Investigation 15.3 Demonstration of pH Curves	239
Exercise (24 – 29)	240
Investigation 15.4 Buffers	241
Exercise (30 – 47)	242
Lab Exercise 15F Mass Percent of Sodium Phosphate	245
Lab Exercise 15G Identifying an Unknown Acid	246
Investigation 15.5 Ammonia Analysis	246
Overview	249
Lab Exercise 15H Interpretation of Results	256

INTRODUCTION TO THE TEACHER

The *Nelson Chemistry Solutions Manual* includes complete solutions to all questions in *Nelson Chemistry* — Exercise questions embedded within each chapter, as well as Overview questions at the end of each chapter. These solutions are presented in the same format as the communication examples in the student book. The way solutions are organized in the Examples and in the *Solutions Manual* for the students to follow provides a model that meets the standard requirements of school and public exams.

The communication skills recommended in Appendix E of the student book are followed in presenting the solutions. Internationally accepted quantity symbols (in italics) and SI unit symbols are used to represent physical quantities and measurements. Commonly used precision and certainty rules are employed to express numerical answers to the correct number of significant digits. The values in the calculation steps for all solutions are always rounded on paper, but never rounded in the calculator.

Communication is receiving an increased emphasis in school and public exams. This trend is consistent with the emphasis placed on communication by the STSC Chemistry Project that led to the production of *Nelson Chemistry*. In the student book, students are introduced to the criteria used to judge scientific communication: international, precise, logical, and simple. The same criteria are used in presenting the solutions in this manual in a clear way, step by step, using international symbols and acceptable scientific language.

As in the student book, the solutions in this manual use a language that reflects a modern view of the nature of science and of scientific problem solving. Concepts are continually tested by their ability to explain or predict observations. The language presents a way of thinking that implicitly infuses a modern view of scientific knowledge into standard, rigorous chemistry content.

Answers to all Overview questions are provided in Appendix A of the student book, with the intention that students can use the Overview questions for independent study prior to tests or exams. These answers are repeated here, with more details added (including calculation steps and italicized teachers' notes for discussion), so that teachers need to look in only one source for all answers and solutions.

Sample reports for all investigations (approximately 50) and lab exercises (approximately 80) are provided. They suggest answers and set models of lab reports to be expected from students. Teachers can use these sample reports as marking schemes, or have students assess their own reports.

This *Solutions Manual* is intended to save teachers' time so that they can use their critical and creative abilities to continually improve their teaching and their students' learning.

UNIT I
MATTER AND ITS DIVERSITY

1 Chemistry, Technology, and Society

INVESTIGATION 1.1

Demonstration of Combustion of Magnesium (page 27)

Problem

What changes occur when magnesium burns?

Evidence

magnesium ribbon:	5 cm of shiny, silver-colored, flexible solid
burning:	bright white light and white smoke produced
product:	dull white powder

Analysis

- Classifications:
  quantitative observation: length 5 cm
  qualitative observations: (all others)
- All observations were visual and chronological.
- A chemical reaction (burning) of magnesium is a possible interpretation of the observations.

According to the evidence gathered in this experiment, the shiny, silvery magnesium burns, producing a bright white light and a white powder.

Exercise (page 29)

1. *Stress that all classification systems are human conveniences to help us organize our knowledge. Discuss the fact that these systems do not always work, and indicate to students that the real world is more complex than any human reconstruction thereof.*
   (a) observation
   (b) observation
   (c) interpretation
   (d) interpretation
2. (a) quantitative
   (b) qualitative
   (c) qualitative
   (d) qualitative
3. (a) theoretical
   (b) empirical
   (c) empirical
   (d) theoretical

4. (a) Empirical knowledge is observable; theoretical knowledge is not.
   (b) Experimental evidence is empirical.
   (c) This knowledge is empirical, qualitative or quantitative, and probably involves interpretations.

Exercise (page 32)

5. The cooking time of a hamburger patty is affected by the mass.
6. The column headings would be as follows.

   | Mass of Patty (g) | Cooking Time (min) |

7. Title: The Effect of Hamburger Mass on Cooking Time
   Vertical axis: Time (min)
   Horizontal axis: Mass (g)

   Refer students to the information on how to prepare tables and graphs in Appendix E, page 548. You might focus on titles, variables, and units.

8. According to my experience, if the mass of the hamburger patty is larger, the cooking time will be longer.
9. *Individual answers; e.g., statements about thickness, temperature of flame, outside temperature, and cover.*
10. Different-sized patties will be cooked over a grill using a meat thermometer to determine when the patty is cooked. The time required will be measured. The manipulated variable is the mass; the responding variable is the time; the controlled variables are diameter of patty, temperature, and rate of cooking.
    Control: design of the cooking technology
11. Based on limited results, it appears that patties with a larger mass require more cooking time.

Exercise (page 34)

12–28 *Answers will vary depending on the school, but it is very important for your students' safety and your legal concerns to cover these questions thoroughly.*

INVESTIGATION 1.2

Classifying Pure Substances (page 37)

Problem

Are water, bluestone, malachite, table salt, and sugar empirically classified as elements or compounds?

Evidence

Substance	Evidence
water	• The cobalt chloride paper turned pink when tested with water.
	• Gas bubbles formed and rose to the surface.
	• A colorless liquid condensed near the top of the flask.
	• The cobalt chloride paper turned pink.
bluestone	• The solid changed from dark blue to almost white.
	• A colorless liquid collected at the top of the test tube. (The cobalt chloride paper turned pink in this liquid.)
malachite	• The green solid slowly turned into a black solid.

CHEMISTRY, TECHNOLOGY, AND SOCIETY

table salt	• No noticeable changes were observed.
sugar	• The white solid melted and turned from a colorless liquid into a brown liquid with bubbles.
	• The liquid turned into a black solid with smoke rising.
	• A burnt smell was detected as some of the solid disappeared.

Analysis

According to the evidence collected, water and table salt are classified as elements since no evidence of decomposition was noted. Bluestone, malachite, and sugar are classified as compounds since they appeared to decompose on heating.

Evaluation

The experimental design of heating to distinguish between elements and compounds is judged to be inadequate even though the problem was answered. A possible flaw in this design is the limitation of the temperature that can be obtained from the heating equipment. Therefore, I am not very confident in the heating design used. A better design would be an electrical decomposition.

The procedure is judged to be inadequate because some compounds may require a higher temperature to decompose than the others. Insufficient heating in steps 1(d) and 4(b) might have significantly affected the results.

The technological skills required in this investigation are judged to be adequate since these skills are minimal and are not likely responsible for any unexpected results obtained.

Based upon my evaluation of the experiment, I am not very certain of the results for water and table salt, but quite certain of those for the other substances. The major source of uncertainty is the suitability of the temperature used.

Overall, the prediction is judged to be inconclusive since only three out of the five substances heated are interpreted to be compounds. The evidence for bluestone, malachite, and sugar as compounds is relatively clear. It is possible that a higher temperature or an electrical decomposition would alter the results for water and table salt. I am still quite confident that the theoretical definition of elements and compounds remains acceptable. However, further tests on water and table salt need to be done to justify this.

Lab Exercise 1A
Decomposition Using Electricity (page 40)

Problem

Are water and table salt classified as elements or compounds?

Analysis

According to the evidence gathered in this experiment, water and table salt are classified as compounds. The evidence clearly indicates that both water and table salt decomposed when electricity was passed through each of them.

Evaluation

The experimental design is judged to be more than adequate since the problem was clearly answered and no obvious flaws exist. Although heating these two substances in Investigation 1.2 did not result in evidence of decomposition, this experimental design involving passing electricity through the liquid state of the pure substance did. The evidence gathered is consistent with water being composed of hydrogen and oxygen and table salt being composed of sodium and chlorine. The experimental design could be extended to include diagnostic tests on each of the products. This would increase the certainty of the results.

Based upon my evaluation of the design, I am quite certain of the results. The only source of uncertainty is the purity of the samples used.

The prediction is judged to be verified because the experimental results were accurately predicted. The theoretical definition of elements and compounds is therefore judged to be acceptable because the prediction based on this definition was verified. Although a limited number of substances were tested, I feel quite confident in this judgment and in using this definition to classify substances.

The focus of this evaluation should be on the experimental design, as a follow-up to Investigation 1.2.

Exercise (page 43)

29. Some examples of STS issues might be global warming, ozone depletion, nuclear wastes, and oil spills.
30. (a) economic
    (b) ecological
    (c) technological
    (d) political
    (e) scientific

    See Appendix D for a more complete list of perspectives that might be taken on an STS issue. The limited list of five perspectives used at this point is gradually extended. A mnemonic used by some teachers for this list is STEEP.
31. (a) Develop new mining techniques and sources of aluminum. Recycle cans to reuse existing aluminum.
    (b) Attach a small piece of iron to the bottom of the can. Consumers separate cans before discarding them into the garbage.
    (c) Design garbage containers that will only accept aluminum cans. Make an effort to seek out appropriate garbage bins.

Overview (page 44)

Review

1. Useful attitudes include open-mindedness, a respect for evidence, and a tolerance of reasonable uncertainty.
2. If the subject of the statement is observable or based on observable objects or changes, then it is empirical. If the subject is non-observable, then it is a theoretical statement.
3. An acceptable law describes, predicts, and is simple.
4. *Answers will vary, but may include* "According to the evidence ...," "Based on the theory ...," "Preliminary results show ...."
5. *Sample answers:*
    (a) oxygen (element), sodium chloride (compound)
    (b) tap water, air (solutions)
    (c) sand and water, milk
6. (a) An element is a substance that cannot be broken down chemically into simpler substances by heat or electricity. According to theory, an element is composed of only one kind of atom.

(b) A compound is a substance that can be decomposed chemically by heat or electricity. According to theory, a compound is composed of two or more kinds of atoms.
7. The invention was the battery; it produces an electric current that is passed through a sample in an attempt to cause it to decompose.
8. Science and technology work together. Sometimes science leads technology, and sometimes technology leads science.
A common misconception presented in many science textbooks is that technology is applied science. Technology can be a result of applying science but need not always be.

Applications

9. From the photograph you cannot observe any odors that may be produced, how much heat is generated, or how fast the reaction occurs.
10. (a) observation, qualitative, empirical
    (b) observation, qualitative, empirical
    (c) interpretation, qualitative, theoretical
    (d) observation, qualitative, empirical
    (e) interpretation, qualitative, theoretical
    (f) observation, quantitative, empirical
11. *The discussion could include stereotyping, prejudice, and bias. Students are likely to relate to music and clothes. A classification might be seen as being useful when tuning in to a radio station that plays your music preference, but might be seen as negative when a famous band member from your class of musical taste dies of a drug overdose and smears the reputation of the rest of the musicians who play the same type of music.*
12. (a) economic
    (b) scientific
    (c) political
    (d) technological, ecological
    (e) scientific
    (f) technological
    (g) ecological, technological
13. A variety of plastic blocks are slowly heated with a weight on top of the block to see if the block is flattened. The manipulated variable is the plastic; the responding variable is the thickness of the block; controlled variables are rate of heating and mass of the weight.
14. Samples of different substances will be heated and an electric current passed through each one. If a sample decomposes, then it is known to be a compound.

Extensions

15. *Students should be encouraged to be creative in coming up with a design that requires minimal human involvement. Point out that while they can crush the steel, aluminum, and plastic containers, they should avoid breaking the glass bottles because often the bottles can be reused. A general approach would be to list the properties of the different materials and identify properties that would allow separation. For example, the magnetism of steel cans allows them to be separated from the other containers using a large magnet. The World Book Encyclopedia is a good source of information on general recycling methods. Your local recycling society can tell you what methods they use.*
16. (a) *The World Book Encyclopedia is a good source of information on the properties and uses of aluminum, steel, glass, and plastic. A table like the following one could be used to list advantages and disadvantages of different materials for making beverage containers.*

Perspective	Advantages	Disadvantages
ALUMINUM		
scientific	• forms self protective coating	• possible link to Alzheimer's disease
technological	• strength allows efficient design • needs no coating	• difficult to separate for recyling
ecological	• can be recycled	• strip mining aluminum ore may be harmful to environment
economic	• recycling is cost effective • lightweight, economical to ship	
political		• aluminum ore must be imported
STEEL		
scientific	• bendable, high m.p.	
technological	• magnetism makes sorting easy	• must be coated
ecological	• decays by rusting	
economic	• relatively inexpensive	
political	• lots of steelworkers' vote	
GLASS		
scientific	• unreactive material	
technological	• transparent, can be colored	• breakable (by sharp blow or sudden temperature change)
ecological	• can be reused	• broken glass a hazard
economic	• relatively inexpensive	
political	• recycling well established	
PLASTIC		
scientific	• application of polymer chemistry	
technological	• transparent, can be colored • can be designed with specific properties	• difficult to sort various types
ecological		• takes a long time to decompose
economic	• lightweight, economical to ship	
political		• need to legislate recycling

(b) *Answers will vary; but students should base their decision on the advantages and disadvantages listed. They should also indicate which advantage they think is the most important.*

17. The experimental design does not make a valid comparison because the two vacuum cleaners do not perform the same task. It is possible that the old vacuum cleaner could also pick up additional dirt if used after the new model. A better experimental design would compare the amounts of dirt picked up by the two vacuum cleaners on two comparable pieces of carpet, with variables such as area, dirtiness, and type of carpet held constant.

2 Elements

Lab Exercise 2A
Testing the Periodic Law (page 50)

Problem

How is atomic volume related to the atomic mass of the elements?

Analysis

According to the evidence gathered for this experiment, the atomic volume increases and decreases in a periodic fashion relative to an increasing atomic mass of the elements. This periodicity is displayed in the graph below.

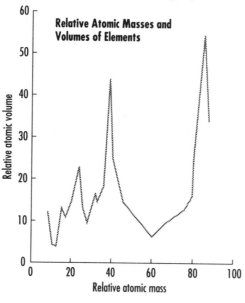

Exercise (page 55)

1. They must be international, logical, precise, and simple.
2. There are approximately five times as many metals as nonmetals.
3. The representative elements are generally considered to be the elements on the left (Groups 1 and 2) and on the right (Groups 13 to 18) in the periodic table.
4. Chemical properties of sulfur and oxygen, such as the chemical formula for the hydrogen compound, are similar.
5. The elements of the noble gas family have similar physical properties (they are all gases) and similar chemical properties (they are all unreactive). This added another family of elements to the Mendeleyev periodic table to illustrate the periodic repetition of properties.

Exercise (page 62)

6. Empirical knowledge is observable, theoretical knowledge is not.
7. Theoretical knowledge is communicated using theoretical descriptions, theoretical hypotheses, theoretical definitions, theories, analogies, and models.

8. A theory must describe, explain, predict, and be simple.
   These criteria will be continually applied to testing theories throughout this textbook.
9. A theory is based on non-observable ideas. A law is based on observable facts.
   A theory is a class of theoretical knowledge, while a law is class of empirical knowledge. Generally, a law is developed before a theory — a theory is created to explain a law.
10. *Refer students to the Glossary (pages 555 – 567) to check their theoretical definitions.*

Exercise (page 64)

11.
	Number of Occupied Energy Levels	Number of Valence Electrons
Be	2	3
Cl	3	7
Kr	4	8
I	5	7
Pb	6	4
As	4	5
Cs	6	1

12. *Diagrams like Figure 2.25 (page 62) should show the number of protons and the arrangement of electrons as follows. Because of space considerations, the electron energy-level models of the elements shown below are presented in a horizontal format, rather than in a vertical format as in the student text.*

   (period 1)
   H $1p^+$, $1e^-$
   He $2p^+$, $2e^-$
   (period 2)
   Li $3p^+$, $2e^-$, $1e^-$
   Be $4p^+$, $2e^-$, $2e^-$
   B $5p^+$, $2e^-$, $3e^-$
   C $6p^+$, $2e^-$, $4e^-$
   N $7p^+$, $2e^-$, $5e^-$
   O $8p^+$, $2e^-$, $6e^-$
   F $9p^+$, $2e^-$, $7e^-$
   Ne $10p^+$, $2e^-$, $8e^-$
   (period 3)
   Na $11p^+$, $2e^-$, $8e^-$, $1e^-$
   Mg $12p^+$, $2e^-$, $8e^-$, $2e^-$
   Al $13p^+$, $2e^-$, $8e^-$, $3e^-$
   Si $14p^+$, $2e^-$, $8e^-$, $4e^-$
   P $15p^+$, $2e^-$, $8e^-$, $5e^-$
   S $16p^+$, $2e^-$, $8e^-$, $6e^-$
   Cl $17p^+$, $2e^-$, $8e^-$, $7e^-$
   Ar $18p^+$, $2e^-$, $8e^-$, $8e^-$
   (period 4)
   K $19p^+$, $2e^-$, $8e^-$, $8e^-$, $1e^-$
   Ca $20p^+$, $2e^-$, $8e^-$, $8e^-$, $2e^-$

13. Reactive elements are believed to have incomplete outer energy levels, whereas the unreactive noble gases are believed to have complete outer electron energy levels.

Exercise (page 66)

14. *See the Glossary for definitions of cation and anion.*
15. *Using the theoretical rule, the charge on the ion will be the difference between the number of electrons in the representative atom and the number of electrons in the nearest noble gas atom.*
16. *Diagrams like Figure 2.26 (page 64) should show the number of protons and the arrangement of electrons as follows.*

    Li $3p^+$, $2e^-$, $1e^-$
    Cl $17p^+$, $2e^-$, $8e^-$, $7e^-$
    Li^+ $3p^+$, $2e^-$
    Cl^- $17p^+$, $2e^-$, $8e^-$, $8e^-$
    K $19p^+$, $2e^-$, $8e^-$, $8e^-$, $1e^-$
    Cl $17p^+$, $2e^-$, $8e^-$, $7e^-$
    K^+ $19p^+$, $2e^-$, $8e^-$, $8e^-$
    Cl^- $17p^+$, $2e^-$, $8e^-$, $8e^-$

Overview (page 67)

Review

1. oxygen, carbon, and hydrogen
2. (a) Dimitri Mendeleyev
   (b) The periodic law (table) was used to successfully predict new elements.
3. (a) Bromine and mercury are liquids at SATP. Helium, nitrogen, oxygen, fluorine, neon, chlorine, argon, krypton, xenon, and radon are gases at SATP.
   (b) The purpose of the staircase line is to separate the metals from the non-metals.
   (c) magnesium, Mg; lead, Pb; fluorine, F
   (d) 1, 8, 13, 14, 17, 29
4. *See Figure 2.11, page 54.*
5. *See Table 2.4, page 59.*
6. Bohr suggested that the properties of the elements can be explained by the arrangement of electrons in specific orbits with certain maximum numbers of electrons (2, 8, 8).
7. Unacceptable theories may be restricted, revised, or replaced.
   An analogy to an old car might be useful; i.e., the car may be restricted to short distances, repaired, or replaced with a newer model.
8. (a) atomic number
   (b) equal to the number of protons (atomic number)
   (c) equal to the last digit of the group number
   (d) equal to the period number
9. (a) Use the theoretical rule that atoms of the representative elements lose or gain electrons to achieve the same electronic arrangement (structure) as the nearest noble gas atom.
   (b) 1+, 2+, 3+, 3–, 2–, 1–
10. Mendeleyev was able to describe all elements using groups with similar properties, such as Group 1. He was able to predict new elements, such as germanium. Finally, the arrangement of rows (periods) and columns (groups) was a simple arrangement.
11. According to the Rutherford model, most of the atom is empty space with a tiny, massive, positively charged nucleus. Only a few alpha particles would pass close enough to the nucleus to be deflected at large angles.

12. Theories are tested by their ability to explain and predict. As new experimental evidence was collected that conflicted with an existing theory, it was revised to account for this new information.
The expression "life-cycle of a theory" might be used in answering this question.

Applications

13. (a) 12p⁺, 12e⁻, 2e⁻
    (b) 13p⁺, 13e⁻, 3e⁻
    (c) 53p⁺, 53e⁻, 7e⁻

14. *Diagrams like Figure 2.25 (page 62) should show the number of protons and the arrangement of electrons as follows.*
    (a) K   19p⁺, 2e⁻, 8e⁻, 8e⁻, 1e⁻
        K⁺   19p⁺, 2e⁻, 8e⁻, 8e⁻
    (b) O   8p⁺, 2e⁻, 6e⁻
        O²⁻ 8p⁺, 2e⁻, 8e⁻
    (c) Cl   17p⁺, 2e⁻, 8e⁻, 7e⁻
        Cl⁻ 17p⁺, 2e⁻, 8e⁻, 8e⁻

15. Noble gases are very unreactive and are thought to have full outer electron energy levels.

16. (a) representative elements
    (b) transition elements, boron, carbon, silicon, and hydrogen

17. (a) sodium ion, Na⁺
    (b) phosphide ion, P³⁻
    (c) sulfide ion, S²⁻

18. *Diagrams like Figure 2.26 (page 64) should show the number of protons and the arrangement of electrons as follows.*
    Mg   12p⁺, 2e⁻, 8e⁻, 2e⁻
    O   8p⁺, 2e⁻, 6e⁻
    Mg²⁺ 12p⁺, 2e⁻, 8e⁻
    O²⁻   8p⁺, 2e⁻, 8e⁻

Extensions

19. The density of gallium can be predicted by averaging the densities of the elements above (Al, 2.70 g/cm³) and below (In, 7.30 g/cm³) gallium in Group 13. Using this method, the predicted density of gallium is 5.00 g/cm³ and the percent difference between the accepted value and the predicted value is 18%. The density of gallium can also be predicted by averaging the densities of the elements to the left (Zn, 7.14 g/cm³) and right (Ge, 5.35 g/cm³) of gallium in Period 4. Using this method, the predicted density of gallium is 6.25 g/cm³ and the percent difference between the accepted value and the predicted value is 5.5%. Averaging these two values gives a predicted density of 5.63 g/cm³, with a percent difference of 4.7%. *Students could also plot a graph of density versus atomic number for the elements in Periods 3, 4, and 5, and use the graph to predict the density of gallium.*

    Based upon the percent differences, the best method, in this case, is the averaging of all neighboring elements.

20. *The paragraph should include the following points:*
    - Empirical knowledge comes from direct observations and leads to descriptions, generalizations and laws, which require the development of empirical concepts.
    - Predictions can be made from generalizations and laws, but they do not explain why a phenomenon occurs.

- Explanations require the development of theoretical knowledge which involves concepts that are not directly observable.
- In the quest for scientific knowledge there is a constant interplay between empirical and theoretical knowledge.
- Theories are developed to explain empirical knowledge, and the theories are tested by designing experiments that attempt to falsify their predictions.
- Scientific concepts are human creations and are restricted representations of the real world.

21. *The report could begin by describing the issue, explaining why the element chosen is in the news. The* CRC Handbook *and the* Merck Index *are good sources for the scientific perspective. Periodicals such as* Discover *and* Scientific American *usually take a scientific and technological perspective. News magazines like* Macleans, Time, *and* Newsweek *often emphasize an economic, ecological, or political perspective.*

Lab Exercise 2B
Testing the Theory of Ions (page 69)

Problem

What is the chemical formula of the compound formed by the reaction of aluminum and fluorine?

Prediction

The formula is predicted to be AlF_3. Each aluminum atom loses three electrons to three fluorine atoms. The aluminum atom forms an Al^{3+} ion and the fluorine atoms form F^- ions with the same number of electrons as the nearest noble gas Ne.

$$\begin{array}{c}
e^- \\
\overset{\frown}{}
\end{array}$$

$$\begin{array}{l} 3e^- \\ 8e^- \\ 2e^- \\ 13p^+ \\ Al \end{array} + 3 \begin{bmatrix} 7e^- \\ 2e^- \\ 9p^+ \\ F \end{bmatrix} \rightarrow \begin{array}{l} 8e^- \\ 2e^- \\ 13p^+ \\ Al^{3+} \end{array} + 3 \begin{bmatrix} 8e^- \\ 2e^- \\ 9p^+ \\ F^- \end{bmatrix}$$

Evaluation

The prediction is judged to be verified because the experimentally determined formula is the same as the predicted formula. The authority used to make the prediction, the restricted quantum mechanics theory of atoms and ions, is judged to be acceptable because the prediction was verified.

UNIT II
CHANGE AND STRUCTURE

3 Compounds

Lab Exercise 3A
Empirical Definitions of Compounds (page 73)

Problem

What properties can be used to develop empirical definitions of ionic and molecular compounds?

Analysis

According to the evidence presented, ionic compounds are solids at SATP and have a high conductivity when dissolved in water. Molecular compounds are solids, liquids, or gases at SATP and form non-conducting solutions.

Lab Exercise 3B
Testing the Classification of Compounds (page 75)

Problem

Of the following compounds, which are ionic and which are molecular?
- pure table salt
- pure sugar
- magnesium nitrate hexahydrate
- citric acid

Prediction

According to the conventional classification of compounds, table salt and magnesium nitrate hexahydrate are ionic (metal-nonmetal combination) and their aqueous solutions should conduct electricity. Sugar and citric acid are molecular (nonmetal-nonmetal combination) and their aqueous solutions should be non-conducting.

Analysis

According to the evidence presented, table salt, magnesium nitrate hexahydrate, and citric acid are ionic because their aqueous solutions conduct electricity. Sugar is molecular because its aqueous solution is non-conducting.

Evaluation

The experimental design is judged to be adequate although its efficiency could be increased without affecting the results and the interpretations. For example, the state of matter and the solubility are not required observations to classify the four substances as ionic or molecular.

> The prediction is judged to be falsified (*could argue inconclusive*) since the prediction for citric acid disagrees with the evidence. The conventional classification system is judged to be unacceptable because the prediction was falsified. The classification method needs to be restricted, revised, or replaced depending on the results obtained from testing other compounds.

Exercise (page 85)

1. (a) sodium oxide
   (b) $CaS_{(s)}$
   (c) potassium nitrate
   (d) $FeCl_{3(s)}$
   (e) mercury(II) oxide
   (f) calcium sulfate-2-water
   (g) $PbO_{2(s)}$
   (h) $Na_2SO_4 \cdot 10H_2O_{(s)}$
   (i) $Al_2O_{3(s)}$
   (j) $Ca_3(PO_4)_{2(s)}$

2. (a) $Cl_{2(g)} + NaOH_{(aq)} \rightarrow NaCl_{(aq)} + H_2O_{(l)} + NaClO_{(aq)}$
   (b) $NaClO_{(aq)} \rightarrow NaCl_{(aq)} + NaClO_{3(aq)}$
   (c) $Na_2OOCCOO_{(aq)} + Ca(OH)_{2(aq)} \rightarrow CaOOCCOO_{(s)} + NaOH_{(aq)}$
   (d) $CoCl_{2(s)} + H_2O_{(l)} \rightarrow CoCl_2 \cdot 6H_2O_{(s)}$

3. (a) $MgO_{(s)}$
   (b) $BaS_{(s)}$
   (c) $ScF_{3(s)}$
   (d) $Fe_2O_{3(s)}$
   (e) $HgCl_{2(s)}$
   (f) $PbBr_{2(s)}$
   (g) $CoI_{2(s)}$

4. (a) ammonium chloride and sodium benzoate $\rightarrow$ ammonium benzoate and sodium chloride
   (b) aluminum nitrate and sodium silicate $\rightarrow$ aluminum silicate and sodium nitrate
   (c) sodium sulfide and water $\rightarrow$ sodium hydrogen sulfide and sodium hydroxide
   (d) nickel(II) oxide and hydrofluoric acid $\rightarrow$ nickel(II) fluoride and water

5. $K^+ \quad Al^{3+} \quad SO_4^{2-} SO_4^{2-}$
   $(1+) + (3+) + 2(2-) = 0$

6. (Discussion)
   Point out to students that the chemical formulas for polyatomic ions are referenced or memorized at this stage in their study of chemistry. Chemists try to work toward theoretical ways of knowing.

Exercise (page 88)

7. (a) $Si_{(s)} + F_{2(g)} \rightarrow SiF_{4(g)}$
   (b) $B_{(s)} + H_{2(g)} \rightarrow B_2H_{4(g)}$
   (c) $C_{12}H_{22}O_{11(aq)} + H_2O_{(g)} \rightarrow C_2H_5OH_{(l)} + CO_{2(g)}$
   (d) $CH_{4(g)} + O_{2(g)} \rightarrow CH_3OH_{(l)}$
   (e) $H_2SO_{4(aq)} + NaOH_{(aq)} \rightarrow H_2O_{(l)} + Na_2SO_{4(aq)}$
   (f) $NH_{3(g)} + HCl_{(g)} \rightarrow NH_4Cl_{(s)}$
   (g) $SO_{2(g)} + H_2O_{(l)} \rightarrow H_2SO_{3(aq)}$

Overview (page 89)

Review

1. (a)

Property	Ionic Compounds	Molecular Compounds
*SATP state	(s) only	(s), (l), or (g)
conductivity†	high	none

(b)

Property	Acids	Bases
*litmus† color	turn red	turn blue

(c) * indicates defining properties; † indicates diagnostic tests.

2. There are two kinds of ions (positive and negative); and the sum of the charges of all the ions is zero.

3. (a) Effective scientific communication is international, logical, precise, and simple.
   (b) International Union of Pure and Applied Chemistry (IUPAC)
   (c) A chemical formula is international, whereas chemical names are not.

4. The kinds of atoms/ions, the number or ratio of the atoms/ions, and the state of matter should be communicated by a chemical formula.

5. Both terms apply to chemical formulas. Chemical formulas can be determined empirically or theoretically.

Applications

6. (a) $NaHSO_{4(s)}$
   (b) $NaOH_{(s)}$
   (c) $CO_{2(g)}$
   (d) $CH_3COOH_{(aq)}$
   (e) $Na_2S_2O_3 \cdot 5H_2O_{(s)}$
   (f) $NaClO_{(s)}$
   (g) $S_{8(s)}$
   (h) $KNO_{3(s)}$
   (i) $H_3PO_{4(aq)}$
   (j) $I_{2(s)}$
   (k) $Al_2O_{3(s)}$
   (l) $KOH_{(s)}$
   (m) $O_{3(g)}$
   (n) $CH_3OH_{(l)}$
   (o) $H_2CO_{3(aq)}$
   (p) $C_3H_{8(g)}$

7. (a) calcium carbonate
   (b) diphosphorus pentaoxide
   (c) magnesium sulfate-7-water (or *magnesium sulfate heptahydrate*)
   (d) dinitrogen oxide
   (e) sodium silicate
   (f) calcium hydrogen carbonate (or *calcium bicarbonate*)
   (g) hydrochloric acid (or *aqueous hydrogen chloride*)
   (h) copper(II) sulfate-5-water (or *copper(II) sulfate pentahydrate*)
   (i) sulfuric acid (or *aqueous hydrogen sulfate*)
   (j) calcium hydroxide
   (k) sulfur trioxide (*not sulfite*)
   (l) tin(II) fluoride

8. (a) Potassium hydroxide and carbonic acid react to form water and potassium carbonate.
   (b) Lead(II) nitrate and ammonium sulfate react to produce lead(II) sulfate and ammonium nitrate.
   (c) Aluminum and iron(II) sulfate react to produce iron and aluminum sulfate.
   (d) Nitrogen dioxide and water react to form nitric acid and nitrogen monoxide.
9. (a) $N_{2(g)} + O_{2(g)} \rightarrow NO_{2(g)}$
   (b) $Fe(CH_3COO)_{3(aq)} + Na_2OOCCOO_{(aq)} \rightarrow Fe_2(OOCCOO)_{3(s)} + NaCH_3COO_{(aq)}$
   (c) $S_{8(s)} + Cl_{2(g)} \rightarrow S_2Cl_{2(l)}$
   (d) $Cu_{(s)} + AgNO_{3(aq)} \rightarrow Ag_{(s)} + Cu(NO_3)_{2(aq)}$
10. (a) $KBr_{(s)}$
    (b) $AgI_{(s)}$
    (c) $PbO_{2(s)}$
    (d) $ZnS_{(s)}$
    (e) $CuO_{(s)}$
    (f) $LiN_{3(s)}$

Extensions

11. *The report could begin by describing applications of the compound chosen to current technology, explaining why the compound is noteworthy. The* CRC Handbook *and the* Merck Index *are good sources for the scientific perspective. Periodicals such as* Discover *and* Scientific American *usually take a scientific and technological perspective. Newspapers and news magazines often emphasize an economic, ecological, social, or political perspective. The* Canadian Encyclopedia *and the* Canada Year Book *are good sources of information on the Canadian chemical industry.*

12. *Encyclopedia articles on IUPAC nomenclature and* Le Système International d'Unités *are good sources of information on communication in the sciences. The IUPAC and SI rules are examples of systems of communication that are international, logical, precise, and simple.*

13. Nitric oxide appears to mediate the control of blood pressure. Research indicates that nitric oxide helps the immune system kill invading parasites that sneak into cells. Under certain conditions it stops cancer cells from dividing. Research also suggests that nitric oxide transmits signals between brain cells ....

Lab Exercise 3C
A Chemical Analysis (page 91)

Problem

Which of the solutions labelled 1, 2, 3, and 4 is $KCl_{(aq)}$, $C_2H_5OH_{(aq)}$, $HCl_{(aq)}$, and $Ba(OH)_{2(aq)}$?

Analysis

According to the evidence gathered in this experiment, the solutions labelled 1, 2, 3, and 4 are $KCl_{(aq)}$, $HCl_{(aq)}$, $C_2H_5OH_{(aq)}$, and $Ba(OH)_{2(aq)}$, respectively. The reasoning is that the evidence indicates an ionic compound, an acid, a molecular compound, and a base, respectively.

4 Chemical Reactions

Exercise (page 98)

1. (a) $2Al_{(s)} + 3CuSO_{4(aq)} \rightarrow 3Cu_{(s)} + Al_2(SO_4)_{3(aq)}$
   (b) 2:3:3:1

Exercise (page 100)

2. scientific
   $2SO_{2(g)} + O_{2(g)} \rightarrow 2SO_{3(g)}$
3. political
   $SO_{3(g)} + H_2O_{(l)} \rightarrow H_2SO_{4(aq)}$
4. technological
   $2CaO_{(s)} + 2SO_{2(g)} + O_{2(g)} \rightarrow 2CaSO_{4(s)}$
5. economic
   $CaO_{(s)} + H_2SO_{3(aq)} \rightarrow H_2O_{(l)} + CaSO_{3(s)}$
6. ecological
   $Al_2(SiO_3)_{3(s)} + 3H_2SO_{4(aq)} \rightarrow 3H_2SiO_{3(aq)} + Al_2(SO_4)_{3(aq)}$

Lab Exercise 4A
The Combustion of Butane (page 101)

Problem

What is the balanced equation for the burning of butane?

Analysis

Based upon the evidence gathered in this investigation, the equation to represent the burning of butane is

$C_4H_{10(g)} + O_{2(g)} \rightarrow CO_{2(g)} + H_2O_{(g)}$

There was no evidence of hydrogen being produced, but there was evidence of carbon dioxide and water vapor being produced. The evidence was not sufficient to indicate how the equation should be balanced.
The reaction of butane and oxygen was not tested for in this experimental design. The problem could not be fully answered with the experimental design used. Critiquing experimental designs is an important skill for students to learn.

Exercise (page 103)

7. (a) formation
   $2Al_{(s)} + 3F_{2(g)} \rightarrow 2AlF_{3(s)}$
   (b) simple decomposition
   $2NaCl_{(s)} \rightarrow 2Na_{(s)} + Cl_{2(g)}$
   (c) complete combustion or formation
   $S_{8(s)} + 8O_{2(g)} \rightarrow 8SO_{2(g)}$
   (d) complete combustion
   $CH_{4(g)} + 2O_{2(g)} \rightarrow CO_{2(g)} + 2H_2O_{(g)}$

(e) simple decomposition
$$2Al_2O_{3(s)} \rightarrow 4Al_{(s)} + 3O_{2(g)}$$
(f) complete combustion
$$C_3H_{8(g)} + 5O_{2(g)} \rightarrow 3CO_{2(g)} + 4H_2O_{(g)}$$
(g) complete combustion or formation
$$2Hg_{(l)} + O_{2(g)} \rightarrow 2HgO_{(s)}$$
(h) simple decomposition
$$2FeBr_{3(s)} \rightarrow 2Fe_{(s)} + 3Br_{2(l)}$$
(i) complete combustion
$$2C_4H_{10(g)} + 13O_{2(g)} \rightarrow 8CO_{2(g)} + 10H_2O_{(g)}$$

INVESTIGATION 4.1

Testing Replacement Reaction Generalizations (page 106)

Problem

What reaction products result when the following substances are mixed?

Prediction

According to the single and double replacement reaction generalizations, the products shown in the following chemical equations should form.

1. $2Al_{(s)}$ + $3CuCl_{2(aq)}$ → $2AlCl_{3(aq)}$ + $3Cu_{(s)}$
   silvery blue   colorless   red-brown

2. $Ba(OH)_{2(aq)}$ + $H_2SO_{4(aq)}$ → $BaSO_{4(s)}$ + $2HOH_{(l)}$
   colorless   colorless   white

3. $2NaBr_{(aq)}$ + $Cl_{2(aq)}$ → $2NaCl_{(aq)}$ + $Br_{2(l)}$
   colorless   colorless colorless   orange in chlorinated hydrocarbon solvent

4. $Zn_{(s)}$ + $CuSO_{4(aq)}$ → $ZnSO_{4(aq)}$ + $Cu_{(s)}$
   silver-grey blue   colorless   red-brown

5. $2HCl_{(aq)}$ + $Mg(OH)_{2(s)}$ → $MgCl_{2(aq)}$ + $2HOH_{(l)}$
   colorless   white   colorless

6. $CaCl_{2(aq)}$ + $Na_2CO_{3(aq)}$ → $CaCO_{3(s)}$ + $2NaCl_{(aq)}$
   colorless   colorless   white   colorless

7. $CoCl_{2(aq)}$ + $2NaOH_{(aq)}$ → $Co(OH)_{2(s)}$ + $2NaCl_{(aq)}$
   pink   colorless   pink   colorless

8. $Ca_{(s)}$ + $2HOH_{(l)}$ → $Ca(OH)_{2(s)}$ + $H_{2(g)}$
   grey   white   colorless gas, pop sound when ignited

9. $NaCH_3COO_{(aq)}$ + $HCl_{(aq)}$ → $NaCl_{(aq)}$ + $CH_3COOH_{(aq)}$
   colorless,   colorless   colorless   colorless, vinegar odor
   odorless

10. $Mg_{(s)}$ + $2HCl_{(aq)}$ → $MgCl_{2(aq)}$ + $H_{2(g)}$
    silver   colorless   colorless   colorless, pop sound when ignited

Evidence/Analysis

	SINGLE REPLACEMENT REACTIONS	
Reaction Number	**Evidence**	**Analysis (products identified)**
1	• Colorless gas bubbles formed initially. • Red-brown solid formed. • Blue solution color changed to colorless.	? $Cu_{(s)}$ $AlCl_{3(aq)}$
3	• Colorless mixture turned yellow-brown and produced an orange color in a chlorinated hydrocarbon.	$Br_{2(aq)}$
4	• Blue solution color changed to colorless. • Red-brown solid formed.	$ZnSO_{4(aq)}$ $Cu_{(s)}$
8	• White precipitate formed. • Colorless gas formed which gave a high-pitched squeal when ignited. • Container became noticeably hotter.	$Ca(OH)_{2(s)}$ $H_{2(g)}$
10	• Colorless gas bubbles formed which produced a pop sound when ignited.	$H_{2(g)}$
	DOUBLE REPLACEMENT REACTIONS	
2	• White precipitate formed.	$BaSO_{4(s)}$
5	• White solid disappeared to form a colorless solution.	$MgCl_{2(aq)}$
6	• White precipitate formed when colorless solutions were mixed.	$CaCO_{3(s)}$
7	• Blue precipitate formed initially. • Final pink-colored precipitate formed.	? $Co(OH)_{2(s)}$
9	• Vinegar odor produced.	$CH_3COOH_{(aq)}$

Evaluation

The experimental design using diagnostic test information to test for predicted products is judged to be adequate to answer the problem. This is the best design available but the overall confidence is low due to the limited evidence obtained.

The procedure is judged to be inadequate because the evidence obtained in most cases was not specific enough to clearly identify the products. An obvious improvement is to extend the procedure to include more detailed diagnostic tests. The technological skills were adequate for the procedure performed.

Based upon my evaluation of this experiment, I am only moderately certain of the experimental results. The major source of uncertainty lies in the use of general color changes and precipitate formation to identify specific products.

The prediction is judged to be verified since all the reactions produced evidence consistent with the predicted products and none were clearly falsified. Unexpected evidence for reactions 1 and 7 suggests that more than one reaction might have occurred. The single and double replacement reaction generalizations are judged to be acceptable since the prediction was verified with no significant conflicting evidence. The certainty of this evaluation could be increased with the improvements discussed.

Exercise (page 107)

8. (a) single replacement
   $Br_{2(l)} + 2NaI_{(aq)} \rightarrow 2NaBr_{(aq)} + I_{2(s)}$
   (b) double replacement (neutralization)
   $H_2SO_{4(aq)} + 2NaOH_{(aq)} \rightarrow 2HOH_{(l)} + Na_2SO_{4(aq)}$

Overview (page 108)

Review

1. The central idea of the kinetic molecular theory is that the smallest entities of a substance are in constant motion.
2. Solids have mostly vibrational motion. Liquids have some vibrational, rotational, and translational motion. Gases have mostly translational motion.
3. Reactant particles must collide with a certain minimum energy and orientation before any rearrangement of atoms or ions occurs.
4. 6.02×10^{23}
5. (a) If a burning splint is inserted into a test tube containing an unknown gas, and a squeal or pop sound is heard, then the gas is likely to be hydrogen.
   (b) If an unknown gas is bubbled through limewater, and the mixture turns cloudy, then the gas is likely to be carbon dioxide.
6. The evidence of conservation of mass supports the idea that atoms are conserved in chemical reactions.
7. Coefficients represent the mole ratio of reactants and products in a chemical reaction. Formula subscripts represent a ratio of ions in an ionic compound or the number of atoms per molecule in a molecular compound.
8. step 1: Write all reactant and product chemical formulas including the states of matter.
   step 2: Begin by balancing the atom or ion present in the greatest number.
   step 3: Repeat step 2 to balance each of the remaining atoms or ions.
   step 4: Check the final reaction equation to ensure that all entities are balanced.
9. formation: elements → compound
   simple decomposition: compound → elements
   complete combustion: substance + oxygen → most common oxides
   single replacement: element + compound → element + compound
   double replacement: compound + compound → compound + compound
10. The type of element (metal or nonmetal) produced in a single replacement reaction is the same as the type of element that reacts.
11. (a) $CO_{2(g)}$
    (b) $H_2O_{(l)}$
    (c) $SO_{2(g)}$
    (d) $Fe_2O_{3(s)}$
12. In general, elements have low solubility in water.
    Most of the exceptions, such as the Group 1 and Group 2 metals and chlorine, involve the element reacting with water rather than dissolving in water.

Applications

13. (a) Two moles of solid nickel(II) sulfide and three moles of oxygen gas react to form two moles of solid nickel(II) oxide and two moles of sulfur dioxide gas.
    2:3:2:2
    (b) Two moles of solid aluminum and three moles of aqueous copper(II) chloride react to produce two moles of aqueous aluminum chloride and three moles of solid copper.
    2:3:2:3

(c) Two moles of liquid hydrogen peroxide react to form two moles of water and one mole of oxygen gas.
2:2:1

14. (a) simple decomposition
$2NaCl_{(s)} \rightarrow 2Na_{(s)} + Cl_{2(g)}$
(b) formation
$4Na_{(s)} + O_{2(g)} \rightarrow 2Na_2O_{(s)}$
(c) single replacement
$2Na_{(s)} + 2HOH_{(l)} \rightarrow H_{2(g)} + 2NaOH_{(aq)}$
(d) double replacement
$AlCl_{3(aq)} + 3NaOH_{(aq)} \rightarrow Al(OH)_{3(s)} + 3NaCl_{(aq)}$
(e) single replacement
$2Al_{(s)} + 3H_2SO_{4(aq)} \rightarrow 3H_{2(g)} + Al_2(SO_4)_{3(aq)}$
(f) complete combustion
$2C_8H_{18(l)} + 25O_{2(g)} \rightarrow 16CO_{2(g)} + 18H_2O_{(g)}$

15. (a) formation
$8Ni_{(s)} + S_{8(s)} \rightarrow 8NiS_{(s)}$
8:1:8
(b) complete combustion
$2C_6H_{6(l)} + 15O_{2(g)} \rightarrow 12CO_{2(g)} + 6H_2O_{(g)}$
2:15:12:6
(c) single replacement
$2K_{(s)} + 2HOH_{(l)} \rightarrow H_{2(g)} + 2KOH_{(aq)}$
2:2:1:2

16. $Cl_{2(g)} + 2KI_{(aq)} \rightarrow I_{2(s)} + 2KCl_{(aq)}$
If a few millilitres of a chlorinated hydrocarbon solvent is added to a test tube containing the reaction mixture, and the solvent layer appears purple, then iodine has likely been formed.

17. (a) $2C_2H_{2(g)} + 5O_{2(g)} \rightarrow 4CO_{2(g)} + 2H_2O_{(g)}$
technological
(b) $MgCl_{2(l)} \rightarrow Mg_{(s)} + Cl_{2(g)}$
scientific
(c) $2Fe_{(s)} + 3CuSO_{4(aq)} \rightarrow Fe_2(SO_4)_{3(aq)} + 3Cu_{(s)}$
economic
(d) $2ZnS_{(s)} + 3O_{2(g)} \rightarrow 2ZnO_{(s)} + 2SO_{2(g)}$
political
(e) $2Pb(C_2H_5)_{4(l)} + 27O_{2(g)} \rightarrow 2PbO_{(s)} + 16CO_{2(g)} + 20H_2O_{(g)}$
ecological

Extensions

18. Smog is a poisonous mixture of smoke and fog in various compositions of chemicals such as sulfur oxides, nitrogen oxides, ozone, hydrocarbons, and other organic compounds. Oxides of sulfur and nitrogen are two main components of smog. Sulfur oxides are produced by fossil fuels that contain traces of sulfur, such as coal or oil. The sulfur oxides react with water to form acids. The following equations represent these reactions.

$S_{8(s)} + 8O_{2(g)} \rightarrow 8SO_{2(g)}$
$S_{8(s)} + 12O_{2(g)} \rightarrow 8SO_{3(g)}$
$SO_{2(g)} + H_2O_{(l)} \rightarrow H_2SO_{3(aq)}$
$SO_{3(g)} + H_2O_{(l)} \rightarrow H_2SO_{4(aq)}$

The main source of nitrogen oxides is the automobile engine. At the high temperature inside an engine, nitrogen burns to form nitrogen monoxide. This compound when released into the air reacts with oxygen to produce nitrogen dioxide, a brown gas responsible for the color of some smog. Nitrogen dioxide also reacts with water to form acids. The following equations summarize these reactions.

$N_{2(g)} + O_{2(g)} \rightarrow 2NO_{(g)}$

$2NO_{(g)} + O_{2(g)} \rightarrow 2NO_{2(g)}$

$2NO_{2(g)} + H_2O_{(l)} \rightarrow HNO_{3(aq)} + HNO_{2(aq)}$

Ecological implications could include a variety of human health problems and defoliation of plants. Solutions may be varied; for example, technological — electric cars, social — car pooling, legal — emission standards.

19. In the contact process for the manufacture of sulfuric acid, $SO_{2(g)}$ is produced by burning sulfur or roasting sulfide ores. The $SO_{2(g)}$ is oxidized to $SO_{3(g)}$ by using $V_2O_{5(s)}$ or $Pt_{(s)}$ as a catalyst.

$$2SO_{2(g)} + O_{2(g)} \xrightarrow{V_2O_{5(s)}} 2SO_{3(g)} + \text{heat}$$

Since the rate of formation of $SO_{3(g)}$ is slow at low temperatures, the temperature is raised to increase the rate of formation. The process is carried out at temperatures of 450°C – 575°C. $SO_{3(g)}$ is absorbed very slowly in water, so the $SO_{3(g)}$ is dissolved in pure $H_2SO_{4(l)}$ to form pyrosulfuric acid:

$SO_{3(g)} + H_2SO_{4(l)} \rightarrow H_2SO_4 \cdot SO_{3(l)}$ (or $H_2S_2O_{7(l)}$)

The $H_2S_2O_{7(l)}$ is then diluted with $H_2O_{(l)}$ to form $H_2SO_{4(l)}$ which is about 98% pure:

$H_2S_2O_{7(l)} + H_2O_{(l)} \rightarrow 2H_2SO_{4(l)}$

Concentrated $H_2SO_{4(l)}$ is frequently used as a dehydrating agent and as a catalyst in reactions to eliminate water molecules. For example:

$$2C_2H_5OH_{(l)} \xrightarrow{H_2SO_{4(l)}} C_2H_5OC_2H_{5(l)} + H_2O_{(l)}$$

Sulfuric acid, like other strong acids, is very corrosive and should be handled with care. When diluting the concentrated acid, always add the acid to the water to dissipate the heat that is generated.

Lab Exercise 4B
Testing Reaction Generalizations (page 109)

Problem
What are the products of the reaction of sodium metal and water?

Prediction
According to the single replacement reaction generalization, the products of the reaction of sodium and water are hydrogen gas and aqueous sodium hydroxide.

$2Na_{(s)} + 2HOH_{(l)} \rightarrow H_{2(g)} + 2NaOH_{(aq)}$

Experimental Design
If a flame is introduced into a small sample of any gas produced, and a squeal or pop is heard, *then* the gas is likely hydrogen.

If red litmus paper is used to check the water (solution) before and after, *and* the red litmus turns blue, *then* sodium hydroxide is likely produced.

If the water is tested by a flame test before and after any reaction, *and* the flame is bright yellow, *then* sodium ions are likely produced.

UNIT III
MATTER AND SYSTEMS

5 Solutions

Lab Exercise 5A
Identification of Solutions (page 114)

Problem

Which of the solutions labelled 1, 2, 3, and 4 is hydrobromic acid, ammonium sulfate, lithium hydroxide, and methanol?

Analysis

According to the litmus and conductivity evidence, solution 1 is methanol, 2 is hydrobromic acid, 3 is ammonium sulfate, and 4 is lithium hydroxide. The evidence is consistent with the solutions being neutral molecular, acidic, neutral ionic, and basic, respectively.

INVESTIGATION 5.1

Chemical Analysis (page 115)

Problem

Which of the white solids labelled 1, 2, 3, and 4 is calcium chloride, citric acid, glucose, and calcium hydroxide?

Experimental Design

A sample of each substance is dissolved in water. A conductivity test and a litmus test are carried out on each aqueous solution. Pure water is tested as a control. Temperature, quantity of solute, and volume of water are controlled variables.

Materials

lab apron
safety glasses
vials of
    calcium chloride $CaCl_{2(s)}$
    citric acid $C_3H_4OH(COOH)_{3(s)}$
    glucose $C_6H_{12}O_{6(s)}$
    calcium hydroxide $Ca(OH)_{2(s)}$
distilled water
50 mL beaker
conductivity tester
red and blue litmus paper
laboratory scoop
stirring rod

Procedure
1. Place about 10 mL of pure water in a clean, 50 mL beaker.
2. Test and record the conductivity.
3. Test with red and blue litmus, and record any color change.
4. For each of the white solids, dissolve a small quantity in 10 mL of water and repeat steps 2 – 3.
5. Samples are disposed of in the sink.

Evidence/Analysis

Solution	Conductivity	Litmus	Analysis
pure water	none	no change	—
1	low	blue turns red	citric acid
2	high	red turns blue	calcium hydroxide
3	none	no change	glucose
4	high	no change	calcium chloride

Exercise (page 119)

1. An electrolyte is a soluble compound whose aqueous solution conducts electricity. According to Arrhenius's theory, solutions formed by electrolytes contain aqueous ions — particles responsible for the conductivity of the solution. Electrolytes may be acids, bases, or neutral ionic compounds.

2. (a) $NaF_{(s)} \rightarrow Na^+_{(aq)} + F^-_{(aq)}$
   (b) $(NH_4)_2SO_{4(s)} \rightarrow 2NH_4^+_{(aq)} + SO_4^{2-}_{(aq)}$
   (c) $HBr_{(g)} \rightarrow H^+_{(aq)} + Br^-_{(aq)}$
   (d) $C_{12}H_{22}O_{11(s)} \rightarrow C_{12}H_{22}O_{11(aq)}$

3. According to Arrhenius's theory, acids dissolve in water and ionize into positive hydrogen ions and negative ions.

4. (a) $Zn_{(s)}, H_2O_{(l)}$
   (b) $Na^+_{(aq)}, Br^-_{(aq)}, H_2O_{(l)}$
   (c) $O_{2(g)}, H_2O_{(l)}$
   (d) $H^+_{(aq)}, NO_3^-_{(aq)}, H_2O_{(l)}$
   (e) $Ca_3(PO_4)_{2(s)}, H_2O_{(l)}$
   (f) $CH_3OH_{(aq)}, H_2O_{(l)}$
   (g) $Al^{3+}_{(aq)}, SO_4^{2-}_{(aq)}, H_2O_{(l)}$
   (h) $K^+_{(aq)}, Cr_2O_7^{2-}_{(aq)}, H_2O_{(l)}$
   (i) $CH_3COOH_{(aq)}, H_2O_{(l)}$
   (j) $S_{8(s)}, H_2O_{(l)}$
   (k) $Cu^{2+}_{(aq)}, SO_4^{2-}_{(aq)}, H_2O_{(l)}$
   (l) $AgCl_{(s)}, H_2O_{(l)}$
   (m) $C_{25}H_{52(s)}, H_2O_{(l)}$

5. (a) $H_3PO_{4(aq)}$
   (b) $H_2SO_{3(aq)}$
   (c) $HNO_{3(aq)}$
   (d) $CH_3COOH_{(aq)}$
   (e) $HI_{(aq)}$

6. (a) aqueous hydrogen sulfate, sulfuric acid
   (b) aqueous hydrogen carbonate, carbonic acid
   (c) aqueous hydrogen sulfide, hydrosulfuric acid
   (d) aqueous hydrogen hypochlorite, hypochlorous acid
   (e) aqueous hydrogen benzoate, benzoic acid

Lab Exercise 5B
Qualitative Analysis (page 120)

Problem

Which of the chemicals numbered 1 to 7 is $KCl_{(s)}$, $Ba(OH)_{2(s)}$, $Zn_{(s)}$, $C_6H_5COOH_{(s)}$, $Ca_3(PO_4)_{2(s)}$, $C_{25}H_{52(s)}$, and $C_{12}H_{22}O_{11(s)}$?

Analysis

According to the evidence, 1 is $C_{12}H_{22}O_{11(s)}$; 2 is $KCl_{(s)}$; 4 is $Ba(OH)_{2(s)}$; 7 is $C_6H_5COOH_{(s)}$; 3, 5, or 6 are $Ca_3(PO_4)_{2(s)}$, $C_{25}H_{52(s)}$, or $Zn_{(s)}$.

Evaluation

The experimental design is judged to be inadequate since the problem could not be completely answered. The design is flawed since solutions 3, 5, and 6 are not distinguishable. A better design could include a simple observation to identify the zinc and measuring the conductivities of both solid and molten forms of the remaining two substances. I am not confident in the given design and am only certain about the identities of four out of the seven chemicals tested.

INVESTIGATION 5.2

The Iodine Clock Reaction (page 121)

Problem

What technological process can be employed to have solution A react with solution B in a time of 20±1 s?

Experimental Design

The solutions are cooled in a systematic trial-and-error approach to produce the desired reaction time. Temperature is the manipulated variable; reaction time is the responding variable; quantity of solute and volume of solution are the controlled variables.

Evidence

Trial	Procedure	Observation
1	1. Measure 5 mL of each solution into separate test tubes. 2. Measure and record the temperature of each solution. 3. Mix solutions back and forth, timing from the moment they contact to the appearance of the blue color.	temperature = 22°C reaction time = 10.4 s
2	4. Use a 400 mL beaker to prepare an ice bath at 12°C. 5. Repeat step 1 and place the test tubes into the ice bath. 6. When the desired temperature is reached repeat step 3.	temperature = 12°C reaction time = 15.7 s
3	7. Repeat steps 4 to 6 using a temperature of 4°C. 8. Do another trial at 4°C.	reaction time = 19.2 s reaction time = 17.8 s
4	9. Repeat steps 4 to 6 using 4°C and replacing the test tubes containing the reaction mixture into the ice bath immediately after mixing. 10. Do another trial by repeating step 9.	reaction time = 20.2 s reaction time = 19.9 s

Analysis

The following procedure produces a reproducible reaction time of 20±1 s.
1. Measure 5 mL of each solution into separate test tubes.
2. Place the solutions into an ice bath at 4°C.
3. When the desired temperature is reached, mix the solutions back and forth. Start the timing when the solutions first come into contact.
4. Immediately replace the test tubes containing the reaction mixture in the ice bath.
5. Stop timing on the first appearance of the blue color.

INVESTIGATION 5.3

Sequential Qualitative Analysis (page 122)

Problem

Are there any lead(II) ions and/or strontium ions present in a sample solution?

Materials

lab apron
safety glasses
sample solution A
$NaCl_{(aq)}$
$Na_2SO_{4(aq)}$
medicine dropper

(3) 50 mL beakers
(2) large test tubes
test tube stopper
filter paper
filter funnel, funnel holder, and stand

Procedure

1. Obtain approximately 10 mL of the sample solution in a clean test tube.
2. Add about 10 mL of $NaCl_{(aq)}$.
3. Stopper the test tube and invert to mix.
4. After the precipitate settles, add a few drops of $NaCl_{(aq)}$ with a medicine dropper.
5. If further precipitation occurs, repeat steps 2 to 4.
6. Filter the mixture and collect the filtrate in a second test tube.
7. Add about 10 mL of $Na_2SO_{4(aq)}$ to the filtrate and note any precipitation.
8. Dispose of all solutions into the labelled waste containers.

Evidence

step 2	• A white precipitate formed.
step 4	• No further precipitation occurred.
step 7	• A white precipitate formed.

Analysis

According to the evidence collected, sample solution A contains both lead(II) ions and strontium ions.

Exercise (page 123)

7. baking powder, solid drain cleaners, sugar, or drink mixes
8. (a) yellow-brown
   (b) colorless

(c) blue
(d) green
9. (a) yellow-red
   (b) blue (halides), green (others)
   (c) yellow
   (d) violet
10. If a magnesium chloride (or any other Group 2 compound) solution is added to the sample, and a precipitate forms, then carbonate ions are likely present.
11. If a sodium chloride solution is added to the sample, and a precipitate forms, then $Tl^+_{(aq)}$ is likely present. If a sodium hydroxide solution is added to the filtrate from the first test, and a precipitate forms, then $Ca^{2+}_{(aq)}$ is likely present. If a sodium sulfate solution is added to the filtrate from the second test, and a precipitate forms, then $Ba^{2+}_{(aq)}$ is likely present.

Lab Exercise 5C
Qualitative Analysis by Color (page 123)

Problem
Which of the solutions labelled 1, 2, 3, and 4 is potassium permanganate, copper(II) sulfate, sodium chloride, and copper(I) nitrate?

Analysis
According to the evidence gathered, solution 1 is $CuNO_{3(aq)}$; 2 is $NaCl_{(aq)}$; 3 is $KMnO_{4(aq)}$; 4 is $CuSO_{4(aq)}$.

Exercise (page 125)

12. $v_{CH_3COOH} = 250 \text{ mL} \times \dfrac{7 \text{ mL}}{100 \text{ mL}} = 0.02 \text{ L}$

13. $m_{NaCl} = 60 \text{ kg} \times \dfrac{25 \text{ kg}}{100 \text{ kg}} = 15 \text{ kg}$

14. $m_{PCBs} = 64 \text{ kg} \times \dfrac{4.0 \text{ mg}}{1 \text{ kg}} = 0.26 \text{ g}$

15. $v_{NH_3} = 0.500 \text{ mol} \times \dfrac{1 \text{ L}}{1.24 \text{ mol}} = 0.403 \text{ L}$

Exercise (page 128)

16. $$v_i C_i = v_f C_f$$
$$v_i \times 17.8 \text{ mol/L} = 2.00 \text{ L} \times 0.200 \text{ mol/L}$$
$$v_{i_{H_2SO_4}} = 22.5 \text{ mL}$$

17. $$v_i C_i = v_f C_f$$
$$1.00 \text{ L} \times 17.4 \text{ mol/L} = v_f \times 0.400 \text{ mol/L}$$
$$v_{f_{CH_3COOH}} = 43.5 \text{ L}$$

18. $$v_i C_i = v_f C_f$$
$$1 \text{ L} \times 72\,786 \text{ ppm} = v_f \times 345 \text{ ppm}$$
$$v_{f_{CO_2}} = 0.2 \text{ kL}$$

19.
$$v_i c_i = v_f c_f$$
$$10.00 \text{ mL} \times c_i = 250.0 \text{ mL} \times 0.274 \text{ g/L}$$
$$c_i = 6.85 \text{ g/L}$$

20. (Discussion) Two examples are pesticides such as DDT concentrated in a food chain, and sulfur oxides dissolving in water in the atmosphere and precipitating as acid rain.

INVESTIGATION 5.4

Solution Preparation (page 128)

Problem

Any practical problem that requires a 500 ppm solution.

Procedure

1. Measure 5.00 g of pure sodium chloride in a weighing boat.
2. Transfer the solid into a clean 100 mL graduated cylinder, rinsing the boat with distilled water into the cylinder.
3. Add 10–20 mL of water, swirl and stir to dissolve the salt.
4. Rinse the stirring rod with water into the cylinder.
5. Add sufficient distilled water to obtain a final volume of 100.0 mL. Use a medicine dropper to set the final volume.
6. Pour the solution into a clean, dry 250 mL beaker and stir with a dry stirring rod.
7. Rinse the 10 mL graduated cylinder with a small quantity of the $NaCl_{(aq)}$ from step 6.
8. Measure 10.0 mL of the $NaCl_{(aq)}$ in the graduated cylinder. Use a medicine dropper to set the final volume.
9. Rinse the 100 mL graduated cylinder with distilled water.
10. Pour the 10.0 mL of $NaCl_{(aq)}$ from the 10 mL graduated cylinder into the 100 mL graduated cylinder.
11. Repeat steps 5 to 10.
12. Repeat steps 5 and 6.

Exercise (page 131)

21. (a) $KCl_{(aq)} \rightarrow K^+_{(aq)} + Cl^-_{(aq)}$
    (b) $Na_2SO_{4(aq)} \rightarrow 2Na^+_{(aq)} + SO_4^{2-}_{(aq)}$
    (c) $Na_3PO_{4(aq)} \rightarrow 3Na^+_{(aq)} + PO_4^{3-}_{(aq)}$

22. (a) $[K^+_{(aq)}] = 0.14$ mol/L, $[NO_3^-_{(aq)}] = 0.14$ mol/L
    (b) $[Ca^{2+}_{(aq)}] = 0.14$ mol/L, $[Cl^-_{(aq)}] = 0.28$ mol/L
    (c) $[NH_4^+_{(aq)}] = 0.42$ mol/L, $[PO_4^{3-}_{(aq)}] = 0.14$ mol/L

23. (a) 10^{-7} mol/L
    (b) 10^{-11} mol/L
    (c) 10^{-2} mol/L
    (d) 10^{-4} mol/L
    (e) 10^{-14} mol/L

24. (a) 3
    (b) 5
    (c) 7
    (d) 10

25. 0.01 or 10⁻² (*to obtain a pH of 7*)

26. (a) $n_{H^+} = 100 \text{ L} \times \dfrac{1 \times 10^{-3} \text{ mol}}{\text{L}} = 0.1 \text{ mol}$

    (b) $n_{H^+} = 100 \text{ L} \times \dfrac{1 \times 10^{-8} \text{ mol}}{\text{L}} = 1 \times 10^{-6} \text{ mol}$

    (c) $n_{H^+} = 100 \text{ L} \times \dfrac{0.05 \text{ mol}}{\text{L}} = 5 \text{ mol}$

27. (Discussion)
    (a) Either yes or no depending on the meaning of "zero."
        Perhaps discuss the statement, "Chemicals aren't toxic, amounts are."
    (b) Yes. It is possible that a sample could have no ions or molecules of a particular chemical.
    (c) No. It is not possible since no instrument is perfect.
    (d) *This is a debatable point and subject to a variety of perspectives. Note that many toxic limits are set as a result of testing on animals.*

28. (Discussion) Both acidic and basic solutions are used for cleaning to improve hygiene, appearance, or efficiency. Some examples of acidic solutions are toilet bowl cleaners and scale removers; and examples of basic solutions are oven, floor, and general purpose cleaners. The risks involved are the corrosive effects that these solutions have when they are spilled on the skin and especially in the eyes. There is a higher risk to small children as they may not be aware of the dangers. In many cases, milder, less concentrated chemicals can be used if cleaning is more frequent and people are willing to do more physical labor. Those cleaners specifically used to improve hygiene and prevent health problems probably have more benefits than risks if handled with care.

Lab Exercise 5D
Qualitative Analysis (page 132)

Problem

Which of the 0.1 mol/L solutions labelled 1, 2, 3, 4, and 5 is $KCl_{(aq)}$, $CaBr_{2(aq)}$, $HCl_{(aq)}$, $CH_3OH_{(aq)}$, and $NaOH_{(aq)}$?

Analysis

According to the evidence obtained, solution 1 is $NaOH_{(aq)}$; 2 is $KCl_{(aq)}$; 3 is $CaBr_{2(aq)}$; 4 is $CH_3OH_{(aq)}$; 5 is $HCl_{(aq)}$.

Evaluation

The experimental design is adequate since the problem was answered. There were no obvious flaws in the design; however, the design can be made more efficient by eliminating the observation for color since this is not needed.

Overall, I am quite confident in the design and the answer. Sources of uncertainty may include measurement errors in the preparation of the solutions and in the instruments used. These would not be a major factor and the level of certainty of the answer is quite high.

INVESTIGATION 5.5

The Solubility of Sodium Chloride in Water (page 133)

Problem
What is the solubility of sodium chloride at room temperature?

Prediction
According to the graph of known empirical solubilities, the solubility of sodium chloride at 21.2°C is 32.02 g/100 mL.

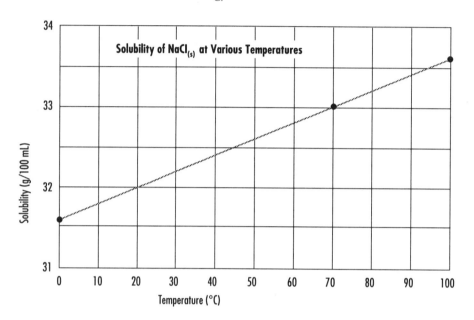

Evidence
temperature of saturated $NaCl_{(aq)}$ = 21.2°C
volume of saturated $NaCl_{(aq)}$ = 10.00 mL
mass of evaporating dish = 152.19 g
mass of evaporating dish and $NaCl_{(s)}$ = 155.23 g
mass of evaporating dish and $NaCl_{(s)}$ (after reheating) = 155.20 g
Some solid splattered out of the evaporating dish after most of the water had evaporated.

Analysis
mass of salt in 10 mL = 155.20 g – 152.19 g = 3.01 g
mass of salt in 100 mL = 3.01 g × 10 = 30.1 g
According to the evidence collected in this experiment, the solubility of sodium chloride at 21.2°C is 30.1 g/100 mL.

Evaluation
The experimental design of crystallization is judged to be adequate to answer the problem with no obvious flaws. An alternative design would be precipitation of a low solubility chloride compound. This alternative is not significantly better than the one used. I am quite confident in the design used.

The procedure is judged to be inadequate because some solid splattered out of the evaporating dish and only one trial was done. The procedure could be improved by covering the evaporating dish (for example, with a watch glass) to prevent splattering, and repeating the procedure for another 10.00 mL of saturated $NaCl_{(aq)}$. The technological skills were adequate for the procedure and equipment used.

Based upon my evaluation of the experiment, I am only moderately certain about my results. The major sources of uncertainty are the splattering of the solid and the dryness of the final solid. With these uncertainties, a percent difference of less than 10% is acceptable.

The percent difference between the experimental value and the predicted value is 6%, which is within the estimated total uncertainty.

$$\% \text{ difference} = \frac{|30.1 \text{ g}/100 \text{ mL} - 32.02 \text{ g}/100 \text{ mL}|}{32.02 \text{ g}/100 \text{ mL}} \times 100 = 6\%$$

The prediction is judged to be verified because the predicted answer agrees with the experimental answer within the allowed percent difference. The percent difference can easily be accounted for by the sources of uncertainty discussed. The graph of empirical solubilities is judged to be an acceptable way of predicting the solubility of sodium chloride. I am quite confident in this judgment.

Exercise (page 136)

29. immiscible with water: mineral oil, gasoline
    miscible with water: methanol, ethanol
30. According to the solubility rules, gases always have a higher solubility in water at lower temperatures. Therefore, more oxygen can dissolve in a cold stream.
31. In hot water, both the detergent and any substance soiling the clothes would have a higher solubility.
    There is also an increase in the rate of reaction at higher temperatures.
32. Carbonated beverages have a higher than atmospheric pressure of carbon dioxide inside their containers. When they are opened, the decrease in pressure makes the dissolved carbon dioxide less soluble and, therefore, escape from the beverages.

Lab Exercise 5E
Solubility and Temperature (page 136)

Problem

How does the temperature of a saturated mixture affect the solubility of potassium nitrate?

Prediction

According to the solubility generalization, as the temperature increases, the solubility of potassium nitrate should increase.

Analysis

According to the evidence, as the temperature increased, the solubility of potassium nitrate also increased.

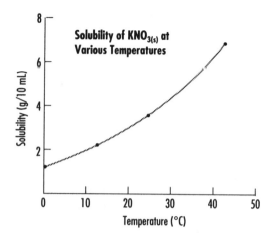

Evaluation

The experimental design is judged to be adequate to answer the problem. The design appears to be the best available. I am confident in this design and certain of the answer obtained. The main source of uncertainty is the dryness of the final solid. However, if the same procedure is followed for each sample, the answer obtained should not be affected.

The prediction based on the solubility generalization is judged to be verified since the experimental result clearly agrees with the predicted result. The solubility generalization is judged to be acceptable because the prediction was verified. Since the trend is obvious on the graph, I feel certain about my judgment. A further test for the solubility generalization is to try the same design using many other solids.

Exercise (page 138)

33. The concept of dynamic equilibrium can be tested by observing the size of crystals in a closed saturated solution and by using a radioactive tracer in any system at equilibrium.

34. $NaCl_{(s)} \rightleftharpoons NaCl_{(aq)}$
    or $NaCl_{(s)} \rightleftharpoons Na^+_{(aq)} + Cl^-_{(aq)}$

35. The theory of dissolving is incorporated into the theory of equilibrium describing saturated solutions. Dissolving is said to continue, even when a solution is saturated.
    The logical consistency of this explanation adds credibility to both theories.

Overview (page 139)

Review

1. conductivity test, litmus test
2. (a) The solute is calcium chloride; the solvent is water.
   (b) The solute is ammonia; the solvent is water.
3. (a) acids, bases, and ionic compounds
   (b) molecular substances
4. Hydrogen ions are responsible for acidic properties and hydroxide ions are responsible for basic properties.
5. Arrhenius studied depression of freezing points and conductivities of solutions.
6. Acids differ from molecular compounds because, when dissolved in water, they produce conducting solutions that turn blue litmus pink. According to

Arrhenius's theory, acids ionize in solution to produce hydrogen ions and negative ions.
7. Solutions make it easy to handle chemicals, to allow chemicals to react, and to control the reactions.
8. A common method of chemical analysis of ions in solution is selective precipitation. For example, adding silver nitrate to a solution containing bromide ions would give a precipitate of silver bromide.
9. (a) 15
   (b) 4
   (c) −1
10. Vinegar is used in making pickles. Ammonia solution is used for cleaning windows. Soft drinks are used for refreshment. Gasoline is used as a fuel for cars.
    Many other examples are possible.
11. The concentrated salt solution in a water softener is necessary for effective operation. A high concentration of hydrogen peroxide for use as a disinfectant would be dangerous.
    Many other examples are possible.
12. Immiscible means that the two liquids do not mix; they form separate layers.
13. According to the solubility rules, the solubility of most solutes in water decreases as the temperature of the solutions drops.
14. If a saturated solution and excess solute are part of a closed system and all observable properties become constant, then a chemical equilibrium exists. According to the theory of dynamic equilibrium, two opposing processes — dissolving and crystallizing — are occurring at the same rate.

Applications

15. (a) If the solution is tested for conductivity, and it conducts electricity, then the solution contains an ionic compound. If it does not conduct, then it contains a molecular compound.
    (b) If the solution is tested with both red and blue litmus paper, and the blue litmus turns red, then the solution contains an acid. If the red litmus turns blue, then a base is present.
    (c) If the freezing point of the solution is measured and the freezing point is less than 0°C, then the solution contains a compound.
16. (a) $Sr(OH)_{2(s)} \rightarrow Sr^{2+}_{(aq)} + 2OH^-_{(aq)}$
    (b) $K_3PO_{4(aq)} \rightarrow 3K^+_{(aq)} + PO_4^{3-}_{(aq)}$
    (c) $HBr_{(g)} \rightarrow H^+_{(aq)} + Br^-_{(aq)}$
    (d) $Mg(CH_3COO)_{2(s)} \rightarrow Mg^{2+}_{(aq)} + 2CH_3COO^-_{(aq)}$
17. (a) $Ca^{2+}_{(aq)}, Cl^-_{(aq)}, H_2O_{(l)}$
    (b) $C_2H_5OH_{(aq)}, H_2O_{(l)}$
    (c) $NH_4^+_{(aq)}, CO_3^{2-}_{(aq)}, H_2O_{(l)}$
    (d) $Cu_{(s)}, H_2O_{(l)}$
    (e) $Pb(OH)_{2(s)}, H_2O_{(l)}$
    (f) $H^+_{(aq)}, SO_4^{2-}_{(aq)}, H_2O_{(l)}$
    (g) $Al^{3+}_{(aq)}, SO_4^{2-}_{(aq)}, H_2O_{(l)}$
    (h) $S_{8(s)}, H_2O_{(l)}$
18.

Test Solution	$Na^+_{(aq)}$	$Li^+_{(aq)}$	$Ca^{2+}_{(aq)}$	$Ni^{2+}_{(aq)}$	$Cu^{2+}_{(aq)}$	$Fe^{3+}_{(aq)}$
Color	none	none	none	green	blue	yellow-brown
Flame Color	yellow	bright red	yellow-red	—	blue or green	—

19. Add an excess of zinc nitrate solution to the unknown solutions to precipitate any sulfide ions present. Filter and test the filtrate for chloride ions by adding silver nitrate solution.
    Many other designs will work.
20. $Sr(OH)_{2(s)}$ or $Ba(OH)_{2(s)}$
21. $c_{NaCl} = \dfrac{3.13 \text{ g}}{20.0 \text{ mL}} \times 100\% = 15.7\%$ W/V
22. $m_{mineral} = \dfrac{440 \text{ mg}}{1000 \text{ mL}} \times 200 \text{ mL} = 88.0 \text{ mg}$
23. $c_{Pb} = \dfrac{0.2 \text{ mg}}{100 \text{ mL}} \times \dfrac{1000 \text{ mL}}{1 \text{ L}} = 2 \text{ mg/L} = 2 \text{ ppm}$

    Therefore, 3 ppm is more concentrated.
24. $m_{CO_3^{2-}} = 225 \text{ mg/L} \times 200 \text{ L} = 45.0 \text{ g}$
25. $m_{Hg} = 5 \text{ mg/L} \times 1.2 \text{ L} = 6 \text{ mg}$
26. $v_{H_2O} = 1 \text{ g} \times \dfrac{1 \text{ L}}{4 \times 10^{-6} \text{ mg}} = 3 \times 10^8 \text{ L}$
27. (a) $Na_2S_{(aq)} \rightarrow 2Na^+_{(aq)} + S^{2-}_{(aq)}$
                           4.48 mol/L 2.24 mol/L
    (b) $Fe(NO_3)_{2(aq)} \rightarrow Fe^{2+}_{(aq)} + 2NO_3^-_{(aq)}$
                             0.44 mol/L 0.88 mol/L
    (c) $K_3PO_{4(aq)} \rightarrow 3K^+_{(aq)} + PO_4^{3-}_{(aq)}$
                             0.525 mol/L 0.175 mol/L

    The calculation of ionic concentrations uses the mole ratio from the dissociation equation. Students should be encouraged to do this calculation by inspection and not by writing a calculation expression.
28. $c_i = \dfrac{5.00 \text{ L} \times 0.125 \text{ mol/L}}{11.6 \text{ mol/L}} = 53.9 \text{ mL}$
29. $c_f = \dfrac{25.0 \text{ mL} \times 2.70 \text{ g/L}}{4.00 \text{ L}} = 16.9 \text{ mg/L}$
30. $m_{KClO_3} = \dfrac{2.16 \text{ g}}{25.0 \text{ mL}} \times 100 \text{ mL} = 8.64 \text{ g}$

    Therefore, the solubility may be expressed as 8.64 g/100 mL.
31. $m_{O_2} = 42 \text{ mg/L} \times 25.0 \text{ L} = 1.1 \text{ g}$
32. According to the solubility rules, some solid sodium carbonate will precipitate from the solution. The reasoning is that the solubility decreases as the temperature decreases. Therefore, the excess sodium carbonate will precipitate until the concentration is the same as the solubility at that temperature.

Extensions

33. The maximum amount of water vapor in the atmosphere decreases with decreasing temperature. When the temperature drops below freezing, the excess water vapor in the atmosphere sublimes onto cold surfaces forming hoarfrost.
34. *Students could form many different hypotheses relating to the release of gas by the cold ice cube. Their hypotheses must be testable. For example:*

Hypothesis
The rate of fizzing of an ice cube increases as the temperature of the soft drink increases.

Experimental Design
Ice cubes are dropped in a soft drink at different temperatures, starting at 0°C for the soft drink, and the rate of fizzing observed.

35. *Students could choose any of several aspects of chemistry and present positive or negative aspects of it. Whichever position they take, it should be examined from a variety of perspectives, e.g., ecological, economic, and technological. For example:*

Use	Positive	Negative
insecticides	• prevent spread of disease	• build up in food chains
herbicides	• improve crop yield	• may harm ecosystem
food additives	• prevent spoilage	• may cause cancer
medicines	• save lives	• side effects
fertilizers	• increase food supplies	• algae in lakes near farms
plastics	• save trees	• not biodegradable
synthetic fibres	• produce low cost clothing	• put trappers, sheep and cotton farmers out of work
cleaning agents	• prevent disease	• pollute the environment
packaging	• facilitates transfer	• produces garbage
steel	• provides strong structures	• rusts and sometimes fails

Lab Exercise 5F
Cation Analysis (page 141)

Problem
What four of five cations are present in the solution sample provided?

Analysis
According to the evidence gathered by this experimental design, the four cations present in the solution are $H^+_{(aq)}$, $Ag^+_{(aq)}$, $Cu^+_{(aq)}$, and $Na^+_{(aq)}$.

The reasoning behind this analysis is that the blue litmus turns red to indicate the hydrogen ion, the potassium acetate produces a precipitate with only silver ions, the potassium chloride produces a white precipitate and removes the green color with only copper(I) ions, and the yellow flame test is characteristic of sodium.

Since only five cations are present and a sulfate precipitate of the fifth cation is produced, there can only be one compound precipitated by the potassium chloride. In other words, lead(II), thalium, and mercury(I) ions are not present. According to chemical references, copper(I) ion forms a green-colored solution but copper(I) chloride solid is white. Note that this flowchart is based on the solubility table on the inside back cover of the student book. Other solubility charts may have other cations that could precipitate.

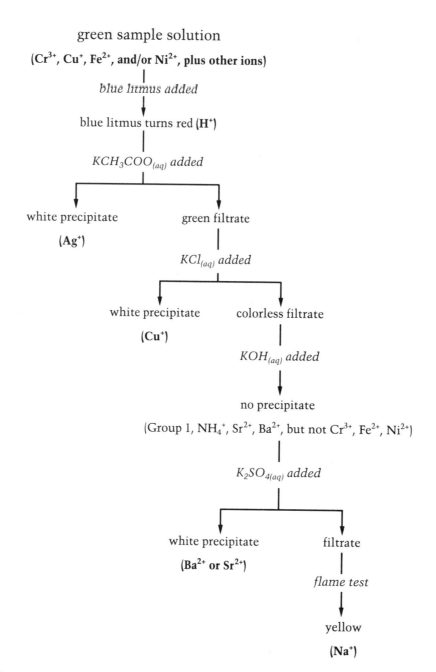

Evaluation

The experiment could be modified to identify a fifth cation by testing the white sulfate precipitate with a flame test to determine whether the other cation is barium or strontium.

6 Gases

Lab Exercise 6A
Pressure and Volume of a Gas (page 144)

Problem

What is the simplest mathematical equation to describe the relationship between the pressure and the volume of a gas?

Analysis

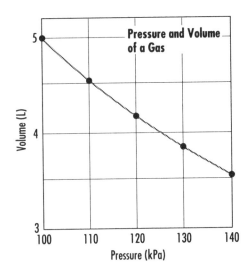

According to the graph, the volume (v) varies inversely with the pressure (p). The simplest mathematical equation to describe the relationship is $pv = k$.

Lab Exercise 6B
Temperature and Volume of a Gas (page 146)

Problem

What is the mathematical relationship between the temperature and the volume of a gas?

Analysis

According to the graph, the volume (v) varies directly with the temperature (t) since the graph is a straight line. The mathematical relationship is $v \propto t$.
Note that the line does not go through the origin.

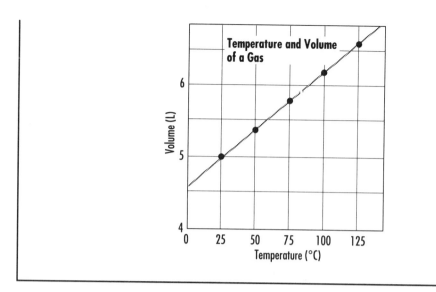

Exercise (page 149)

1. $\dfrac{v_1}{T_1} = \dfrac{v_2}{T_2}$

   $v_2 = \dfrac{v_1 T_2}{T_1} = \dfrac{0.10 \text{ L} \times 371 \text{ K}}{298 \text{ K}} = 0.12 \text{ L}$

2. (a) $p_2 = \dfrac{p_1 T_2}{T_1} = \dfrac{225 \text{ kPa} \times 318 \text{ K}}{291 \text{ K}} = 246 \text{ kPa}$

   (b) $\dfrac{p_1 v_1}{T_1} = \dfrac{p_2 v_2}{T_2}$

   $v_2 = \dfrac{p_1 v_1 T_2}{p_2 T_1} = \dfrac{225 \text{ kPa} \times 27 \text{ L} \times 298 \text{ K}}{100 \text{ kPa} \times 291 \text{ K}} = 62 \text{ L}$

3. $p_1 v_1 = p_2 v_2$

   $v_2 = \dfrac{p_1 v_1}{p_2} = \dfrac{100 \text{ kPa} \times 600 \text{ mL}}{250 \text{ kPa}} = 240 \text{ mL}$

4. $\dfrac{p_1 v_1}{T_1} = \dfrac{p_2 v_2}{T_2}$

   $v_2 = \dfrac{p_1 v_1 T_2}{p_2 T_1} = \dfrac{100 \text{ kPa} \times 5.00 \text{ L} \times 308 \text{ K}}{90 \text{ kPa} \times 293 \text{ K}} = 5.8 \text{ L}$

5. $\dfrac{p_1}{T_1} = \dfrac{p_2}{T_2}$

   $T_2 = \dfrac{p_2 T_1}{p_1} = \dfrac{250 \text{ kPa} \times 308 \text{ K}}{150 \text{ kPa}} = 513 \text{ K or } 240°\text{C}$

6. $\dfrac{p_1 v_1}{T_1} = \dfrac{p_2 v_2}{T_2}$

   $T_2 = \dfrac{p_2 v_2 T_1}{p_1 v_1} = \dfrac{35.0 \text{ atm} \times 23.0 \text{ mL} \times 313 \text{ K}}{1.00 \text{ atm} \times 500 \text{ mL}} = 504 \text{ K or } 231°\text{C}$

7. $\dfrac{p_1V_1}{T_1} = \dfrac{p_2V_2}{T_2}$

$V_2 = \dfrac{p_1V_1T_2}{p_2T_1} = \dfrac{800 \text{ kPa} \times 1.0 \text{ L} \times 298 \text{ K}}{100 \text{ kPa} \times 303 \text{ K}} = 7.9 \text{ L}$

8. No, all gases have similar properties and follow the same gas laws.
9. The amount of gas did not change when the pressure, volume, or temperature changes.

Exercise (page 153)

10. (a) According to the kinetic molecular theory, as the temperature increases, the average speed of the gas particles increases. If the volume is kept constant, then faster moving particles will collide more often with the sides of the container. More collisions means a greater pressure.
    (b) According to the kinetic molecular theory, as the volume decreases, the particles are confined to a smaller space and will collide more often with each other and the walls of the container. More collisions means a greater pressure.
    (c) According to the kinetic molecular theory, perfume molecules collide randomly with air molecules and slowly move throughout the air in the room.
    (d) A bullet moves in a straight line over a long distance before it hits its target. A gas molecule only moves a very short distance before colliding and changing its direction.

11. $V_{H_2} = 7.50 \text{ mol} \times \dfrac{24.8 \text{ L}}{1 \text{ mol}} = 186 \text{ L}$

12. $n_{SO_2} = 50 \text{ mL} \times \dfrac{1 \text{ mol}}{24.8 \text{ L}} = 2.0 \text{ mmol}$

13. $V_{Ne} = 2.25 \text{ mol} \times \dfrac{22.4 \text{ L}}{1 \text{ mol}} = 50.4 \text{ L}$

14. $n_{O_2} = (0.20 \times 20.0 \text{ L}) \times \dfrac{1 \text{ mol}}{22.4 \text{ L}} = 0.18 \text{ mol}$

Exercise (page 157)

15. (a) 200.59 g/mol
    (b) 48.00 g/mol
    (c) 44.01 g/mol
    (d) 17.04 g/mol
    (e) 34.08 g/mol
    (f) 92.02 g/mol
    (g) 67.45 g/mol

16. Molar mass converts between mass and amount in moles. Using the mole ratio from a balanced chemical equation, it is then possible to determine the masses of reactants and products.

17. (a) $n_{C_8H_{18}} = 50 \text{ g} \times \dfrac{1 \text{ mol}}{114.26 \text{ g}} = 0.44 \text{ mol}$

    (b) $n_{CH_3OH} = 70.0 \text{ g} \times \dfrac{1 \text{ mol}}{32.05 \text{ g}} = 2.18 \text{ mol}$

(c) $n_{Cl_2} = 500 \text{ mg} \times \dfrac{1 \text{ mol}}{70.90 \text{ g}} = 7.05 \text{ mmol}$

18. (a) $m_{CCl_3F_3} = 2.50 \text{ mol} \times \dfrac{120.91 \text{ g}}{1 \text{ mol}} = 302 \text{ g}$

(b) $m_{Rn} = 15 \text{ mmol} \times \dfrac{222 \text{ g}}{1 \text{ mol}} = 3.3 \text{ g}$

(c) $m_{O_2} = 20.0 \text{ kmol} \times \dfrac{32.00 \text{ g}}{1 \text{ mol}} = 640 \text{ kg}$

19. $n_{CO_2} = 0.13 \text{ g} \times \dfrac{1 \text{ mol}}{44.01 \text{ g}} = 0.0030 \text{ mol}$

$v_{CO_2} = 0.0030 \text{ mol} \times \dfrac{24.8 \text{ L}}{1 \text{ mol}} = 73 \text{ mL}$

20. $n_{NO_2} = 1.00 \text{ Mg} \times \dfrac{1 \text{ mol}}{46.01 \text{ g}} = 0.0217 \text{ Mmol}$

$v_{NO_2} = 0.0217 \text{ Mmol} \times \dfrac{24.8 \text{ L}}{1 \text{ mol}} = 0.539 \text{ ML}$

21. $n_{CO_2} = 1.00 \text{ Mg} \times \dfrac{1 \text{ mol}}{44.01 \text{ g}} = 0.0227 \text{ Mmol}$

$v_{CO_2} = 0.0227 \text{ Mmol} \times \dfrac{22.4 \text{ L}}{1 \text{ mol}} = 0.509 \text{ ML}$

22. $n_{O_2} = 1.9 \text{ kL} \times \dfrac{1 \text{ mol}}{24.8 \text{ L}} = 0.077 \text{ kmol}$

$m_{O_2} = 0.077 \text{ kmol} \times \dfrac{32.00 \text{ g}}{1 \text{ mol}} = 2.5 \text{ kg}$

23. $n_{H_2O} = 1.00 \text{ L} \times \dfrac{1 \text{ mol}}{24.8 \text{ L}} = 0.0403 \text{ mol}$

$m_{H_2O} = 0.0403 \text{ mol} \times \dfrac{18.02 \text{ g}}{1 \text{ mol}} = 0.727 \text{ g}$

Exercise (page 159)

24. $pv = nRT$

$n_{CH_4} = \dfrac{pv}{RT} = \dfrac{210 \text{ kPa} \times 500 \text{ mL}}{8.31 \text{ kPa}\cdot\text{L}/(\text{mol}\cdot\text{K}) \times 308 \text{ K}} = 41 \text{ mmol}$

25. $pv = nRT$

$p = \dfrac{nRT}{v} = \dfrac{30 \text{ mol} \times 8.31 \text{ kPa}\cdot\text{L}/(\text{mol}\cdot\text{K}) \times 313 \text{ K}}{50 \text{ L}}$

$p_{air} = 1.6 \text{ MPa}$

26. $n_{O_2} = 50 \text{ kg} \times \dfrac{1 \text{ mol}}{32.00 \text{ g}} = 1.6 \text{ kmol}$

$pv = nRT$

$$V = \frac{nRT}{p} = \frac{1.6 \text{ kmol} \times 8.31 \text{ kPa•L/(mol•K)} \times 398 \text{ K}}{150 \text{ kPa}}$$

$V_{O_2} = 34$ kL

27. $n_{NH_3} = 10.5 \text{ g} \times \dfrac{1 \text{ mol}}{17.04 \text{ g}} = 0.616$ mol

    $pv = nRT$

    $T_{NH_3} = \dfrac{pv}{nR} = \dfrac{85.0 \text{ kPa} \times 30.0 \text{ L}}{0.616 \text{ mol} \times 8.31 \text{ kPa•L/(mol•K)}} = 498$ K or 225°C

28. $pv = nRT$ and $n = \dfrac{m}{M}$

    Therefore, $pv = \dfrac{mRT}{M}$

    and $M = \dfrac{mRT}{pv}$

29. $M = \dfrac{mRT}{pv} = \dfrac{1.25 \text{ g} \times 8.31 \text{ kPa•L/(mol•K)} \times 273 \text{ K}}{100 \text{ kPa} \times 1.00 \text{ L}}$

    $M_{gas} = 28.4$ g/mol

30. React a known mass of zinc with an excess of hydrochloric acid and collect the hydrogen gas by the displacement of water in a graduated cylinder.

INVESTIGATION 6.1

Determining the Molar Mass of a Gas (page 160)

Problem

What is the molar mass of butane?

Prediction

According to the chemical formula, $C_4H_{10(g)}$, the molar mass is 58.14 g/mol.

Evidence

initial mass of butane lighter = 18.41 g
final mass of butane lighter = 16.60 g
volume of gas collected = 830 mL
ambient temperature = 24.6°C
ambient pressure (corrected) = 93.6 kPa

Analysis

$pv = nRT$

$n_{C_4H_{10}} = \dfrac{93.6 \text{ kPa} \times 830 \text{ mL}}{8.31 \text{ kPa•L/(mol•K)} \times 297.6 \text{ K}} = 31.4$ mmol

$M_{C_4H_{10}} = \dfrac{(18.41 - 16.60) \text{ g}}{0.0314 \text{ mol}} = 57.6$ g/mol

According to the evidence collected in this experiment, the molar mass of butane is 57.6 g/mol.

Evaluation

The experimental design involved the collection of butane gas and the measurement of the mass of butane liquid. This design is judged to be adequate to answer the problem. There are no obvious flaws, and there seem to be no better alternatives. I am quite confident in this design.

The procedure was adequate, although additional trials that provide more evidence would definitely increase the certainty of the answer. The procedure was modified using a one-litre graduated cylinder, and a sink instead of a large bucket. This change is not expected to make a significant difference. The measurements of the volume (steps 4 and 5) and the final dry mass of the butane lighter (step 7) could have significantly affected the results. It was difficult to ascertain that the lighter was perfectly dry.

Technological skills were relatively simple and adequate. After an initial unsuccessful attempt to invert the full cylinder of water, plastic wrap was used to cover the open end of the cylinder to successfully invert it.

Based upon my evaluation, I am moderately certain of the results. The major sources of uncertainty are measurements and the dryness of the lighter. These uncertainties might affect the answer by approximately 5%.

The accuracy of the prediction is shown by the calculation below.

$$\% \text{ difference} = \frac{|57.6 \text{ g/mol} - 58.14 \text{ g/mol}|}{58.14 \text{ g/mol}} \times 100 = 0.9\%$$

The prediction is judged to be verified since the experimental value agrees very well with the predicted value. The small percent difference of 0.9% can easily be accounted for by the sources of uncertainty in this experiment. The chemical formula is judged to be acceptable since the prediction was verified. I am quite confident in this judgment.

Overview (page 161)

Review

1. (a) The volume of a gas sample varies inversely with the pressure.
   (b) The volume of a gas sample varies directly with the temperature.
   (c) The volume of a gas sample varies directly with the temperature and inversely with the pressure.
2. The chemical properties vary considerably from the unreactive noble gases to the very reactive halogens. All gases have very similar physical properties.
3. The product of the pressure and volume of a gas is equal to the product of the amount of gas, universal gas constant, and absolute temperature.
4. The behavior of gases becomes similar to an ideal gas as temperature increases and pressure decreases.
5. A law is empirical, for example, the volume-temperature relationship is a law. A theory is based on non-observable ideas such as the random, colliding motion of molecules.
6. Avogadro's idea is theoretical since it is based on a non-observable concept. Molecules in a gas cannot be seen or counted directly.
7. (a) 44.02 g/mol
   (b) 44.11 g/mol
8. (a) $n_{Ne} = 14 \text{ g} \times \dfrac{1 \text{ mol}}{20.18 \text{ g}} = 0.69 \text{ mol}$

(b) $n_{UF_6} = 598 \text{ mg} \times \dfrac{1 \text{ mol}}{352.03 \text{ g}} = 1.70 \text{ mol}$

(c) $n_{SO_2} = 29.8 \text{ kg} \times \dfrac{1 \text{ mol}}{64.06 \text{ g}} = 0.465 \text{ kmol}$

9. (a) $m_{Br_2} = 26 \text{ mol} \times \dfrac{159.80 \text{ g}}{1 \text{ mol}} = 4.2 \text{ kg}$

(b) $m_{Kr} = 8.34 \text{ μmol} \times \dfrac{83.80 \text{ g}}{1 \text{ mol}} = 699 \text{ μg}$

(c) $m_{SO_3} = 2.7 \text{ kmol} \times \dfrac{80.06 \text{ g}}{1 \text{ mol}} = 0.22 \text{ Mg}$

10. (a) $n_{CO} = 5.1 \text{ L} \times \dfrac{1 \text{ mol}}{24.8 \text{ L}} = 0.21 \text{ mol}$

(b) $n_{F_2} = 20.7 \text{ mL} \times \dfrac{1 \text{ mol}}{22.4 \text{ L}} = 0.924 \text{ mmol}$

(c) $n_{NO_2} = 90 \text{ kL} \times \dfrac{1 \text{ mol}}{24.8 \text{ L}} = 3.6 \text{ kmol}$

11. (a) $v_{H_2} = 500 \text{ mol} \times \dfrac{24.8 \text{ L}}{1 \text{ mol}} = 12.4 \text{ kL}$

(b) $v_{H_2S} = 56 \text{ kmol} \times \dfrac{24.8 \text{ L}}{1 \text{ mol}} = 1.4 \text{ ML}$

Applications

12. $n_{Ar} = 4.2 \text{ kg} \times \dfrac{1 \text{ mol}}{39.95 \text{ g}} = 0.11 \text{ kmol}$

$v_{Ar} = 0.11 \text{ kmol} \times \dfrac{24.8 \text{ L}}{1 \text{ mol}} = 2.6 \text{ kL}$

13. $\dfrac{p_1 v_1}{T_1} = \dfrac{p_2 v_2}{T_2}$

$\dfrac{500 \text{ mL} \times 1.50 \text{ atm}}{(273 + 24) \text{ K}} = \dfrac{250 \text{ mL} \times 3.50 \text{ atm}}{T_2}$

$T_2 = \dfrac{250 \text{ mL} \times 3.50 \text{ atm} \times 297 \text{ K}}{500 \text{ mL} \times 1.50 \text{ atm}} = 347 \text{ K or } 74°C$

14. A low pressure system refers to an air mass with a pressure lower than normal; a high pressure system refers to higher than normal atmospheric air pressure.

15. (a) $p_1 v_1 = p_2 v_2$

$v_2 = 300 \text{ L} \times \dfrac{p_1}{p_2} = 300 \text{ L} \times \dfrac{1}{2} = 150 \text{ L}$

(b) $\dfrac{v_1}{T_1} = \dfrac{v_2}{T_2}$

$v_2 = 300 \text{ L} \times \dfrac{333 \text{ K}}{303 \text{ K}} = 330 \text{ L}$

(c) $pv = nRT$

$v_{CO_2} = \dfrac{1 \text{ mol} \times 8.31 \text{ kPa·L/(mol·K)} \times 295 \text{ K}}{94.0 \text{ kPa}} = 26.1 \text{ L}$

$V_{CO_2} = 26.1 \text{ L/mol}$

(d) A measured volume of a cold soft drink will be gently warmed to drive off the carbon dioxide gas that is collected by the downward displacement of water.
Alternative design: Limewater will be added to the soft drink until no further precipitation occurs. The mixture will be filtered and the mass of precipitate measured.

16. $p_1v_1 = p_2v_2$

$v_2 = 28.8 \text{ L} \times \dfrac{100 \text{ kPa}}{350 \text{ kPa}} = 8.23 \text{ L}$

17. $\dfrac{p_1v_1}{T_1} = \dfrac{p_2v_2}{T_2}$

$p_2 = \dfrac{600 \text{ kPa} \times 10.0 \text{ kL} \times 383 \text{ K}}{18.0 \text{ kL} \times 423 \text{ K}} = 302 \text{ kPa}$

18. $\dfrac{p_1}{T_1} = \dfrac{p_2}{T_2}$

$p_2 = p_1 \times \dfrac{473 \text{ K}}{373 \text{ K}} = 1.27 \, p_1$

Therefore, a 27% increase in pressure is obtained.

19. (a) 6 mol of $O_{2(g)}$ is consumed for each 6 mol of $CO_{2(g)}$ and 6 mol of $H_2O_{(g)}$ produced. Therefore, the ratio of gas consumed to gas produced is 6:12 or 1:2.

    (b) Assuming sufficient air (oxygen) is available, the first reaction produces greater leavening because 12 volumes of gas are produced, compared with 4 volumes in the second reaction.

20. $v_{CO_2} = 1.0 \text{ g} \times \dfrac{1 \text{ mol}}{44.01 \text{ g}} \times \dfrac{24.8 \text{ L}}{1 \text{ mol}} = 0.56 \text{ L}$

21. $n_{H_2O} = 1.0 \text{ g} \times \dfrac{1 \text{ mol}}{18.02 \text{ g}} = 0.055 \text{ mol}$

    $pv = nRT$

    $v_{H_2O} = \dfrac{0.055 \text{ mol} \times 8.31 \text{ kPa·L/(mol·K)} \times 371 \text{ K}}{103 \text{ kPa}} = 1.7 \text{ L}$

22. $pv = nRT$

$$V_{Cl_2} = \frac{26.5 \text{ kmol} \times 8.31 \text{ kPa·L/(mol·K)} \times 308 \text{ K}}{400 \text{ kPa}} = 170 \text{ kL}$$

23. $pv = nRT$

$$n_{Br_2} = \frac{60 \text{ kPa} \times 18.8 \text{ L}}{8.31 \text{ kPa·L/(mol·K)} \times 413 \text{ K}} = 0.33 \text{ mol}$$

24. (a) $pv = nRT$

$$n_{gas} = \frac{102.2 \text{ kPa} \times 1.25 \text{ L}}{8.31 \text{ kPa·L/(mol·K)} \times 296.4 \text{ K}} = 0.0519 \text{ mol}$$

$$M_{gas} = \frac{9.31 \text{ g} - 7.02 \text{ g}}{0.0519 \text{ mol}} = 44.2 \text{ g/mol}$$

(b) The gas may be $CO_{2(g)}$ (44.01 g/mol), but this is not very certain since other possibilities such as $N_2O_{(g)}$ exist. A diagnostic test would increase the certainty.

25. $pv = nRT$

$$V_{CO_2} = \frac{1 \text{ mol} \times 8.31 \text{ kPa·L/(mol·K)} \times 1073 \text{ K}}{7500 \text{ kPa}} = 1.19 \text{ L}$$

$V_{CO_2} = 1.19$ L/mol

Alternatively,

$$V_{CO_2} = 24.8 \text{ L/mol} \times \frac{101.3 \text{ kPa}}{7500 \text{ kPa}} \times \frac{1073 \text{ K}}{298 \text{ K}} = 1.21 \text{ L/mol}$$

The slight difference in the two answers are the result of the use of different, rounded constants in the calculation. Either answer is acceptable.

Extensions

26. *Answers will vary. A sample is provided below.*

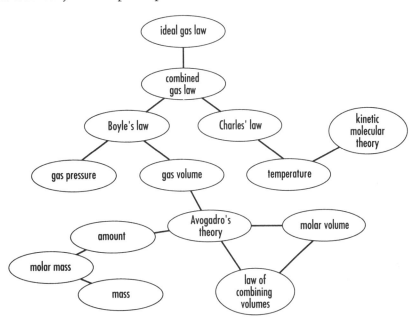

27. When a gas expands, it is doing work. The energy for this work comes from the kinetic energy of the molecules of the gas. The molecules' loss of kinetic energy is reflected by a drop in the temperature of the gas. Conversely, when a gas is compressed, work is done on it. Some of the work goes into increasing the kinetic energy of the gas molecules, which is reflected by an increase in the temperature of the gas.

28. $$\frac{p_1 v_1}{T_1} = \frac{p_2 v_2}{T_2}$$

$$v_2 = \frac{1.00 \text{ m}^3 \times 102 \text{ kPa} \times 285 \text{ K}}{96 \text{ kPa} \times 250 \text{ K}} = 1.21 \text{ m}^3$$

29. Combustion products from automobiles and industrial processes are warmer (and consequently lighter) than the surrounding air. Normally, these combustion products rise, carrying the polluted air away from ground level, and drawing fresh air in to replace it. When a temperature inversion occurs, warm air at higher altitudes moves on top of colder air on the ground. Under these conditions the warm polluted air is not able to rise through the warm air aloft, so it remains trapped under a layer of warm air.

30.

    (a) At high pressures, a real gas has a larger volume than an ideal gas. This is explained by the real volume occupied by the molecules of the gas.
    (b) At low temperatures, a real gas has a smaller volume than an ideal gas. This is explained by the intermolecular (van der Waals) forces between the molecules of the gas.
    Have students with graphing calculators input van der Waals equation of state for a real gas and van der Waals constants to plot actual graphs.

    $$(p + \frac{n^2 a}{v^2})(V - nb) = nRT$$

    For nitrogen, a = 1.390 L^2•atm/mol^2, b = 0.039 13 L/mol.

31. *The following are some possible designs for testing gas laws:*
    Boyle's law: A pressure gauge is used to measure the gas pressure in the cylinder of a pump, as the gas is compressed. The change in volume is estimated by measuring the distance the piston has moved into the cylinder. (Remember that the pressure gauge measures how much the pressure is above atmospheric pressure.)
    Charles' law: The piston of a pump is moved halfway down the cylinder and then the valve is closed. The volume of trapped gas is measured when the pump is placed in pails of water at various temperatures. The change in volume is estimated by measuring the distance the piston has moved into the cylinder.

32. *Some topics that could be discussed are: availability of resource and consumer stations, cost, safety, environmental impact, and engine performance. Several perspectives will be required to adequately discuss the benefits and risks. This is a good opportunity to use the Decision-Making Model (Appendix D, page 538). The evaluation (step 5) suggests one method of comparing compressed gases with gasoline.*

    A lot of information is available on methane and propane, particularly from Energy, Mines and Resources Canada. Information on hydrogen is most readily available from popular science magazines.

Lab Exercise 6C
Analyzing Gas Samples (page 163)

Problem
What mass of sulfur dioxide gas is present in a 20.00 L sample of air?

Analysis

$$n_{SO_2} = \frac{pv}{RT} = \frac{100 \text{ kPa} \times 0.26 \text{ L}}{8.31 \text{ kPa} \cdot \text{L}/(\text{mol} \cdot \text{K}) \times 298 \text{ K}} = 0.010 \text{ mol}$$

$$m_{SO_2} = 0.010 \text{ mol} \times \frac{64.06 \text{ g}}{1 \text{ mol}} = 0.67 \text{ g}$$

According to the evidence collected in this investigation, the mass of sulfur dioxide present in the 20.00 L sample of air is 0.67 g.

Chemical Reaction Calculations

Lab Exercise 7A
Chemical Analysis Using a Graph (page 165)

Problem

What mass of lead(II) nitrate is present in 20.0 mL of a solution?

Analysis

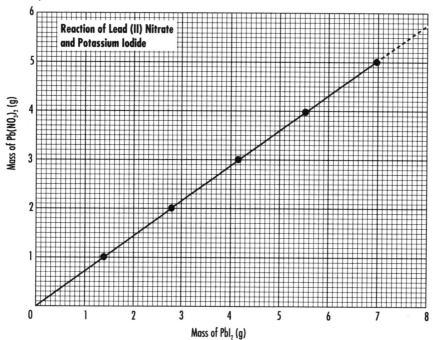

According to the evidence collected and the graph, the mass of lead(II) nitrate in solution 1 is 3.15 g and in solution 2 is 5.40 g.

INVESTIGATION 7.1

Analysis of Sodium Carbonate Solution (page 166)

Problem

What mass of sodium carbonate is present in a 50.0 mL sample of a solution?

Materials

lab apron	wash bottle of distilled water
safety glasses	100 mL graduated cylinder
$Na_2CO_{3(aq)}$	stirring rod
$CaCl_{2(aq)}$	medicine dropper
100 mL beaker	filter paper
250 mL beaker	filter funnel, funnel holder, and stand
400 mL beaker	centigram balance
paper towel	watch glass (optional)

Procedure
1. Measure 50 mL of $Na_2CO_{3(aq)}$ using the graduated cylinder and pour into a clean 250 mL beaker.
2. Slowly add, with stirring, about 50 mL of $CaCl_{2(aq)}$.
3. Add a few extra drops of $CaCl_{2(aq)}$ when the top of the mixture becomes clear.
4. If any cloudiness is visible, add more $CaCl_{2(aq)}$ and then repeat step 3. Repeat until no cloudiness is observed.
5. Measure the mass of a piece of filter paper.
6. Filter the mixture and discard the filtrate in the sink.
7. Dry the filter paper and precipitate overnight on a labelled folded paper towel or watch glass.
8. Measure the mass of filter paper plus dried precipitate.

Evidence
- Clear liquid turned cloudy on adding a few drops of $CaCl_{2(aq)}$.
- Some white precipitate stuck to the sides of the beaker.

volume of $Na_2CO_{3(aq)}$ = 50.0 mL
total volume of $CaCl_{2(aq)}$ added = 100 mL
mass of filter paper = 0.78 g
mass of filter paper and precipitate = 2.07 g

Analysis
mass of $CaCO_{3(s)}$ precipitate = 2.07 g – 0.78 g = 1.29 g

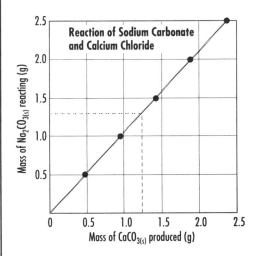

According to the evidence collected and the graph, the mass of $Na_2CO_{3(s)}$ present in the 50.0 mL sample of sodium carbonate solution is estimated to be 1.37 g.

Evaluation

The filtration design is judged to be adequate because the problem was answered with no obvious flaws. This appears to be the best design and I am very confident in it.

The procedure was adequate because sufficient evidence was obtained. However, additional trials would improve the certainty of the answer. Collecting and drying the precipitate, if not done correctly, would have a significant effect on the results.

The technological skills are judged to be adequate because they were easy to perform. It was not possible to remove all of the precipitate from the sides of the beaker.

Overall, I am reasonably certain about my result. The major sources of uncertainty are measurement errors and the precipitate left on the sides of the beaker. I estimate that these uncertainties might account for about 5% difference.

INVESTIGATION 7.2

Decomposing Malachite (page 167)

Problem
How is the mass of copper(II) oxide formed from the decomposition of malachite related to the mass of malachite reacted?

Prediction
My hypothesis is that the law of conservation of mass and the balanced chemical equation can be used to determine the mass of copper(II) oxide that will be formed. According to this hypothesis, the mass of copper(II) oxide will be less than the mass of malachite used. A guess would be one-half the mass of malachite. The reasoning is that if carbon dioxide and water vapor escape, then the mass of what is left will be less. Since the copper(II) oxide makes up one-half of the total products (based on the coefficients), the mass of copper(II) oxide should be one-half as well.

$$Cu(OH)_2 \cdot CuCO_{3(s)} \rightarrow 2CuO_{(s)} + CO_{2(g)} + H_2O_{(g)}$$

Procedure
1. Measure and record the mass of a clean crucible.
2. Transfer a small sample, about one teaspoonful, of malachite into the crucible.
3. Measure and record the total mass.
4. Using a burner adjusted to the hottest flame, heat the sample until the sample is completely black. Use a glass stirring rod to mix the contents and to break up any lumps.
5. Shut off the burner and allow the crucible to cool.
6. Measure and record the final mass of the crucible and contents.

Evidence
- The green powder changed to a black powder.

mass of crucible = 27.24 g
mass of crucible + malachite = 29.36 g
mass of crucible + product = 28.76 g

Analysis

CLASS RESULTS FOR DECOMPOSING MALACHITE	
Mass of Malachite (g)	Mass of Copper(II) Oxide (g)
2.12	1.52
2.65	1.94
1.21	0.85
2.35	1.69
1.75	1.24
2.02	1.41
1.38	0.98
1.15	0.83
1.43	1.03

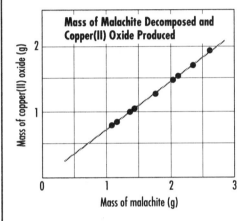

According to the evidence collected and displayed on the graph, the mass of copper(II) oxide varies directly with the mass of malachite. From the slope of the graph, the ratio of the mass of copper(II) oxide to the mass of malachite is found to be approximately 0.72.

Evaluation

The experimental design of decomposition by heating is judged to be adequate to answer the problem. This design seems simple and should give a highly certain result.

The procedure is judged to be adequate as long as the sample was heated carefully until it was completely black. The technological skills were also adequate because the results from several groups, as seen on the graph, were quite consistent.

Based upon my evaluation of this experiment, I am very certain of the results. The major source of uncertainty is the mass measurement and this should only contribute to a difference of about 1%.

The percent difference between the experimental value and the predicted value is 44%, which is significantly higher than the expected 1%.

$$\% \text{ difference} = \frac{|0.72 - 0.50|}{0.50} \times 100 = 44\%$$

> The prediction is judged to be falsified since the predicted answer does not agree with the experimental answer. The hypothesis is unacceptable because the prediction was falsified. The mass of copper(II) oxide product is less than malachite, but not as little as one-half of the initial mass of malachite. Because of the consistency of the class results, I am certain of this judgment.

Exercise (page 168)

1. The purpose of a quantitative analysis is to determine the quantity of a substance in a sample.
2. In quantitative analysis, a graph is first prepared using known values of two substances involved in a reaction. Then a measured value of one substance is located on the line and the value of the other substance is read on the corresponding axis.
3. The manipulated variable is the mass of lead(II) nitrate and the responding variable is the mass of lead(II) iodide.

The answers to questions 4 and 5 will vary slightly since these answers are obtained by reading the graph produced in Investigation 7.2.

4. (a) 0.18 g
   (b) 0.37 g
   (c) 0.55 g
   (d) 0.95 g
5. (a) 0.34 g
   (b) 0.69 g
   (c) 1.04 g
   (d) 1.84 g
6. The ratio of the mass of copper(II) oxide to the mass of malachite is a constant, approximately equal to 0.72.

   $$\frac{0.72}{1} = \frac{m}{400 \text{ g}}$$

   $m_{CuO} = 0.29$ kg

7. | Science | Technology |
   |---|---|
   | international | localized |
   | more theoretical | more empirical |
   | emphasizes ideas | emphasizes methods and materials |
   | natural products and processes | manufactured products and processes |

8. (a) technological
   (b) scientific
   (c) scientific
   (d) technological
   (e) scientific
   (f) technological

Exercise (page 171)

9. $8Zn_{(s)} + S_{8(s)} \rightarrow 8ZnS_{(s)}$
   25 g   m
   65.38 g/mol   256.48 g/mol

   $n_{Zn} = 25 \text{ g} \times \dfrac{1 \text{ mol}}{65.38 \text{ g}} = 0.38 \text{ mol}$

   $n_{S_8} = 0.38 \text{ mol} \times \dfrac{1}{8} = 0.048 \text{ mol}$

   $m_{S_8} = 0.048 \text{ mol} \times \dfrac{256.48 \text{ g}}{1 \text{ mol}} = 12 \text{ g}$

   According to the stoichiometric method, 12 g of sulfur are required to react with 25 g of zinc.

10. $2Al_2O_{3(s)} \rightarrow 4Al_{(s)} + 3O_{2(g)}$
    100 g   m
    101.96 g/mol   26.98 g/mol

    $n_{Al_2O_3} = 100 \text{ g} \times \dfrac{1 \text{ mol}}{101.96 \text{ g}} = 0.981 \text{ mol}$

    $n_{Al} = 0.981 \text{ mol} \times \dfrac{4}{2} = 1.96 \text{ mol}$

    $m_{Al} = 1.96 \text{ mol} \times \dfrac{26.98 \text{ g}}{1 \text{ mol}} = 52.9 \text{ g}$

    According to the stoichiometric method, 52.9 g of aluminum can be produced from 100 g of aluminum oxide.

11. $C_3H_{8(g)} + 5O_{2(g)} \rightarrow 3CO_{2(g)} + 4H_2O_{(g)}$
    10.0 g   m
    44.11 g/mol   32.00 g/mol

    $n_{C_3H_8} = 10.0 \text{ g} \times \dfrac{1 \text{ mol}}{44.11 \text{ g}} = 0.227 \text{ mol}$

    $n_{O_2} = 0.227 \text{ mol} \times \dfrac{5}{1} = 1.13 \text{ mol}$

    $m_{O_2} = 1.13 \text{ mol} \times \dfrac{32.00 \text{ g}}{1 \text{ mol}} = 36.3 \text{ g}$

    According to the stoichiometric method, 36.3 g of oxygen is required to completely burn 10.0 g of propane.

12. $2NaCl_{(aq)} + Pb(NO_3)_{2(aq)} \rightarrow PbCl_{2(s)} + 2NaNO_{3(aq)}$
    2.57 g   m
    58.44 g/mol   278.10 g/mol

    $n_{NaCl} = 2.57 \text{ g} \times \dfrac{1 \text{ mol}}{58.44 \text{ g}} = 0.0440 \text{ mol}$

    $n_{PbCl_2} = 0.0440 \text{ mol} \times \dfrac{1}{2} = 0.0220 \text{ mol}$

    $m_{PbCl_2} = 0.0220 \text{ mol} \times \dfrac{278.10 \text{ g}}{1 \text{ mol}} = 6.11 \text{ g}$

According to the stoichiometric method, 6.11 g of lead(II) chloride precipitate is produced from 2.57 g of sodium chloride.

13. $2Al_{(s)} + 3H_2SO_{4(aq)} \rightarrow 3H_{2(g)} + Al_2(SO_4)_{3(aq)}$
   2.73 g   m
   26.98 g/mol   2.02 g/mol

$n_{Al} = 2.73 \text{ g} \times \dfrac{1 \text{ mol}}{26.98 \text{ g}} = 0.101 \text{ mol}$

$n_{H_2} = 0.101 \text{ mol} \times \dfrac{3}{2} = 0.152 \text{ mol}$

$m_{H_2} = 0.152 \text{ mol} \times \dfrac{2.02 \text{ g}}{1 \text{ mol}} = 0.307 \text{ g}$

According to the stoichiometric method, 0.307 g of hydrogen is produced from 2.73 g of aluminum.

14. $2KOH_{(aq)} + Cu(NO_3)_{2(aq)} \rightarrow Cu(OH)_{2(s)} + 2KNO_{3(aq)}$
    2.67 g   m
    56.11 g/mol   97.57 g/mol

$n_{KOH} = 2.67 \text{ g} \times \dfrac{1 \text{ mol}}{56.11 \text{ g}} = 0.0476 \text{ mol}$

$n_{Cu(OH)_2} = 0.0476 \text{ mol} \times \dfrac{1}{2} = 0.0238 \text{ mol}$

$m_{Cu(OH)_2} = 0.0238 \text{ mol} \times \dfrac{97.57 \text{ g}}{1 \text{ mol}} = 2.32 \text{ g}$

According to the stoichiometric method, 2.32 g of copper(II) hydroxide precipitate is produced from 2.67 g of potassium hydroxide.

Lab Exercise 7B
Testing Gravimetric Stoichiometry (page 172)

Problem

What mass of lead is produced by the reaction of 2.13 g of zinc with an excess of lead(II) nitrate in solution?

Prediction

According to the stoichiometric method, 6.75 g of lead is produced by the reaction of 2.13 g of zinc. The reasoning is shown below.

$Zn_{(s)} + Pb(NO_3)_{2(aq)} \rightarrow Pb_{(s)} + Zn(NO_3)_{2(aq)}$
2.13 g   m
65.38 g/mol   207.20 g/mol

$n_{Zn} = 2.13 \text{ g} \times \dfrac{1 \text{ mol}}{65.38 \text{ g}} = 0.0326 \text{ mol}$

$n_{Pb} = 0.0326 \text{ mol} \times \dfrac{1}{1} = 0.0326 \text{ mol}$

$m_{Pb} = 0.0326 \text{ mol} \times \dfrac{207.20 \text{ g}}{1 \text{ mol}} = 6.75 \text{ g}$

Analysis

mass of lead = 7.60 g − 0.92 g = 6.68 g
According to the evidence collected, the mass of lead precipitate is 6.68 g.

Evaluation

The experimental design of filtration is judged to be adequate since the problem was answered. There seem to be no better alternative designs. I am quite confident in this design because there are no apparent flaws. The major sources of uncertainty are the measurements and the purity or dryness of the precipitate. Since these uncertainties should be small, probably less than 5%, I am very certain of the experimental results.

The percent difference between the experimental result and the predicted value is:

$$\% \text{ difference} = \frac{|6.68 \text{ g} - 6.75 \text{ g}|}{6.75 \text{ g}} \times 100 = 1\%$$

The percent difference was found to be a low of 1% and is well within the estimated total percent uncertainty. Therefore, the prediction is judged to be verified since the small difference can easily be accounted for by the sources of uncertainty. The method of gravimetric stoichiometry is acceptable because the prediction was verified.

Lab Exercise 7C
Designing and Testing Stoichiometry (page 172)

Problem

What is the mass of precipitate formed when 2.98 g of sodium phosphate in solution reacts with an excess of aqueous calcium nitrate?

Prediction

According to the stoichiometric method, 2.82 g of precipitate is formed when 2.98 g of sodium phosphate in solution reacts with an excess of aqueous calcium nitrate. This prediction is based on the following calculations.

$2Na_3PO_{4(aq)} + 3Ca(NO_3)_{2(aq)} \rightarrow Ca_3(PO_4)_{2(s)} + 6NaNO_{3(aq)}$
2.98 g m
163.94 g/mol   310.18 g/mol

$$n_{Na_3PO_4} = 2.98 \text{ g} \times \frac{1 \text{ mol}}{163.94 \text{ g}} = 0.0182 \text{ mol}$$

$$n_{Ca_3(PO_4)_2} = 0.0182 \text{ mol} \times \frac{1}{2} = 0.00909 \text{ mol}$$

$$m_{Ca_3(PO_4)_2} = 0.00909 \text{ mol} \times \frac{310.18 \text{ g}}{1 \text{ mol}} = 2.82 \text{ g}$$

Experimental Design

Calcium nitrate solution is added to a measured mass of sodium phosphate dissolved in water. Sufficient calcium nitrate solution is added until no further precipitation occurs. The precipitate is then separated by filtration, dried and weighed.

Materials

lab apron	filter paper
safety glasses	filter funnel, funnel holder, and stand
$Na_3PO_{4(s)}$	laboratory scoop
$Ca(NO_3)_{2(aq)}$	centigram balance
100 mL beaker	stirring rod with spatula
250 mL beaker	wash bottle of distilled water
medicine dropper	

Procedure

1. Measure 2.98 g of $Na_3PO_{4(s)}$ in a clean, dry 250 mL beaker.
2. Dissolve the solid in about 50 mL of distilled water.
3. Slowly add about 50 mL of $Ca(NO_3)_{2(aq)}$ with stirring.
4. Let the precipitate settle until a clear layer is seen at the top.
5. Add a few drops of $Ca(NO_3)_{2(aq)}$ down the side of the beaker.
6. If any cloudiness is visible, add more $Ca(NO_3)_{2(aq)}$ and repeat steps 4 and 5 until no further precipitation occurs.
7. Measure the mass of a piece of filter paper.
8. Set up the filtration apparatus and filter the mixture.
9. Dry the filter paper and precipitate.
10. Measure the mass of dried filter paper and precipitate.
11. Dispose of the solutions in the sink and the solids in the waste basket.

Analysis

mass of precipitate = 3.82 g – 0.93 g = 2.89 g
According to the evidence collected, the mass of precipitate is 2.89 g.

Evaluation

The experimental design of filtration is judged to be adequate to answer the problem with no obvious flaws. An alternative to part of the design would be to test the filtrate for excess calcium nitrate. However, this might improve the efficiency, but not the results. I am confident in this design since it appears to be the best available. The major uncertainties are mass measurements, washing of the precipitate to remove all traces of solution, and the drying of the precipitate. These uncertainties might affect the results by 5%. Overall, I am very certain of the experimental results.

The percent difference is only 2%, and is less than the estimated difference of 5%. Since the predicted answer agrees with the experimental answer and the difference can be accounted for by the uncertainties discussed, the prediction is judged to be verified. The method of stoichiometry is acceptable since the prediction was verified.

INVESTIGATION 7.3

Testing the Stoichiometric Method (page 174)

Problem

What is the mass of precipitate produced by the reaction of 2.00 g of strontium nitrate in solution with excess (2.56 g) copper(II) sulfate in solution?

Prediction

According to the stoichiometric method, 1.74 g of strontium sulfate precipitate will be produced from the reaction of 2.00 g of strontium nitrate with excess copper(II) sulfate. The reasoning is shown below.

$$Sr(NO_3)_{2(aq)} + CuSO_{4(aq)} \rightarrow SrSO_{4(s)} + Cu(NO_3)_{2(aq)}$$
2.00 g　　　　　　　　　　　　　　m
211.64 g/mol　　　　　　　　　　183.68 g/mol

$$n_{Sr(NO_3)_2} = 2.00 \text{ g} \times \frac{1 \text{ mol}}{211.64 \text{ g}} = 0.009\,45 \text{ mol}$$

$$n_{SrSO_4} = 0.009\,45 \text{ mol} \times \frac{1}{1} = 0.009\,45 \text{ mol}$$

$$m_{SrSO_4} = 0.009\,45 \text{ mol} \times \frac{183.68 \text{ g}}{1 \text{ mol}} = 1.74 \text{ g}$$

Experimental Design

Strontium nitrate is dissolved in water and reacted with a solution containing excess copper(II) sulfate. The precipitate is separated by filtration and then dried and weighed.

Materials

lab apron	wash bottle of distilled water
safety glasses	laboratory scoop
$Sr(NO_3)_{2(s)}$	centigram balance
$CuSO_4 \cdot 5H_2O_{(s)}$	stirring rod
150 mL beaker	filter paper
250 mL beaker	filter funnel, funnel holder, and stand
400 mL beaker	watch glass

Procedure

1. Obtain 2.56 g of $CuSO_4 \cdot 5H_2O_{(s)}$ in a clean, dry 150 mL beaker.
2. Dissolve the solid in approximately 50 mL of distilled water.
3. Obtain 2.00 g of $Sr(NO_3)_{2(s)}$ in a clean, dry 250 mL beaker.
4. Dissolve the solid in approximately 50 mL of distilled water.
5. While stirring, slowly pour the $CuSO_{4(aq)}$ into the $Sr(NO_3)_{2(aq)}$ and record the observations.
6. Measure and record the mass of a piece of filter paper.
7. Filter the mixture.
8. Set the filter paper and precipitate aside to dry overnight.
9. Measure and record the mass of filter paper and precipitate.

Evidence

mass of 250 mL beaker = 102.41 g
mass of 250 mL beaker + $Sr(NO_3)_{2(s)}$ = 104.41 g
mass of 150 mL beaker = 68.83 g
mass of 150 mL beaker + $CuSO_4 \cdot 5H_2O_{(s)}$ = 71.39 g
mass of filter paper = 0.90 g
mass of filter paper and precipitate = 2.66 g
- The blue copper(II) sulfate solution mixed with the colorless strontium nitrate solution to produce a white precipitate and a blue solution.

> *Analysis*
> mass of precipitate = 2.66 g – 0.90 g = 1.76 g
> According to the evidence collected in this experiment, the mass of the strontium sulfate precipitate from the reaction of strontium nitrate and copper(II) sulfate solutions was determined to be 1.76 g.
>
> *Evaluation*
> The filtration design is adequate to answer the problem and I am confident that this design is the best. The procedure was also adequate because sufficient evidence was collected, and there are no obvious improvements to be made. Technological skills were adequate as long as care was taken in filtering and washing the precipitate.
> Overall, I am very certain of my results. The major uncertainty is likely caused by the number of mass measurements made. This might account for a 5% difference.
> The percent difference was found to be 1%, which is well within the estimated 5% uncertainty.
>
> $$\% \text{ difference} = \frac{|1.76 \text{ g} - 1.74 \text{ g}|}{1.74 \text{ g}} \times 100 = 1\%$$
>
> Based on the very low percent difference, the prediction is judged to be verified. The stoichiometric method is acceptable because the prediction was verified. I am quite confident in this judgment.

Exercise (page 174)

15. $Pb(NO_3)_{2(aq)} + 2KBr_{(aq)} \rightarrow PbBr_{2(s)} + 2KNO_{3(aq)}$

    m 3.65 g

    331.22 g/mol   367.00 g/mol

    $n_{PbBr_2} = 3.65 \text{ g} \times \dfrac{1 \text{ mol}}{367.00 \text{ g}} = 0.009\,95 \text{ mol}$

    $n_{Pb(NO_3)_2} = 0.009\,95 \text{ mol} \times \dfrac{1}{1} = 0.009\,95 \text{ mol}$

    $m_{Pb(NO_3)_2} = 0.009\,95 \text{ mol} \times \dfrac{331.22 \text{ g}}{1 \text{ mol}} = 3.29 \text{ g}$

    Using evidence and the method of stoichiometry, the mass of lead(II) nitrate in the sample is 3.29 g.

16. $2AgNO_{3(aq)} + Na_2CrO_{4(aq)} \rightarrow Ag_2CrO_{4(s)} + 2NaNO_{3(aq)}$

    m 2.89 g

    169.88 g/mol   331.74 g/mol

    $n_{Ag_2CrO_4} = 2.89 \text{ g} \times \dfrac{1 \text{ mol}}{331.74 \text{ g}} = 0.008\,71 \text{ mol}$

    $n_{AgNO_3} = 0.008\,71 \text{ mol} \times \dfrac{2}{1} = 0.0174 \text{ mol}$

    $m_{AgNO_3} = 0.0174 \text{ mol} \times \dfrac{169.88 \text{ g}}{1 \text{ mol}} = 2.96 \text{ g}$

    According to the methods of filtration and stoichiometry, 2.96 g of silver nitrate was present in the original solution.

17. mass of zinc reacted = 50.0 g − 38.5 g = 11.5 g

$2CrCl_{3(aq)} + 3Zn_{(s)} \rightarrow 2Cr_{(s)} + 3ZnCl_{2(aq)}$
m   11.5 g
158.35 g/mol 65.38 g/mol

$n_{Zn} = 11.5 \text{ g} \times \dfrac{1 \text{ mol}}{65.38 \text{ g}} = 0.176 \text{ mol}$

$n_{CrCl_3} = 0.176 \text{ mol} \times \dfrac{2}{3} = 0.117 \text{ mol}$

$m_{CrCl_3} = 0.117 \text{ mol} \times \dfrac{158.35 \text{ g}}{1 \text{ mol}} = 18.6 \text{ g}$

The sample contained 18.6 g of chromium(III) chloride according to the method of stoichiometry.

Lab Exercise 7D
Testing a Chemical Process (page 175)

Problem

What is the mass of sodium silicate in a 25.0 mL sample of the solution used in a chemical process?

Analysis

mass of precipitate = 9.45 g − 0.98 g = 8.47 g

$3Na_2SiO_{3(aq)} + 2Fe(NO_3)_{3(aq)} \rightarrow Fe_2(SiO_3)_{3(s)} + 6NaNO_{3(aq)}$
m   8.47 g
122.07 g/mol   339.97 g/mol

$n_{Fe_2(SiO_3)_3} = 8.47 \text{ g} \times \dfrac{1 \text{ mol}}{339.97 \text{ g}} = 0.0249 \text{ mol}$

$n_{Na_2SiO_3} = 0.0249 \text{ mol} \times \dfrac{3}{1} = 0.0747 \text{ mol}$

$m_{Na_2SiO_3} = 0.0747 \text{ mol} \times \dfrac{122.07 \text{ g}}{1 \text{ mol}} = 9.12 \text{ g}$

According to the evidence given and the stoichiometric method, the mass of sodium silicate in the sample is 9.12 g.

Evaluation

The experimental design is adequate to answer the problem but it is not the best design available. Using a prepared graph relating the mass of precipitate to the mass of sodium silicate would be a more efficient way of determining the answer. An alternative design is to crystallize the sample solution of sodium silicate by evaporating the solvent to obtain the mass of solid present. I am quite confident in the design used since it does not appear to have any flaws. The color of the filtrate suggests that an excess of iron(III) nitrate was added.

Major uncertainties are the usual ones associated with this design: accuracy of measuring instruments, complete washing of the precipitate, and the purity of the final precipitate. All of these uncertainties might account for a 5% difference.

The percent difference between the experimental result and the average expected result is approximately 42% as shown below.

$$\% \text{ difference} = \frac{|9.12 \text{ g} - 6.45 \text{ g}|}{6.45 \text{ g}} \times 100 = 42\%$$

This percent difference is far greater than would be reasonably expected based on the total percent uncertainty for this type of experiment. Therefore, the prediction is clearly falsified. The process is not operating as expected and it should be revised or adjusted.

Exercise (page 178)

18. $2CH_3OH_{(l)} + 3O_{2(g)} \rightarrow 2CO_{2(g)} + 4H_2O_{(g)}$
    15 g v (SATP)
    32.05 g/mol   24.8 L/mol

    $n_{CH_3OH} = 15 \text{ g} \times \dfrac{1 \text{ mol}}{32.05 \text{ g}} = 0.47 \text{ mol}$

    $n_{O_2} = 0.47 \text{ mol} \times \dfrac{3}{2} = 0.70 \text{ mol}$

    $v_{O_2} = 0.70 \text{ mol} \times \dfrac{24.8 \text{ L}}{1 \text{ mol}} = 17 \text{ L}$

    According to the method of stoichiometry, 17 L of oxygen at SATP is needed to completely burn 15 g of methanol.

19. $C_3H_{8(g)} + 5O_{2(g)} \rightarrow 3CO_{2(g)} + 4H_2O_{(g)}$
    m 50 L (SATP)
    44.11 g/mol   24.8 L/mol

    $n_{O_2} = 50 \text{ L} \times \dfrac{1 \text{ mol}}{24.8 \text{ L}} = 2.0 \text{ mol}$

    $n_{C_3H_8} = 2.0 \text{ mol} \times \dfrac{1}{5} = 0.40 \text{ mol}$

    $m_{C_3H_8} = 0.40 \text{ mol} \times \dfrac{44.11 \text{ g}}{1 \text{ mol}} = 18 \text{ g}$

    According to the method of stoichiometry, 18 g of propane can be burned using 50 L of oxygen gas at SATP.

20. $2H_{2(g)} + O_{2(g)} \rightarrow 2H_2O_{(g)}$
    v 300 L
    40°C, 150 kPa   40°C, 150 kPa

    $v_{H_2} = 300 \text{ L} \times \dfrac{2}{1} = 600 \text{ L}$

    According to the law of combining volumes, 600 L of hydrogen can be burned using 300 L of oxygen.

21. $2NaCl_{(l)} \rightarrow 2Na_{(s)} + Cl_{2(g)}$
    100 kg v (SATP)
    22.99 g/mol   24.8 L/mol

$$n_{Na} = 100 \text{ kg} \times \frac{1 \text{ mol}}{22.99 \text{ g}} = 4.35 \text{ kmol}$$

$$n_{Cl_2} = 4.35 \text{ kmol} \times \frac{1}{2} = 2.17 \text{ kmol}$$

$$v_{Cl_2} = 2.17 \text{ kmol} \times \frac{24.8 \text{ L}}{1 \text{ mol}} = 53.9 \text{ kL}$$

According to the method of stoichiometry, 53.9 kL of chlorine at SATP is produced along with 100 kg of sodium metal.

22. $CH_{4(g)} + 2O_{2(g)} \rightarrow CO_{2(g)} + 2H_2O_{(g)}$
    2.00 ML   v (SATP)
    0°C, 120 kPa   24.8 L/mol

$$n_{CH_4} = \frac{pv}{RT} = \frac{120 \text{ kPa} \times 2.00 \text{ ML}}{8.31 \text{ kPa·L/(mol·K)} \times 273 \text{ K}} = 0.106 \text{ Mmol}$$

$$n_{O_2} = 0.106 \text{ Mmol} \times \frac{2}{1} = 0.212 \text{ Mmol}$$

$$v_{O_2} = 0.212 \text{ Mmol} \times \frac{24.8 \text{ L}}{1 \text{ mol}} = 5.25 \text{ ML}$$

According to the stoichiometric method, 5.25 ML of oxygen at SATP is required to burn 2.00 ML of methane at 0°C and 120 kPa.

INVESTIGATION 7.4

The Universal Gas Constant (page 178)

Problem

What is the experimental value of the universal gas constant?

Prediction

According to the data table on the inside back cover of the student book, the value of the universal gas constant is 8.31 kPa·L/(mol·K).

Evidence

mass of magnesium = 0.08 g
volume of gas = 87.5 mL
ambient temperature = 26.8°C
ambient pressure = 93.6 kPa

Analysis

$Mg_{(s)} + 2HCl_{(aq)} \rightarrow H_{2(g)} + MgCl_{2(aq)}$
0.08 g   87.5 mL
24.31 g/mol   26.8°C, 93.6 kPa

$$n_{Mg} = 0.08 \text{ g} \times \frac{1 \text{ mol}}{24.31 \text{ g}} = 3 \text{ mmol}$$

$$n_{H_2} = 3 \text{ mmol} \times \frac{1}{1} = 3 \text{ mmol}$$

$$pv = nRT$$

$$R = \frac{pv}{nT} = \frac{93.6 \text{ kPa} \times 87.5 \text{ mL}}{3 \text{ mmol} \times 299.8 \text{ K}} = 8 \text{ kPa}\cdot\text{L}/(\text{mol}\cdot\text{K})$$

According to the evidence gathered, the experimental value of the universal gas constant is 8 kPa·L/(mol·K).

Evaluation

The experimental design involving a reaction to produce a gas and the collection of the gas is judged to be adequate to answer the problem. A better design would be to use a known mass of gas (for example, from a butane lighter, Investigation 6.1). This would eliminate uncertainties associated with the purity of the magnesium. I am only moderately confident in the design that was used.

The procedure used is judged to be inadequate for several reasons. The magnesium should be cleaned and its purity should be known. A larger graduated cylinder is required so that a greater mass of magnesium can be used. One significant digit for the mass of magnesium recorded is inadequate. You also need to be sure that the copper wire does not affect the results. The technological skills were adequate for the procedure used.

Based upon this evaluation, I am only moderately certain of the results. The major sources of uncertainty are related to the measurements made (especially the very small mass of magnesium, the oxide coating on the magnesium, and the purity of the magnesium). It is not possible to estimate how much the total of these uncertainties would be.

The percent difference was found to be 0.1%, which is quite remarkable considering the possible uncertainties discussed above.

$$\% \text{ difference} = \frac{|8 \text{ kPa}\cdot\text{L}/(\text{mol}\cdot\text{K}) - 8.31 \text{ kPa}\cdot\text{L}/(\text{mol}\cdot\text{K})|}{8.31 \text{ kPa}\cdot\text{L}/(\text{mol}\cdot\text{K})} \times 100 = 0.1\%$$

The prediction is judged to be verified since there is a close agreement between the predicted answer and the experimental answer. The difference can easily be accounted for by the many uncertainties in this experiment. As a result, the value of the gas constant is considered acceptable.

Exercise (page 181)

23. $H_2SO_{4(aq)} + 2NH_{3(aq)} \rightarrow (NH_4)_2SO_{4(aq)}$
    50.0 mL   24.4 mL
    C   2.20 mol/L

$$n_{NH_3} = 24.4 \text{ mL} \times \frac{2.20 \text{ mol}}{1 \text{ L}} = 53.7 \text{ mmol}$$

$$n_{H_2SO_4} = 53.7 \text{ mmol} \times \frac{1}{2} = 26.8 \text{ mmol}$$

$$C_{H_2SO_4} = \frac{26.8 \text{ mmol}}{50.0 \text{ mL}} = 0.537 \text{ mol/L}$$

The concentration of sulfuric acid is 0.537 mol/L according to the method of stoichiometry.

24. $3Ca(OH)_{2(aq)} + Al_2(SO_4)_{3(aq)} \rightarrow 3CaSO_{4(s)} + 2Al(OH)_{3(s)}$

   $v 25.0 \text{ mL}$

   $0.0250 \text{ mol/L} \quad 0.125 \text{ mol/L}$

   $n_{Al_2(SO_4)_3} = 25.0 \text{ mL} \times \dfrac{0.125 \text{ mol}}{1 \text{ L}} = 3.13 \text{ mmol}$

   $n_{Ca(OH)_2} = 3.13 \text{ mmol} \times \dfrac{3}{1} = 9.38 \text{ mmol}$

   $v_{Ca(OH)_2} = 9.38 \text{ mmol} \times \dfrac{1 \text{ L}}{0.0250 \text{ mol}} = 375 \text{ mL}$

   According to the stoichiometric method, 375 mL of calcium hydroxide solution is required to react with the aluminum sulfate solution.

25. $3Na_2CO_{3(aq)} + 2FeCl_{3(aq)} \rightarrow Fe_2(CO_3)_{3(s)} + 6NaCl_{(aq)}$

   $v 75.0 \text{ mL}$

   $0.250 \text{ mol/L} \quad 0.200 \text{ mol/L}$

   $n_{FeCl_3} = 75.0 \text{ mL} \times \dfrac{0.200 \text{ mol}}{1 \text{ L}} = 15.0 \text{ mmol}$

   $n_{Na_2CO_3} = 15.0 \text{ mmol} \times \dfrac{3}{2} = 22.5 \text{ mmol}$

   $v_{Na_2CO_3} = 22.5 \text{ mmol} \times \dfrac{1 \text{ L}}{0.250 \text{ mol}} = 90.0 \text{ mL}$

   The minimum volume of sodium carbonate solution needed is 90.0 mL according to the stoichiometric method shown above.

Lab Exercise 7E
Testing Solution Stoichiometry (page 182)

Problem

What mass of precipitate is produced by the reaction of 20.0 mL of 0.210 mol/L sodium sulfide with an excess quantity of aluminum nitrate solution?

Prediction

According to the stoichiometric method, 210 mg of precipitate is produced by the reaction of 20.0 mL of 0.210 mol/L sodium sulfide with excess aluminum nitrate solution. This prediction is based on the following calculation.

$3Na_2S_{(aq)} + 2Al(NO_3)_{3(aq)} \rightarrow Al_2S_{3(s)} + 6NaNO_{3(aq)}$

$20.0 \text{ mL} m$

$0.210 \text{ mol/L} 150.14 \text{ g/mol}$

$n_{Na_2S} = 20.0 \text{ mL} \times \dfrac{0.210 \text{ mol}}{1 \text{ L}} = 4.20 \text{ mmol}$

$n_{Al_2S_3} = 4.20 \text{ mmol} \times \dfrac{1}{3} = 1.40 \text{ mmol}$

$m_{Al_2S_3} = 1.40 \text{ mmol} \times \dfrac{150.14 \text{ g}}{1 \text{ mol}} = 210 \text{ mg or } 0.210 \text{ g}$

Analysis
mass of precipitate = 1.17 g − 0.97 g = 0.20 g
According to the evidence collected, 0.20 g of precipitate was produced.

Evaluation
The experimental design is judged to be adequate because the problem can be answered with no apparent flaws. There does not appear to be a better design that can be used to answer the problem. The positive test for excess aluminum nitrate gives me confidence in this design.

Uncertainty of the accuracy of the measuring instruments and possible uncertainties in the procedure of washing and drying the precipitate would probably affect the results by about 5%. Based upon the evaluation of this experiment, I am very certain of the results.

$$\% \text{ difference} = \frac{|0.20 \text{ g} - 0.210 \text{ g}|}{0.210 \text{ g}} \times 100 = 5\%$$

The percent difference is 5%, the same as the estimated uncertainty. Since the predicted answer closely agrees with the experimental answer and the difference can be accounted for by the uncertainties discussed, the prediction is judged to be verified. The method of stoichiometry used in this experiment is acceptable because the prediction was verified. I am very certain of this judgment.

Lab Exercise 7F
Determining a Solution Concentration (page 182)

Problem
What is the molar concentration of silver nitrate in the solution to be recycled?

Analysis
mass of $Ag_2SO_{4(s)}$ = 6.74 g − 1.27 g = 5.47 g

$2AgNO_{3(aq)} + Na_2SO_{4(aq)} \rightarrow Ag_2SO_{4(s)} + 2NaNO_{3(aq)}$
100 mL   5.47 g
C 311.80 g/mol

$$n_{Ag_2SO_4} = 5.47 \text{ g} \times \frac{1 \text{ mol}}{311.80 \text{ g}} = 0.0175 \text{ mol}$$

$$n_{AgNO_3} = 0.0175 \text{ mol} \times \frac{2}{1} = 0.0351 \text{ mol}$$

$$C_{AgNO_3} = \frac{0.0351 \text{ mol}}{0.100 \text{ L}} = 0.351 \text{ mol/L}$$

According to the evidence, the molar concentration of silver nitrate is 0.351 mol/L.

Lab Exercise 7G
Solution and Gas Stoichiometry (page 183)

Problem

What volume of carbon dioxide gas at 100 kPa and 35°C is produced by the complete reaction of 50 mL of a 0.200 mol/L baking soda solution with excess hydrochloric acid?

Prediction

According to the method of stoichiometry, 0.26 L of carbon dioxide is produced from the reaction of baking soda and hydrochloric acid. The reasoning is shown below.

$NaHCO_{3(aq)} + HCl_{(aq)} \rightarrow CO_{2(g)} + NaCl_{(aq)} + H_2O_{(l)}$
50 mL $\quad\quad\quad\quad\quad\quad\quad\quad$ v
0.200 mol/L $\quad\quad\quad\quad\quad\quad$ 35°C, 100 kPa

$n_{NaHCO_3} = 50 \text{ mL} \times \dfrac{0.200 \text{ mol}}{1 \text{ L}} = 10 \text{ mmol}$

$n_{CO_2} = 10 \text{ mmol} \times \dfrac{1}{1} = 10 \text{ mmol}$

$v_{CO_2} = \dfrac{nRT}{p} = \dfrac{10 \text{ mmol} \times 8.31 \text{ kPa·L/(mol·K)} \times 308 \text{ K}}{100 \text{ kPa}} = 0.26 \text{ L}$

Exercise (page 184)

26. (a) $CaCO_{3(s)} \rightarrow CaO_{(s)} + CO_{2(g)}$
    (b) $CO_{2(g)} + NH_{3(aq)} + H_2O_{(l)} \rightarrow NH_4HCO_{3(aq)}$
    (c) $NH_4HCO_{3(aq)} + NaCl_{(aq)} \rightarrow NH_4Cl_{(aq)} + NaHCO_{3(s)}$
    (d) $2NaHCO_{3(s)} \rightarrow Na_2CO_{3(s)} + H_2O_{(g)} + CO_{2(g)}$
    (e) $CaO_{(s)} + H_2O_{(l)} \rightarrow Ca(OH)_{2(s)}$
    (f) $Ca(OH)_{2(s)} + 2NH_4Cl_{(aq)} \rightarrow 2NH_{3(aq)} + CaCl_{2(aq)} + 2H_2O_{(l)}$
    (g) $CaCO_{3(s)} + 2NaCl_{(aq)} \rightarrow Na_2CO_{3(aq)} + CaCl_{2(aq)}$
27. $CaCO_{3(s)} + 2NaCl_{(aq)} \rightarrow Na_2CO_{3(aq)} + CaCl_{2(aq)}$
28. $CaCO_{3(s)}$, $NaCl_{(s)}$. These materials are obtained from limestone and salt, which occur naturally.
29. The primary product is sodium carbonate (soda ash) and the by-product is calcium chloride.
30. Sodium hydrogen carbonate is an intermediate that can be sold as baking soda. The relative amounts that are processed may need to be adjusted and less primary product may be obtained.
31. Additional natural resources are water and fuel.
32. (Discussion) The use of inexpensive raw materials makes the Solvay process economical. Saleable by-products and intermediates also help.

33. (a) $CaCO_{3(s)} + 2NaCl_{(aq)} \rightarrow Na_2CO_{3(aq)} + CaCl_{2(aq)}$
    10.0 kL   m
    5.40 mol/L   105.99 g/mol

$n_{NaCl} = 10.0 \text{ kL} \times \dfrac{5.40 \text{ mol}}{1 \text{ L}} = 54.0 \text{ kmol}$

$n_{Na_2CO_3} = 54.0 \text{ kmol} \times \dfrac{1}{2} = 27.0 \text{ kmol}$

$m_{Na_2CO_3} = 27.0 \text{ kmol} \times \dfrac{105.99 \text{ g}}{1 \text{ mol}} = 2.86 \text{ Mg}$

According to the method of stoichiometry, 2.86 Mg of soda ash can be produced.

(b) $CaCO_{3(s)} + 2NaCl_{(aq)} \rightarrow Na_2CO_{3(aq)} + CaCl_{2(aq)}$
    500 kg   v
    100.09 g/mol   5.40 mol/L

$n_{CaCO_3} = 500 \text{ kg} \times \dfrac{1 \text{ mol}}{100.09 \text{ g}} = 5.00 \text{ kmol}$

$n_{NaCl} = 5.00 \text{ kmol} \times \dfrac{2}{1} = 9.99 \text{ kmol}$

$v_{NaCl} = 9.99 \text{ kmol} \times \dfrac{1 \text{ L}}{5.40 \text{ mol}} = 1.85 \text{ kL}$

According to the stoichiometric method, 1.85 kL of brine solution is required.

(c) $CaCO_{3(s)} + 2NaCl_{(aq)} \rightarrow Na_2CO_{3(aq)} + CaCl_{2(aq)}$
              350 L   v
              2.00 mol/L   3.00 mol/L

$n_{NaCl} = 350 \text{ L} \times \dfrac{2.00 \text{ mol}}{1 \text{ L}} = 700 \text{ mol}$

$n_{CaCl_2} = 700 \text{ mol} \times \dfrac{1}{2} = 350 \text{ mol}$

$v_{CaCl_2} = 350 \text{ mol} \times \dfrac{1 \text{ L}}{3.00 \text{ mol}} = 117 \text{ L}$

According to the method of stoichiometry, 117 L of calcium chloride solution can be produced.

34. (a) $NH_4HCO_{3(aq)} + NaCl_{(aq)} \rightarrow NH_4Cl_{(aq)} + NaHCO_{3(s)}$
        v   4.00 kL
        0.700 mol/L   3.00 mol/L

$n_{NaCl} = 4.00 \text{ kL} \times \dfrac{3.00 \text{ mol}}{1 \text{ L}} = 12.0 \text{ kmol}$

$n_{NH_4HCO_3} = 12.0 \text{ kmol} \times \dfrac{1}{1} = 12.0 \text{ kmol}$

$v_{NH_4HCO_3} = 12.0 \text{ kmol} \times \dfrac{1 \text{ L}}{0.700 \text{ mol}} = 17.1 \text{ kL}$

According to the method of stoichiometry, 17.1 kL of ammonium hydrogen carbonate solution is required.

(b) (Discussion) In the solubility table, high solubility means a solubility greater than or equal to 0.1 mol/L, but this does not mean that the compound is soluble to an unlimited extent. Under the conditions in the Solvay process, the quantity of sodium hydrogen carbonate produced exceeds its solubility.

35. (Discussion) The primary reason is likely economics. A larger production usually means a lower unit cost.

Exercise (page 187)

36. $n_{CoCl_2 \cdot 2H_2O} = 2.00 \text{ L} \times \dfrac{25.0 \text{ mmol}}{1 \text{ L}} = 50.0 \text{ mmol}$

$m_{CoCl_2 \cdot 2H_2O} = 50.0 \text{ mmol} \times \dfrac{165.87 \text{ g}}{1 \text{ mol}} = 8.29 \text{ g}$

Procedure
1. Obtain 8.29 g of $CoCl_2 \cdot 2H_2O_{(s)}$ in a clean, dry 250 mL beaker.
2. Dissolve the solid in about 100 mL of pure water.
3. Transfer the solution into a 2.00 L volumetric flask.
4. Add pure water until the volume reaches 2.00 L.
5. Stopper the flask and mix the contents thoroughly by repeatedly inverting the flask.

37. $n_{(NH_4)_2OOCCOO} = 100 \text{ mL} \times \dfrac{0.250 \text{ mol}}{1 \text{ L}} = 25.0 \text{ mmol}$

$m_{(NH_4)_2OOCCOO} = 25.0 \text{ mmol} \times \dfrac{124.12 \text{ g}}{1 \text{ mol}} = 3.10 \text{ g}$

38. $n_{KMnO_4} = 500.0 \text{ mL} \times \dfrac{75 \text{ mmol}}{1 \text{ L}} = 0.038 \text{ mol}$

$m_{KMnO_4} = 0.038 \text{ mol} \times \dfrac{158.04 \text{ g}}{1 \text{ mol}} = 5.9 \text{ g}$

39. $n_{NaOH} = 500 \text{ mL} \times \dfrac{10.0 \text{ mol}}{1 \text{ L}} = 5.00 \text{ mol}$

$m_{NaOH} = 5.00 \text{ mol} \times \dfrac{40.00 \text{ g}}{1 \text{ mol}} = 200 \text{ g}$

Exercise (page 189)

40. $v_i C_i = v_f C_f$

$v_i \times \dfrac{5.00 \text{ mol}}{1 \text{ L}} = 2.00 \text{ L} \times \dfrac{0.500 \text{ mol}}{1 \text{ L}}$

$v_{i_{H_2SO_4}} = 200 \text{ mL}$

Procedure
1. Add about 1.00 L of pure water to a clean 2.00 L volumetric flask.
2. Measure 200 mL of the commercial stock solution using a 250 mL graduated cylinder.
3. Transfer the solution into the volumetric flask
4. Add pure water until the final volume is reached.
5. Stopper the flask and mix the solution thoroughly.

41.
$$v_i C_i = v_f C_f$$
$$250 \text{ mL} \times \frac{15.4 \text{ mol}}{1 \text{ L}} = v_f \times \frac{2.5 \text{ mol}}{1 \text{ L}}$$
$$v_f = 1.5 \text{ L}$$

42. (a) HNO_3
$$v_i C_i = v_f C_f$$
$$5.00 \text{ mL} \times \frac{0.0500 \text{ mol}}{1 \text{ L}} = 100.0 \text{ mL} \times C_f$$
$$C_f = 2.50 \text{ mmol/L}$$

(b) $CuSO_4$
$$n_{CuSO_4} = 10.0 \text{ mL} \times \frac{2.50 \text{ mmol}}{1 \text{ L}} = 25.0 \text{ μmol}$$
$$m_{CuSO_4} = 25.0 \text{ μmol} \times \frac{159.61 \text{ g}}{1 \text{ mol}} = 3.99 \text{ mg}$$

(c) No, it is not possible to measure this mass on a centigram balance or on a milligram balance.

43. (a) The final concentration is $\frac{1}{25}$ the original concentration.

(b) To perform the same reaction, about 25 times the initial volume would be required.

Lab Exercise 7H
Titration Analysis of Vinegar (page 191)

Problem
What is the concentration of acetic acid, $CH_3COOH_{(aq)}$, in a sample of vinegar?

Analysis

$CH_3COOH_{(aq)} + NaOH_{(aq)} \rightarrow NaCH_3COO_{(aq)} + HOH_{(l)}$
10.00 mL   13.0 mL
C   0.202 mol/L

$$n_{NaOH} = 13.0 \text{ mL} \times \frac{0.202 \text{ mol}}{1 \text{ L}} = 2.63 \text{ mmol}$$

$$n_{CH_3COOH} = 2.63 \text{ mmol} \times \frac{1}{1} = 2.63 \text{ mmol}$$

$$C_{CH_3COOH} = \frac{2.63 \text{ mmol}}{10.00 \text{ mL}} = 0.263 \text{ mol/L}$$

According to the titration evidence, the molar concentration of acetic acid in a sample of vinegar is 0.263 mol/L.

INVESTIGATION 7.7

Titration Analysis of Hydrochloric Acid (page 192)

Problem
What is the molar concentration of a given hydrochloric acid solution?

Evidence
mass of $Na_2CO_{3(s)}$ used to prepare 100.0 mL of 0.100 mol/L solution = 1.06 g

	VOLUME OF $HCl_{(aq)}$ REQUIRED TO REACT WITH 10.00 mL OF 0.100 mol/L $Na_2CO_{3(aq)}$		
Trial	Final Buret Reading (mL)	Initial Buret Reading (mL)	Volume of $HCl_{(aq)}$ Added (mL)
1	16.7	0.2	16.5
2	32.7	16.7	16.0
3	48.6	32.7	15.9
4	17.2	1.2	16.0

Analysis

best average volume of $HCl_{(aq)}$ = $\dfrac{16.0 \text{ mL} + 15.9 \text{ mL} + 16.0 \text{ mL}}{3}$ = 16.0 mL

$$Na_2CO_{3(aq)} + 2HCl_{(aq)} \rightarrow 2NaCl_{(aq)} + CO_{2(g)} + H_2O_{(l)}$$
10.00 mL   16.0 mL
0.100 mol/L   C

$n_{Na_2CO_3} = 10.00 \text{ mL} \times \dfrac{0.100 \text{ mol}}{1 \text{ L}}$ = 1.00 mmol

$n_{HCl} = 1.00 \text{ mmol} \times \dfrac{2}{1}$ = 2.00 mmol

$C_{HCl} = \dfrac{2.00 \text{ mmol}}{16.0 \text{ mL}}$ = 0.125 mol/L

According to the evidence collected in this experiment, the molar concentration of the given hydrochloric acid solution is 0.125 mol/L.

Evaluation
The titration design is judged to be adequate because the problem was answered. This appears to be a good design and I have much confidence in it. The procedure is judged to be adequate because sufficient evidence was obtained. There are no apparent improvements that need to be made. The preparation of the sodium carbonate solution, if not done correctly, would have significantly affected the results. The technological skills were adequate as evidenced by the agreement of the titration volumes in trials 2 to 4.

Based upon my evaluation of this experiment, I am very confident in my results. The main sources of uncertainty are the common measurement errors. I would expect these uncertainties to be less than 5% in total.

Lab Exercise 7I
Filtration Analysis (page 193)

Problem

What is the molar concentration of the hydrochloric acid in a solution of kettle-scale remover?

Materials

lab apron
safety glasses
kettle-scale remover
1.05 mol/L $Pb(NO_3)_{2(aq)}$
$KI_{(aq)}$
50 mL beaker
250 mL beaker
25 mL graduated cylinder
filter paper
filter funnel, funnel holder, and stand
medicine dropper
stirring rod with spatula
wash bottle of distilled water
centigram balance

Analysis

mass of precipitate = 9.71 g − 0.89 g = 8.82 g

$$2HCl_{(aq)} + Pb(NO_3)_{2(aq)} \rightarrow PbCl_{2(s)} + 2HNO_{3(aq)}$$
$$25 \text{ mL} \qquad\qquad\qquad\qquad 8.82 \text{ g}$$
$$C \qquad\qquad\qquad\qquad\qquad 278.10 \text{ g/mol}$$

$$n_{PbCl_2} = 8.82 \text{ g} \times \frac{1 \text{ mol}}{278.10 \text{ g}} = 0.0317 \text{ mol}$$

$$n_{HCl} = 0.0317 \text{ mol} \times \frac{2}{1} = 0.0634 \text{ mol}$$

$$C_{HCl} = \frac{0.0634 \text{ mol}}{0.025 \text{ L}} = 2.5 \text{ mol/L}$$

According to the evidence given, the molar concentration of hydrochloric acid in a solution of kettle-scale remover is 2.5 mol/L.

Lab Exercise 7J
Gas Volume Analysis (page 194)

Problem

What is the molar concentration of the hydrochloric acid in a solution of kettle-scale remover?

Materials

lab apron
safety glasses
kettle-scale remover
zinc metal
large plastic tray
thermometer
barometer
1000 mL graduated cylinder
18 × 150 mm test tube
1-hole rubber stopper with glass tube
rubber tubing

Analysis

$$Zn_{(s)} + 2HCl_{(aq)} \rightarrow H_{2(g)} + ZnCl_{2(aq)}$$
$$\qquad\qquad 25 \text{ mL} \qquad\quad 822 \text{ mL}$$
$$\qquad\qquad C \qquad\qquad\quad 22.6°C, 98.7 \text{ kPa}$$

$$n_{H_2} = \frac{pv}{RT} = \frac{98.7 \text{ kPa} \times 822 \text{ mL}}{8.31 \text{ kPa·L/mol·K} \times 295.6 \text{ K}} = 33.0 \text{ mmol}$$

$$n_{HCl} = 33.0 \text{ mmol} \times \frac{2}{1} = 66.1 \text{ mmol}$$

$$C_{HCl} = \frac{66.1 \text{ mmol}}{25 \text{ mL}} = 2.6 \text{ mol/L}$$

According to the evidence given, the molar concentration of hydrochloric acid in a solution of kettle-scale remover is 2.6 mol/L.

Lab Exercise 7K
Titration Analysis (page 194)

Problem
What is the molar concentration of the hydrochloric acid in a solution of kettle-scale remover?

Materials
lab apron
safety glasses
kettle-scale remover
$Ba(OH)_{2(aq)}$
(2) 100 mL beakers
250 mL waste beaker
10 mL volumetric pipet
pipet bulb

meniscus finder
wash bottle of distilled water
(2) 250 mL Erlenmeyer flasks
small funnel
50 mL buret
buret clamp
bromothymol blue indicator
ring stand

Analysis
Best average volume of titrant = 13.7 mL

$$2HCl_{(aq)} + Ba(OH)_{2(aq)} \rightarrow BaCl_{2(aq)} + 2H_2O_{(l)}$$
10.00 mL   13.7 mL
C   0.974 mol/L

$$n_{Ba(OH)_2} = 13.7 \text{ mL} \times \frac{0.974 \text{ mol}}{1 \text{ L}} = 13.3 \text{ mmol}$$

$$n_{HCl} = 13.3 \text{ mmol} \times \frac{2}{1} = 26.6 \text{ mmol}$$

$$C_{HCl} = \frac{26.6 \text{ mmol}}{10.00 \text{ mL}} = 2.66 \text{ mol/L}$$

According to the evidence given, the molar concentration of hydrochloric acid in a solution of kettle-scale remover is 2.67 mol/L.

Exercise (page 195)

44. (a) The titration design in Lab Exercise 7K was the best because it was the most efficient, involved multiple trials, and gave the most precise answer.
    (b) An alternative and simpler design is to measure the mass of zinc metal before and after the reaction. The mass of zinc reacted can be used to calculate the concentration of the hydrochloric acid.
45. (a) The question cannot be answered. Ammonia is a gas at SATP and will escape from the solution when it is heated.

(b) The question cannot be answered. Heating will not decompose water into its elements.

(c) The final solid that is collected will contain both products as well as excess copper(II) sulfate. An answer can be obtained but it would be completely unreliable.

Overview (page 196)

Review

1. Chemical science is international in scope and technology is more localized. The approach in science is more theoretical, whereas in technology it is more empirical.

2. Technology is evaluated on the basis of simplicity, reliability, efficiency, and cost.

3. (a) The limiting reagent is the sample.
   (b) Excess reagents are used to make certain that all of the sample has reacted.

4. (a) gravimetric, gas, and solution stoichiometry
   (b) mass, volume, volume
   (c) molar mass, molar volume, molar concentration

5. Stoichiometry calculations may be found in the Prediction and Analysis sections.

6. The balanced equation provides the mole ratio of the two chemicals being considered.

7. crystallization, filtration, gas collection, titration

8. (a) volumetric flask
   (b) 10 mL graduated pipet
   (c) 10 mL volumetric pipet
   (d) Erlenmeyer flask

9. (a) $n_{H_2SO_4} = 10 \text{ mL} \times \dfrac{0.350 \text{ mol}}{1 \text{ L}} = 3.5 \text{ mmol}$

   (b) $n_{NaOH} = 15.0 \text{ kg} \times \dfrac{1 \text{ mol}}{40.00 \text{ g}} = 0.375 \text{ kmol}$

   (c) $n_{CH_4} = 10 \text{ L} \times \dfrac{1 \text{ mol}}{24.8 \text{ L}} = 0.40 \text{ mol}$

   (d) $pv = nRT$
   $$v_{NH_3} = \dfrac{5.1 \text{ mol} \times 8.31 \text{ kPa} \cdot \text{L}/(\text{mol} \cdot \text{K}) \times 303 \text{ K}}{1100 \text{ kPa}} = 12 \text{ L}$$

   (e) $v_{HCl} = 2.13 \text{ Mmol} \times \dfrac{1 \text{ L}}{6.0 \text{ mol}} = 0.36 \text{ ML}$

   (f) $m_{K_2Cr_2O_7} = 15 \text{ mmol} \times \dfrac{294.20 \text{ g}}{1 \text{ mol}} = 4.4 \text{ g}$

Applications

10. $n_{Na_2OOCCOO} = 250.0 \text{ mL} \times \dfrac{0.375 \text{ mol}}{1 \text{ L}} = 93.8 \text{ mmol}$

    $m_{Na_2OOCCOO} = 93.8 \text{ mmol} \times \dfrac{134.00 \text{ g}}{1 \text{ mol}} = 12.6 \text{ g}$

11. (a) $n_{KHC_4H_4O_6} = 100.0 \text{ mL} \times \dfrac{0.150 \text{ mol}}{1 \text{ L}} = 15.0 \text{ mmol}$

$$m_{KHC_4H_4O_6} = 15.0 \text{ mmol} \times \frac{188.19 \text{ g}}{1 \text{ mol}} = 2.82 \text{ g}$$

(b) 1. Obtain the 2.82 g of potassium hydrogen tartrate in a clean, dry 100 mL beaker.
2. Dissolve the solid using about 50 mL of pure water.
3. Transfer the solution into a clean 100 mL volumetric flask.
4. Add pure water to the calibration line.
5. Stopper and mix the solution.

12. $$v_i C_i = v_f C_f$$
$$v_i \times 14.6 \text{ mol/L} = 500 \text{ mL} \times 1.25 \text{ mol/L}$$
$$v_i = 42.8 \text{ mL}$$
H_3PO_4

13. (a) $$v_i C_i = v_f C_f$$
$$v_i \times 0.400 \text{ mol/L} = 100.0 \text{ mL} \times 0.100 \text{ mol/L}$$
$$v_i = 25.0 \text{ mL}$$
$K_2Cr_2O_7$

(b) 1. Add approximately 50 mL of pure water to a 100 mL volumetric flask.
2. Measure 25.00 mL of potassium dichromate solution using a pipet.
3. Transfer the solution slowly, with mixing, into the volumetric flask.
4. Add pure water to the calibration line.
5. Stopper and mix the solution.

14. $2CuO_{(s)} + C_{(s)} \rightarrow 2Cu_{(s)} + CO_{2(g)}$
500 kg m
79.55 g/mol   12.01 g/mol

$$n_{CuO} = 500 \text{ kg} \times \frac{1 \text{ mol}}{79.55 \text{ g}} = 6.29 \text{ kmol}$$

$$n_C = 6.29 \text{ kmol} \times \frac{1}{2} = 3.14 \text{ kmol}$$

$$m_C = 3.14 \text{ kmol} \times \frac{12.01 \text{ g}}{1 \text{ mol}} = 37.7 \text{ kg}$$

15. $2C_8H_{18(l)} + 25O_{2(g)} \rightarrow 16CO_{2(g)} + 18H_2O_{(g)}$
692 g m
114.26 g/mol   44.01 g/mol

$$n_{C_8H_{18}} = 692 \text{ g} \times \frac{1 \text{ mol}}{114.26 \text{ g}} = 6.06 \text{ mol}$$

$$n_{CO_2} = 6.06 \text{ mol} \times \frac{16}{2} = 48.5 \text{ mol}$$

$$m_{CO_2} = 48.5 \text{ mol} \times \frac{44.01 \text{ g}}{1 \text{ mol}} = 2.13 \text{ kg}$$

16. $AgNO_{3(aq)} + NaOH_{(aq)} \rightarrow AgOH_{(s)} + NaNO_{3(aq)}$
10.00 mL m
0.500 mol/L   124.88 g/mol

$$n_{AgNO_3} = 10.00 \text{ mL} \times \frac{0.500 \text{ mol}}{1 \text{ L}} = 5.00 \text{ mmol}$$

$$n_{AgOH} = 5.00 \text{ mmol} \times \frac{1}{1} = 5.00 \text{ mmol}$$

$$m_{AgOH} = 5.00 \text{ mmol} \times \frac{124.88 \text{ g}}{1 \text{ mol}} = 624 \text{ mg}$$

17. $3HCl_{(aq)} \;+\; Al(OH)_{3(s)} \;\rightarrow\; AlCl_{3(aq)} + 3H_2O_{(l)}$
    v   912 mg
    0.10 mol/L   78.01 g/mol

$$n_{Al(OH)_3} = 912 \text{ mg} \times \frac{1 \text{ mol}}{78.01 \text{ g}} = 11.7 \text{ mmol}$$

$$n_{HCl} = 11.7 \text{ mmol} \times \frac{3}{1} = 35.1 \text{ mmol}$$

$$v_{HCl} = 35.1 \text{ mmol} \times \frac{1 \text{ L}}{0.10 \text{ mol}} = 0.35 \text{ L}$$

18. $SO_{3(g)} + H_2O_{(l)} \;\rightarrow\; H_2SO_{4(aq)}$
    10.0 Mg   7.00 kL
    80.06 g/mol   C

$$n_{SO_3} = 10.0 \text{ Mg} \times \frac{1 \text{ mol}}{80.06 \text{ g}} = 0.125 \text{ Mmol}$$

$$n_{H_2SO_4} = 0.125 \text{ Mmol} \times \frac{1}{1} = 0.125 \text{ Mmol}$$

$$C_{H_2SO_4} = \frac{125 \text{ kmol}}{7.00 \text{ kL}} = 17.8 \text{ mol/L}$$

19. $2KOH_{(aq)} \;+\; H_2SO_{4(aq)} \;\rightarrow\; K_2SO_{4(aq)} \;+\; 2HOH_{(l)}$
    9.44 mL   10.00 mL
    50.6 mmol/L   C

$$n_{KOH} = 9.44 \text{ mL} \times \frac{50.6 \text{ mmol}}{1 \text{ L}} = 478 \text{ μmol}$$

$$n_{H_2SO_4} = 478 \text{ μmol} \times \frac{1}{2} = 239 \text{ μmol}$$

$$C_{H_2SO_4} = \frac{239 \text{ μmol}}{10.00 \text{ mL}} = 23.9 \text{ mmol/L}$$

20. $2H_2O_{2(aq)} \;\rightarrow\; 2H_2O_{(l)} + O_{2(g)}$
    v   500 mL
    0.88 mol/L   SATP

$$n_{O_2} = 500 \text{ mL} \times \frac{1 \text{ mol}}{24.8 \text{ L}} = 20.2 \text{ mmol}$$

$$n_{H_2O_2} = 20.2 \text{ mmol} \times \frac{2}{1} = 40.3 \text{ mmol}$$

$$v_{H_2O_2} = 40.3 \text{ mmol} \times \frac{1 \text{ L}}{0.88 \text{ mol}} = 46 \text{ mL}$$

21. $CH_{4(g)}$ + $2H_2O_{(g)}$ → $4H_{2(g)}$ + $CO_{2(g)}$
       1.0 t   v
       18.02 g/mol   25°C, 120 kPa

$n_{H_2O} = 1.0 \text{ Mg} \times \dfrac{1 \text{ mol}}{18.02 \text{ g}} = 0.055 \text{ Mmol}$

$n_{H_2} = 0.055 \text{ Mmol} \times \dfrac{4}{2} = 0.11 \text{ Mmol}$

$pv = nRT$

$v_{H_2} = \dfrac{0.11 \text{ Mmol} \times 8.31 \text{ kPa·L/(mol·K)} \times 298 \text{ K}}{120 \text{ kPa}} = 2.3 \text{ ML}$

22. $2H_2O_{(l)}$ → $2H_{2(g)}$ + $O_{2(g)}$
                    v   52 kL
                    25°C, 120 kPa   25°C, 120 kPa

$v_{H_2} = 52 \text{ kL} \times \dfrac{2}{1} = 104 \text{ kL}$

23. $n_{HOOCCOOH} = (85.97 - 84.56) \text{ g} \times \dfrac{1 \text{ mol}}{90.04 \text{ g}} = 15.7 \text{ mmol}$

$C_{HOOCCOOH} = \dfrac{15.7 \text{ mmol}}{10.00 \text{ mL}} = 1.57 \text{ mol/L}$

24. A measured volume of oxalic acid will be reacted with an excess of zinc. The hydrogen gas will be collected and its volume, temperature, and pressure measured.

25. (Design 1) A measured volume of sodium hydroxide will be reacted with an excess of magnesium chloride. The mixture will be filtered to determine the mass of the precipitate.
    (Design 2) A measured volume of sodium hydroxide will be titrated with a standard solution of hydrochloric acid using bromothymol blue indicator to determine the equivalence point.

26. (a) The experimental design is inadequate since litmus is an acid-base indicator and the reactants are neither acids nor bases. The problem cannot be answered.
    (b) The industrial design appears adequate to answer the question since the product silver sulfate has low solubility.
    (c) The industrial design is inadequate since the use of the lead(II) compound is unwarranted. Many other substances could be used that are less toxic or harmful if spilled.
    (d) The experimental design is inadequate since the concentration of the hydrochloric acid is not precisely known and concentrated hydrochloric acid is not a primary standard. The question could be answered, but with a low level of certainty.

27. *The list could include nomenclature rules, chemical formula theories and rules, states of matter and solubility generalizations, conservation of atoms/ions idea, concept of molar mass and molar volume, mass-amount conversions, molar concentration and volume conversions, ideal gas law, mole ratio concept, certainty and precision rules, international rules for symbols of elements, quantities and numbers. Concept maps will vary. A sample concept map is shown below.*

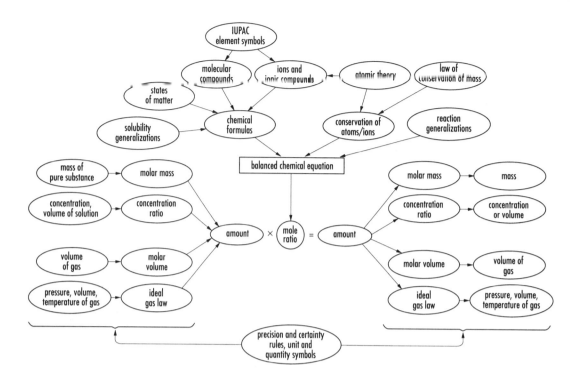

Extensions

28. $Zn_{(s)}$ + $2HNO_{3(aq)}$ → $Zn(NO_3)_{2(aq)}$ + $2H_{2(g)}$
    25.0 g   100 mL
    65.38 g/mol   0.197 mol/L

$$n_{Zn} = 25.0 \text{ g} \times \frac{1 \text{ mol}}{65.38 \text{ g}} = 0.382 \text{ mol}$$

$$n_{HNO_3} = 0.100 \text{ L} \times \frac{0.197 \text{ mol}}{1 \text{ L}} = 0.0197 \text{ mol}$$

Therefore, nitric acid is the limiting reagent.

$$n_{H_2} = 0.0197 \text{ mol} \times \frac{2}{2} = 0.0197 \text{ mol}$$

$$pv = nRT$$

$$v_{H_2} = \frac{0.0197 \text{ mol} \times 8.31 \text{ kPa·L/(mol·K)} \times 298 \text{ K}}{101.3 \text{ kPa}} = 0.482 \text{ L}$$

29. *The experiment should be designed to prevent deception of the audience. For example, by distracting the audience and substituting a bent spoon for an unbent one, or bending the spoon by hand.*

30. *The specific procedures for testing and treating water vary somewhat from one community to another. Information on your community's water supply can be obtained from your local water treatment plant. The following information was obtained from the E.L. Smith water treatment plant in Edmonton.*
    Water is routinely tested for alkalinity (by titration), pH (by pH meter), turbidity (by transmission of light), organic content (by color and ultraviolet light), and hardness (by precipitation of calcium and magnesium compounds). Water is also given a number of non-routine tests as the need arises. For example, total organic content, presence of pesticides and herbicides.

The treatment of water involves several examples of large-scale stoichiometry. Aluminum sulfate is added to the untreated river water. The aluminum ions react with hydroxide ions in the water to produce a gelatinous precipitate of aluminum hydroxide. As the precipitate settles to the bottom, it carries suspended material with it.

Powdered activated charcoal is added to remove organic compounds that give the water an unpleasant odor and taste. Chlorine dioxide is added to kill bacteria. Calcium hydroxide is added to precipitate calcium carbonate. Carbon dioxide gas is added to the water to lower the pH. Ammonia is added to the water early in the process, to ensure thorough mixing before chlorine gas is added. The ammonia and chlorine react to produce chloramine, a disinfectant.

31. The mass of a headache tablet is measured. The tablet is dissolved in methanol to make about 100 mL of solution. The solution is titrated with sodium hydroxide solution using a phenolphthalein indicator. The mass of ASA in the tablet is calculated by stoichiometry from the volume of sodium hydroxide solution used. Finally the filler content is determined by subtracting the mass of ASA from the mass of the original tablet.

32. *The federal and provincial governments provide current information on the training and requirements involved in beginning many different careers, including analytical chemist.*
    In general, analytical chemist requires graduation from high school at the university entrance level, with emphasis on science, plus completion of a two-year community college program in analytical technology, or a four-year university degree program specializing in chemistry.

Lab Exercise 7L
Chemical Analysis (page 199)

Problem
What is the concentration of oxalic acid, $HOOCCOOH_{(aq)}$, in a rust-removing solution?

Analysis
$HOOCCOOH_{(aq)} + 2NaOH_{(aq)} \rightarrow 2HOH_{(l)} + Na_2OOCCOO_{(aq)}$
C 0.0161 mol/L
10.00 mL   13.4 mL

$$n_{NaOH} = 13.4 \text{ mL} \times \frac{0.0161 \text{ mol}}{1 \text{ L}} = 0.216 \text{ mmol}$$

$$n_{HOOCCOOH} = 0.216 \text{ mmol} \times \frac{1}{2} = 0.108 \text{ mmol}$$

$$C_{HOOCCOOH} = \frac{0.108 \text{ mmol}}{10.00 \text{ mL}} = 0.0108 \text{ mol/L}$$

Before dilution, the molar concentration of the oxalic acid solution was 100 times 0.0108 mol/L, or 1.08 mol/L.

$$n_{HOOCCOOH} = 100 \text{ mL} \times \frac{1.08 \text{ mol}}{1 \text{ L}} = 108 \text{ mmol}$$

$$m_{HOOCCOOH} = 108 \text{ mmol} \times \frac{90.04 \text{ g}}{1 \text{ mol}} = 9.71 \text{ g}$$

$$c_{HOOCCOOH} = \frac{9.71 \text{ g}}{100 \text{ mL}} \times 100 = 9.71\% \text{ W/V}$$

According to the evidence gathered in this experiment, the concentration of the oxalic acid in the rust-removing solution is 9.71% W/V.

Evaluation

Titration is a good choice as the experimental design for this analysis. Crystallization is an alternative design which is more efficient, but provides a certainty of only two significant digits compared with three from a titration. I am very confident in the titration design.

The accuracy of the prediction is measured by a percent difference of 3%, calculated as follows:

$$\% \text{ difference} = \frac{|9.71\% - 10\%|}{10\%} \times 100 = 3\%$$

The prediction is verified because the accuracy of the prediction, as reflected by a percent difference of 3%, is very high and the percent difference can easily be accounted for by measurement uncertainties. The labelling is judged to be acceptable because the prediction is verified. I am quite confident about this judgment since a less than 5% difference is acceptable for a cleaning agent.

UNIT IV
CHEMICAL DIVERSITY AND SYSTEMS OF BONDING

8 Chemical Bonding

INVESTIGATION 8.1

Activity Series (page 203)

Problem
What is the order of reactivity (activity series) of calcium, copper, iron, lead, magnesium, and zinc metals with dilute hydrochloric acid?

Experimental Design
Approximately equal size, clean pieces of each metal are placed into separate equal volumes of hydrochloric acid. The manipulated variable is the metal; the responding variable is the rate of reaction; the controlled variables are the size of each piece of metal, temperature, volume of the acid, and concentration of the acid.

Materials

lab apron	(6) large test tubes
safety glasses	(6) 1-hole stoppers
equal size pieces of $Ca_{(s)}$, $Cu_{(s)}$, $Fe_{(s)}$, $Pb_{(s)}$, $Mg_{(s)}$, and $Zn_{(s)}$	test tube rack
	medicine dropper
1.0 mol/L $HCl_{(aq)}$	250 mL beaker
steel wool	

Procedure

1. Clean each metal piece with steel wool to remove the oxide coating.
2. Obtain about 200 mL of $HCl_{(aq)}$.
3. Label six test tubes with the six metal symbols.
4. Carefully pour $HCl_{(aq)}$ into each test tube to a depth of one-third. Make sure all test tubes have the same volume of acid.
5. Quickly add a different metal to each test tube and immediately insert a 1-hole stopper when each metal is added.
6. Observe immediately and after 10 min, and record the relative rate of bubbling (on a scale of 1 for fastest to 6 for slowest).
7. Dispose of the liquids in the sink while running cold water.
8. Solids may be rinsed and recycled, if possible, or disposed of in the waste basket.

Evidence

Metal Sample	Observations
calcium	• relative rate = 1 • very rapid bubbling • test tube was quite hot
copper	• relative rate = 6 • no evidence of reaction observed

iron	• relative rate = ? • might have been a few very tiny bubbles
lead	• relative rate = ? • a few bubbles of gas on surface
magnesium	• relative rate = 2 • metal bubbled rapidly while staying at the surface • test tube was slightly warm
zinc	• relative rate = 3 • large, noticeable bubbles of gas on the surface after 5 min

Analysis

According to the evidence gathered in this experiment, calcium is the most reactive, followed by magnesium and then zinc. There was so little reaction of iron and lead that it was not possible to rank these metals. Copper appears to be the least reactive.

Evaluation

The reaction with hydrochloric acid design appears to be adequate because it provided an answer to the problem. There are no obvious flaws with the design. Some difficulties were encountered with the procedure: some of the metals (such as calcium) were not easy to clean with steel wool; the less reactive metals would require a higher concentration of $HCl_{(aq)}$, or a longer reaction time, or both. For these reasons, the procedure used was inadequate to gather sufficient evidence for analysis. The technological skills were simple and, therefore, adequate. I am quite confident in the results for the three most reactive metals, but not certain of the order for the less reactive metals.

Lab Exercise 8A
Metal and Nonmetal Oxides (page 203)

Problem

How does the acidic or basic nature of the oxides of elements vary from left to right across a period of the periodic table?

Analysis

The tests indicate that the oxides of sodium and magnesium are basic, the oxide of aluminum appears to be both acidic and basic, and the oxides of silicon, phosphorus, sulfur, and chlorine are acidic. Therefore, the oxides of the elements from left to right across Period 3 change from basic (metal oxides) to acidic (nonmetal oxides).

Exercise (page 205)

1. Elements are classified into families in the periodic table according to their chemical properties.
2. (a) Francium would be expected to be the most reactive metal.
   (b) Fluorine would be expected to be the most reactive nonmetal.
3. The noble gases are classified as a chemical family because of their extremely low reactivity.

4. The staircase line separates the metals from the nonmetals.
5. When a metal reacts with a nonmetal, an ionic compound is formed. Ionic compounds are non-conducting, crystalline solids at SATP, have relatively high melting points, and conduct electricity in their molten and aqueous states.
6. Silver, platinum, and gold occur naturally in their element form. These metals are not very reactive so they occupy a position near the bottom of an activity series.

Exercise (page 208)

7. (a) A bonding electron is a single, unpaired electron in a valence orbital of an atom.
   (b) A lone pair is a pair of electrons in a filled valence orbital of an atom.
8. (a) Electronegativity is a number that represents the relative ability of an atom to attract a pair of bonding electrons in its valence level.
   (b) A covalent bond is the simultaneous attraction of two nuclei for a shared pair of bonding electrons.
   (c) An ionic bond is the simultaneous attraction between positive and negative ions.

9.
Element Symbol	Electronegativity	Group Number	Valence Electrons	Electron Dot Diagram	Bonding Electrons	Lone Pair Electrons
Na	0.9	1	1	Na·	1	0
Mg	1.2	2	2	·Mg·	2	0
Al	1.5	13	3	·Al·	3	0
Si	1.8	14	4	·Si·	4	0
P	2.1	15	5	:P·	3	1
S	2.5	16	6	:S·	2	2
Cl	3.0	17	7	:Cl·	1	3
Ar	–	18	8	:Ar:	0	4

10. The electron dot diagrams for metal atoms have vacant valence orbitals and no lone pairs, while those for nonmetals have lone pairs and no vacant valence orbitals.
11. Electronegativities generally decrease from top to bottom within a group and increase from left to right within a period.
12. Hydrogen is an exception to almost every rule or generalization about elements. Hydrogen is unique in having only one valence orbital (in its ground state).

Exercise (page 212)

13. Ionic compounds have low chemical reactivity compared with the elements from which they are formed.
    Some ionic compounds are so stable that they were once thought to be elements.
14. Nonmetals have much higher electronegativities than metals. When a metal and a nonmetal react, the valence electrons of the metal are removed or pulled away by the nonmetal forming positive and negative ions with stable octets.
15. The hardness of ionic compounds and their high melting and boiling points suggest that ionic bonds are strong.

16. K· + ·C̈l: → K⁺ [:C̈l:]⁻
    0.8   3.0

17. ·Ca· + ·Ö: → Ca²⁺ [:Ö:]²⁻
                 calcium oxide

18. $Cl_{2(g)} + 2e^- \rightarrow 2Cl^-_{(s)}$
    $Mg_{(s)} \rightarrow Mg^{2+}_{(s)} + 2e^-$
    _____
    $Mg_{(s)} + Cl_{2(g)} \rightarrow MgCl_{2(s)}$

19. $3[O_{2(g)} + 4e^- \rightarrow 2O^{2-}_{(s)}]$
    $4[Al_{(s)} \rightarrow Al^{3+}_{(s)} + 3e^-]$
    _____
    $4Al_{(s)} + 3O_{2(g)} \rightarrow 2Al_2O_{3(s)}$

20. Based only on differences in electronegativity, francium fluoride (FrF) and cesium fluoride (CsF) would be expected to be the most strongly ionic of all binary compounds.

21. Both gold and selenium have an electronegativity of 2.4. It is not possible to predict which atom will have the greater attraction for the electrons. This problem indicates that the rules and theories for ionic compounds are restricted.

22. The chemical formula of an ionic compound indicates the simplest ratio of ions in the ionic crystal. The chemical formula of a molecular compound indicates the actual number of atoms in a molecule of the substance.

23. (Enrichment) *This assignment could be completed over the remainder of the chapter and include the identification of ionic, molecular, and acid categories. To reduce the complexity, natural products (e.g., corn flour) or non-specific ingredients (e.g., food coloring) could be omitted from the list as long as the product name is clearly identified. Encourage the use of chemical reference books in the library to help identify specific chemicals.*

Exercise (page 214)

24. Examples of molecular elements include H_2, N_2, O_2, P_4, and S_8. Examples of molecular compounds include HCl, H_2O, and CH_4.

25. :F̈:F̈:   :C̈l:C̈l: :B̈r:B̈r: :Ï:Ï: :Ät:Ät:

    The similarity in the electron dot diagrams of the halogens is consistent with the similarity in their chemical properties.

26. The electron dot diagram for a hydrogen molecule shows that each atom has the same number of valence electrons as helium, the nearest noble gas.

27. :N:::N: N≡N

28. (a) :S̈::S̈:

    (b) *Shape cannot be predicted.*

29. (Enrichment) One possibility is predicted to be:

    P = P
    | |
    P = P

When compared with the known structure (phosphorus atoms at the corners of a tetrahedron), the prediction is shown to be false. The molecular theory is too limited to explain this molecule.

Exercise (page 218)

30. In one mole of the pollutant compound,

$$m_N = \frac{30.4}{100} \times 92.0 \text{ g} = 28.0 \text{ g}$$

$$n_N = 28.0 \text{ g} \times \frac{1 \text{ mol}}{14.01 \text{ g}} = 2.00 \text{ mol}$$

$$m_O = \frac{69.6}{100} \times 92.0 \text{ g} = 64.0 \text{ g}$$

$$n_O = 64.0 \text{ g} \times \frac{1 \text{ mol}}{16.00 \text{ g}} = 4.00 \text{ mol}$$

According to the evidence, the compound is N_2O_4, dinitrogen tetraoxide.

31. In one mole of ethane,

$$m_C = \frac{79.8}{100} \times 30.1 \text{ g} = 24.0 \text{ g}$$

$$n_C = 24.0 \text{ g} \times \frac{1 \text{ mol}}{12.01 \text{ g}} = 2.00 \text{ mol}$$

$$m_H = \frac{20.2}{100} \times 30.1 \text{ g} = 6.08 \text{ g}$$

$$n_H = 6.08 \text{ g} \times \frac{1 \text{ mol}}{1.01 \text{ g}} = 6.02 \text{ mol}$$

According to the evidence, the molecular formula for ethane is C_2H_6.

32. (Discussion) Three ways of knowing the chemical formulas for molecular compounds are by memorizing, by reference, and by empirical evidence from, for example, combustion analysis and spectrometry.

A theoretical way of knowing molecular formulas is presented next.

Lab Exercise 8B
Chemical Analysis (page 219)

Problem

What is the molecular formula of the unknown carbohydrate?

Analysis

In one mole of the compound tested,

$$m_C = \frac{40.0}{100} \times 180.2 \text{ g} = 72.1 \text{ g}$$

$$n_C = 72.1 \text{ g} \times \frac{1 \text{ mol}}{12.01 \text{ g}} = 6.00 \text{ mol}$$

$$m_H = \frac{6.8}{100} \times 180.2 \text{ g} = 12 \text{ g}$$

$$n_H = 12 \text{ g} \times \frac{1 \text{ mol}}{1.01 \text{ g}} = 12 \text{ mol}$$

$$m_O = \frac{53.2}{100} \times 180.2 \text{ g} = 95.9 \text{ g}$$

$$n_O = 95.9 \text{ g} \times \frac{1 \text{ mol}}{16.00 \text{ g}} = 5.99 \text{ mol}$$

According to the evidence, the molecular formula of the unknown carbohydrate is $C_6H_{12}O_6$.

Exercise (page 221)

33. The bonding capacity of hydrogen is one and of sulfur is two. Hydrogen can only form one bond and sulfur can form two bonds.

34. The molecular formula CH_3OH is a better representation of the structure of the molecule than CH_4O. CH_3OH indicates that one of the hydrogen atoms is bonded to the oxygen atom.

35. (a) hydrogen chloride   H:Cl: H—Cl

    (b) ammonia   H:N:H   H—N—H
                                 H |
                                                H

    (c) hydrogen sulfide   H:S:H   H—S—H

    (d) carbon dioxide   :O::C::O: O=C=O

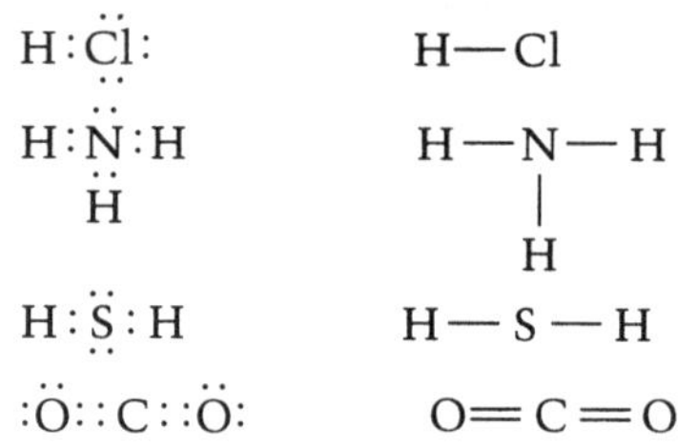

36. (a)  H—O—O—H

    (b) 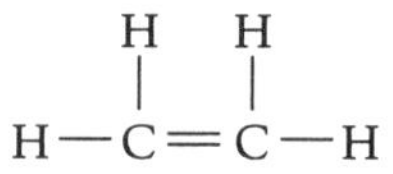 H—C=C—H with H's on each C

    (c) H—C≡N

    (d) 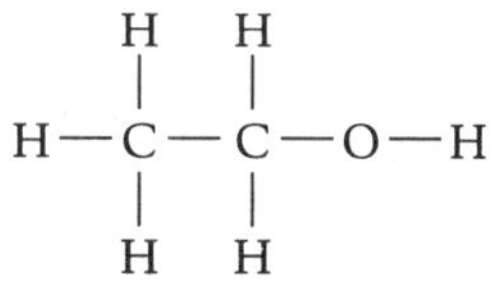 H—C—C—O—H with H's on carbons

    (e) 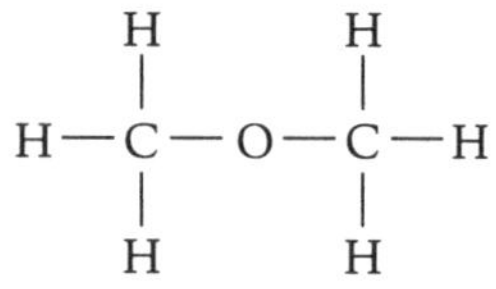 H—C—O—C—H with H's on carbons

    (f) 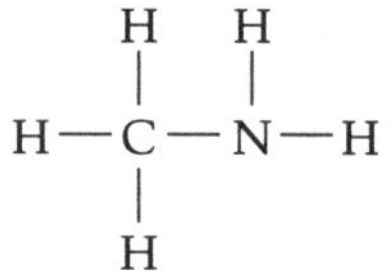 H—C—N—H with H's

Note how the structural diagrams and molecular formulas are related in (d), (e), and (f).

37. (Discussion) Explanations from theories should be logical, consistent, and simple.

INVESTIGATION 8.2

Molecular Models (page 221)

Problem

How can theory, represented by molecular models, explain the following series of chemical reactions that have occurred in a laboratory?

Evidence/Analysis

(a)
$$CH_4 + Cl-Cl \rightarrow CH_3Cl + H-Cl$$

It appears that a Cl atom in a Cl_2 molecule had been exchanged with one of the H atoms in CH_4.

(b)
$$H_2C=CH_2 + Cl-Cl \rightarrow CH_2Cl-CH_2Cl$$

The Cl_2 molecule had been added to the carbon atoms. One of the electron pairs between the carbon atoms was used to form two new bonds with the Cl atoms.

(c)
$$H_2N-NH_2 + O=O \rightarrow N\equiv N + 2\ H_2O$$

There was no simple rearrangements that could change the reactant models to the product models. The reactant models had to be disassembled completely.

(d)
$$CH_3-CH_2-CH_2-OH \rightarrow CH_3-CH=CH_2 + H_2O$$

The removal of the O—H from the end carbon and H from the middle carbon in the reactant model formed H_2O and left with the carbon an "extra" pair of electrons to complete the double bond in the product.

(e)
$$HCOOH + H-C(H)_2-OH \rightarrow HCOO-CH + H_2O$$

The two reactant models were joined by removing the O—H from one and the H from the other. It is not clear from this explanation which reactant model lost the O—H.

Evaluation

All proposed explanations are logical and simple, except for part (c), because a simple rearrangement can be made to change the reactant models into product models. All product models are consistent with molecular theory because they all can be constructed following the theoretical rules discussed previously. All reactions, except those represented by parts (c) and (d), are consistent with a collision of reactant molecules. The reaction represented by part (c) may involve a series of several reactions and not just a single rearrangement. Overall, the explanation is acceptable.

INVESTIGATION 8.3

Evidence for Double Bonds (page 222)

Problem

Which of the common substances supplied contain molecules with double covalent bonds between carbon atoms?

Evidence

Substance Tested	Observation
cyclohexane	• yellow-brown color very slowly faded
cyclohexene	• color of bromine instantly disappeared
mineral oil	• yellow-brown color very slowly faded
paint thinner	• yellow-brown color very slowly faded
liquid paraffin	• yellow-brown color very slowly faded
soybean oil	• color of bromine instantly disappeared
corn oil	• color of bromine instantly disappeared
peanut oil	• color of bromine instantly disappeared

Analysis

According to the evidence collected, a carbon-carbon double bond is likely present in molecules present in cyclohexene, soybean oil, corn oil, and peanut oil; a carbon-carbon double bond is not likely present in molecules present in cyclohexane, mineral oil, paint thinner, and liquid paraffin.

Exercise (page 223)

38. (a) IBr   iodine bromide   :Ï:B̈r: I—Br

    (b) PCl_3 phosphorus trichloride   :C̈l:P̈:C̈l: Cl—P—Cl
                                           :C̈l: |
                                                                Cl

    (c) OCl_2 oxygen dichloride   :C̈l:Ö:C̈l: Cl—O—Cl

    (d) CS_2 carbon disulfide   :S̈::C::S̈:   S=C=S

    (e) SO   sulfur oxide   :S̈::Ö: S=O

39. (Extension) The first four predictions are clearly supported by empirical evidence. However, the empirical evidence for SO is uncertain (could be S_2O_2). *Note that SO_2 and SO_3 are the more common oxides.*

40. (Discussion) The theory used to predict molecular formulas is judged to be acceptable because most of the formulas predicted in question 38 are verified. Additional predictions based on this theory should be made and checked to further test this theory.

Lab Exercise 8C
Predicting Molecular Formulas (page 224)

Problem

What are the molecular formula and the chemical name of the simplest compound formed when oxygen reacts with fluorine?

Prediction

According to the molecular theory presented in this chapter, the simplest compound is OF_2 as shown by the following electron dot diagram.

:F̈:Ö:F̈:

Analysis

In one mole of the compound,

$m_O = \dfrac{29.5}{100} \times 54.0 \text{ g} = 15.9 \text{ g}$

$n_O = 15.9 \text{ g} \times \dfrac{1 \text{ mol}}{16.00 \text{ g}} = 0.996 \text{ mol}$

$m_F = \dfrac{70.5}{100} \times 54.0 \text{ g} = 38.1 \text{ g}$

$n_F = 38.1 \text{ g} \times \dfrac{1 \text{ mol}}{19.00 \text{ g}} = 2.00 \text{ mol}$

According to the evidence, the molecular formula is OF_2.

Evaluation

The prediction is judged to be verified because the empirical molecular formula is the same as the predicted formula. The molecular theory used to make the prediction is judged to be acceptable because the prediction was verified. This judgment is even more certain if it is combined with other successful predictions made previously (question 38).

INVESTIGATION 8.4

Evidence for Polar Molecules (page 227)

Problem

Which of the various substances contain polar molecules?

Evidence/Analysis

The following liquids were found to be attracted to both charged strips:
$H_2O_{(l)}$, $CH_3OH_{(l)}$, $C_2H_5OH_{(l)}$, $C_2H_4(OH)_{2(l)}$, $(CH_3)_2CO_{(l)}$, $C_2H_4Cl_{2(l)}$
According to the evidence, these liquids contain polar molecules.
  The following liquids were found to be unaffected by either charged strip:
$CCl_{4(l)}$, $C_2Cl_{4(l)}$, $C_5H_{12(l)}$, $C_6H_{14(l)}$
According to the evidence, these liquids do not contain polar molecules; that is, the molecules are non-polar.

Lab Exercise 8D
London Forces (page 229)

Problem

What is the relationship between the boiling points of a family of hydrogen compounds and the number of electrons per molecule?

Prediction

According to the theory of London forces, the boiling points of the members of a family of hydrogen compounds should increase as the number of electrons per molecule increases.

Analysis

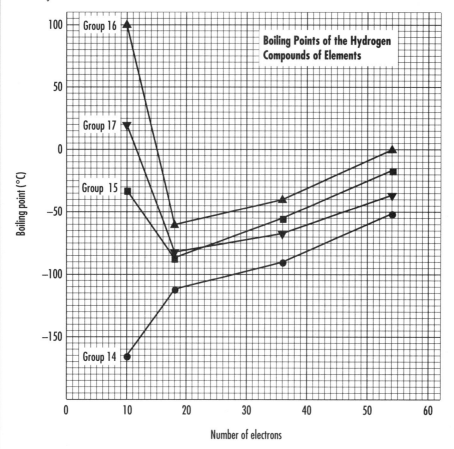

According to the evidence, the boiling point generally increases with increasing number of electrons, with some obvious exceptions such as NH_3, H_2O, and HF.

Evaluation

The experimental design is judged to be adequate because the problem was answered with no obvious flaws in the design. It is important that the same kind of compound, a hydrogen compound, is controlled. A more extensive design using other series of compounds could be used. However, the experimental results for this design appear to be quite certain.

Three out of the sixteen compounds tested (about 19%) differed substantially from the prediction. For this reason the prediction is judged to be inconclusive. The London force theory remains acceptable, but further tests need to be done to increase the certainty of this evaluation.

Synthesis

More families of compounds of Groups 14 to 17 elements need to be investigated, especially compounds of nitrogen, oxygen, and fluorine. It may be that hydrogen compounds of the same elements are exceptions and the London force theory has to be restricted or revised depending on the results of further tests.

INVESTIGATION 8.5

Hydrogen Bond Formation (page 230)

Problem

Are additional hydrogen bonds formed when water and glycerol are mixed?

Prediction

According to the hydrogen bond theoretical definition, water and glycerol should form hydrogen bonds between their molecules because each compound contains O—H bonds as shown in Figure 8.29, page 230.

Procedure

1. Set the nested polystyrene cups inside the 250 mL beaker.
2. Measure 30 mL of distilled water with a graduated cylinder and pour into the polystyrene cups.
3. Measure and record the temperature of the water.
4. Measure 30 mL of glycerol in a graduated cylinder.
5. Dry the thermometer and measure the temperature of the glycerol.
6. Quickly pour the glycerol into the water, cover and insert thermometer.
7. Swirl the mixture and note any temperature change.
8. Record the maximum temperature reached.
9. Dispose of solutions in the sink.
10. Rinse and dry apparatus and repeat steps 2 to 9.

Evidence

	Trial 1	Trial 2
initial temperature of 30 mL of water (°C)	24.2	24.5
initial temperature of 30 mL of glycerol (°C)	25.0	25.2
final temperature of the mixture (°C)	30.0	30.2

Analysis

	Trial 1	Trial 2
average initial temperature (°C)	24.6	24.9
temperature change (°C)	5.4	5.3

The obvious increase in temperature in both trials suggests that hydrogen bonds are formed when water and glycerol are mixed.

Evaluation

The experimental design is judged to be adequate because the problem was answered in a simple way. The procedure was also adequate although it might have been better to measure the temperature of the glycerol first and then pour the water into the glycerol. Glyerol does not pour very quickly. Technological skills were adequate as shown by the close agreement between the two trials. Overall, I am very certain of the results. The major source of uncertainty is the mixing of the two liquids.

The prediction is verified since an obvious temperature increase occurred, suggesting the formation of hydrogen bonds. However, the evidence does not rule out the possibility of the formation of other types of bonds. Because the prediction is considered to be verified, the hydrogen bond concept is judged to be acceptable. I am reasonably confident in this judgment.

Overview (page 231)

Review

1. (a) Chemical reactivity increases with increasing atomic size in Groups 1 and 2.
   (b) Chemical reactivity decreases with increasing atomic size in Groups 16 and 17.
   (c) Within period 3, chemical reactivity decreases from sodium to silicon and then increases from phosphorus to chlorine. Argon is very unreactive.
   (d) All of the elements in Group 18 have very low reactivity.
2. Metals react with nonmetals to form ionic compounds containing ionic bonds. Nonmetals react with other nonmetals to form molecular compounds.
3. Eight
4. An ionic compound is a pure substance formed from metals and nonmetals; is a hard, crystalline solid at SATP with a high melting and boiling point; and conducts electricity in molten and aqueous states.
5. Ionic compounds are neutral, three-dimensional structures of oppositely charged ions, held together by the simultaneous attraction of positive and negative ions.
6. The electronegativities of the representative metals are lower than the electronegativities of the representative nonmetals.

7. (a) $\text{Mg} + \ddot{\text{S}}: \rightarrow \text{Mg}^{2+}[:\ddot{\text{S}}:]^{2-}$

   (b) $:\ddot{\text{Cl}} + \cdot\text{Al}\cdot + \cdot\ddot{\text{Cl}}: \rightarrow \text{Al}^{3+}[:\ddot{\text{Cl}}:]^{-}_3$

   $:\ddot{\text{Cl}}:$

8.

Atom	Bonding Electrons	Lone Pairs
(a) ·Ca·	2	0
(b) ·Al·	3	0
(c) ·Ge·	4	0
(d) :N·	3	1
(e) :S·	2	2
(f) :Br·	1	3
(g) :Ne:	0	4

9. (a) $:N:::N: + :\ddot{I}:\ddot{I}: \rightarrow :\ddot{I}:N:\ddot{I}:$
   $\qquad\qquad\qquad\qquad\qquad\quad :\ddot{I}:$

   $N\equiv N + I-I \rightarrow I-N-I$
   $\qquad\qquad\qquad\qquad\quad |$
   $\qquad\qquad\qquad\qquad\quad I$

   (b) $H:\ddot{O}:\ddot{O}:H \rightarrow H:\ddot{O}:H + :\ddot{O}::\ddot{O}:$

   $H-O-O-H \rightarrow H-O-H + O=O$

10. The numbers in a molecular formula indicate the actual number of atoms of each element in the molecule. The numbers in an ionic formula indicate the ratio of the ions in the ionic crystal.

11. The idea of double or triple covalent bonds explains empirical molecular formulas such as $O_{2(g)}$ and $N_{2(g)}$ without changing the assumptions of covalent bonding.
12. The rapid reaction of some substances with bromine is evidence that a double or triple carbon-carbon bond is present in the substance.
13. Photosynthesis in green plants and the simple decomposition of water are examples of endothermic chemical changes.
14. The combustion of gasoline in a car engine and the metabolism of fats and carbohydrates in the human body are examples of exothermic chemical changes.
15. The boiling point of a substance is an indication of the strength of its intermolecular forces. The higher the boiling point, the stronger the intermolecular forces.
16. London forces act between all molecules. An example of a substance where London forces are the only type of intermolecular force is iodine, $I_{2(s)}$. Dipole-dipole forces are present in liquid hydrogen chloride, $HCl_{(l)}$. Hydrogen bonds are present in water, $H_2O_{(l)}$.

Applications

17. (a) positive and negative ions
    (b) nonmetal atoms
    (c) all molecules (solid and liquid states)
    (d) polar molecules
    (e) molecules containing H—F, H—O, or H—N bonds.
18. (a)
$$2K_{(s)} \rightarrow 2K^+_{(s)} + 2e^-$$
$$Br_{2(l)} + 2e^- \rightarrow 2Br^-_{(s)}$$
$$\overline{2K_{(s)} + Br_{2(l)} \rightarrow 2KBr_{(s)}}$$
   (b)
$$2Sr_{(s)} \rightarrow 2Sr^{2+}_{(s)} + 4e^-$$
$$O_{2(g)} + 4e^- \rightarrow 2O^{2-}_{(s)}$$
$$\overline{2Sr_{(s)} + O_{2(g)} \rightarrow 2SrO_{(s)}}$$

19. (a) fluorine, chlorine, bromine, iodine
    (b) reduction
    (c) The most reactive halogen, fluorine, has the greatest tendency to gain electrons in a reduction half-reaction. In fluorine there are fewer inner electrons that shield the electrons from the attraction of the nucleus.
    (d) The order of the activity series in Group 17 is completely consistent with the electronegativity values. As the reactivity decreases, the electronegativity decreases.
20. (a) Na_2O, MgO, Al_2O_3 are classified as ionic. SiO_2, P_2O_5, SO_2, Cl_2O are classified as molecular.
    (b) Na_2O (2.6), MgO (2.3), Al_2O_3 (2.0), SiO_2 (1.7), P_2O_5 (1.4), SO_2 (1.0), Cl_2O (0.5)
    (c) The larger electronegativity differences correspond to the ionic compounds and the smaller electronegativity differences correspond to the molecular compounds.
21. The high melting and boiling points of ionic compounds are due to the strong, simultaneous forces of attraction between the positive and negative ions.
22. In one mole of nicotine,
$$m_C = \frac{74.0}{100} \times 162.24 \text{ g} = 120 \text{ g}$$

$$n_C = 120 \text{ g} \times \frac{1 \text{ mol}}{12.01 \text{ g}} = 10.0 \text{ mol}$$

$$m_H = \frac{8.7}{100} \times 162.24 \text{ g} = 14 \text{ g}$$

$$n_H = 14 \text{ g} \times \frac{1 \text{ mol}}{1.01 \text{ g}} = 14 \text{ mol}$$

$$m_N = \frac{17.3}{100} \times 162.24 \text{ g} = 28.1 \text{ g}$$

$$n_N = 28.1 \text{ g} \times \frac{1 \text{ mol}}{14.01 \text{ g}} = 2.00 \text{ mol}$$

According to the evidence gathered, the molecular formula for nicotine is predicted to be $C_{10}H_{14}N_2$.

23. (a) H—P—H PH_3,   phosphorus trihydride
          |
          H

    (b)   Cl $SiCl_4$,   silicon tetrachloride
           |
       Cl—Si—Cl
           |
          Cl

    (c) O=C=O CO_2,   carbon dioxide

    (d) F—B—F BF_3,   boron trifluoride (violates octet rule)
          |
          F

24. Both intermolecular forces and covalent bonds are explained as a simultaneous attraction of opposite charges. Covalent bonds involve shared electrons in overlapping orbitals but intermolecular forces do not.

Extensions

25. The bond lengths given in Table 8.6 indicate that as the number of bonds between two atoms increases (single, double, triple), the bond length decreases.

26. Energy is absorbed in order to break the H—H and Cl—Cl bonds, and energy is released when the H—Cl bonds form. Overall, more energy is released than is absorbed, resulting in an exothermic reaction.

27.
(a) H:N̈:H + H:C̈l: → [H:N̈:H]⁺ [:C̈l:]⁻
    H   H

(b) H:Ö:H + H:Ö:H → [H:Ö:H]⁺ + [:Ö:H]⁻
                      H

(c)
                            H F
H:N̈:H + :F̈:B:F̈: → H:N:B:F
   H   :F̈:   H F

28. The molecules of "saturated fats" contain only single bonds between carbon atoms, while the molecules of "unsaturated fats" contain some double bonds between the carbon atoms. Many margarine containers list the mass of polyunsaturates, monounsaturates, and saturates in a typical 10 g mass of margarine.

29.
```
    H H H H              H H H
    | | | |              | | |
H—C—C—C—C—H        H—C—C—C—H
    | | | |              |   |
    H H H H              H   H
                       H—C—H
                         |
                         H
```

In mass spectrometry, the mass of the molecule is accurately determined, as well as the masses of the molecular fragments that form when the molecule is broken apart. From this information, the various clusters of atoms that make up the molecule can be determined. Infrared spectrometry determines the types of bonds (e.g., H—C, C—C) in the molecule. Nuclear magnetic resonance indicates the relative number of CH_3, CH_2, and CH groups in a molecule.

Lab Exercise 8E
Bonding Theory (page 233)

Problem
What compound forms in the reaction of phosphorus and fluorine?

Prediction
According to molecular theory, the simplest compound is PF_3 as shown by the following electron dot diagram.
The reasoning behind this prediction is that Lewis theory dictates that phosphorus atoms have three bonding electrons and that fluorine atoms have one. Therefore, three fluorine atoms are expected to have been used to pair all electrons and provide the octet required by this theory.

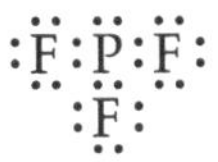

:F̈:P̈:F̈:
 :F̈:

Analysis
In one mole of the compound,

$$m_P = \frac{24.5}{100} \times 126 \text{ g} = 30.9 \text{ g}$$

$$n_P = 30.9 \text{ g} \times \frac{1 \text{ mol}}{30.97 \text{ g}} = 0.997 \text{ mol}$$

$$m_F = \frac{75.5}{100} \times 126 \text{ g} = 95.1 \text{ g}$$

$$n_F = 95.1 \text{ g} \times \frac{1 \text{ mol}}{19.00 \text{ g}} = 5.01 \text{ mol}$$

According to the evidence, the molecular formula is PF_5.

Evaluation
The prediction is judged to be falsified because the predicted answer does not agree with the experimental answer. Experimental uncertainties would not likely account for the difference. It is quite certain that the molecular theory being tested is unacceptable in this case because of the falsified prediction. The theory needs to be restricted or revised to improve its predictive power.
This anomaly is common to the period 3 elements but not common to the period 2 elements. It appears that elements such as phosphorus and sulfur can have all of their valence electrons available as bonding electrons, that is, 5 and 6 electrons, respectively. After discovering a number of these unexpected results, the pattern of period 2 versus period 3 becomes apparent, and the bonding theory is revised to explain and then predict when the higher bonding capacity will occur.

9 Organic Chemistry

Exercise (page 237)

1. Organic chemistry is the study of the molecular compounds of carbon. Originally organic chemistry was only associated with compounds obtained from living organisms.
2. Carbon atoms can bond together to form chains, rings, spheres, sheets, and tubes. Carbon atoms can form any combination of single, double, or triple covalent bonds.
3. oxygen (2), sulfur (2), phosphorus (3), nitrogen (3), halogens (1)
4.

INVESTIGATION 9.1

Models of Organic Compounds (page 238)

Problem
What are the structures of the isomers of C_4H_{10}, $C_2H_3Cl_3$, C_2H_6O, and C_2H_7N?

Evidence/Analysis
C_4H_{10}

100   CHEMICAL DIVERSITY AND SYSTEMS OF BONDING

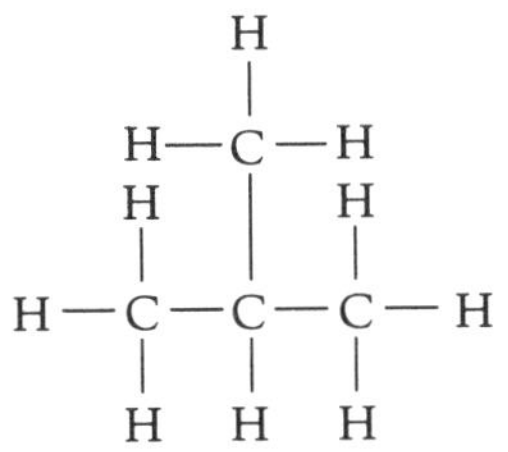

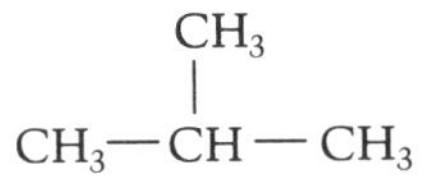

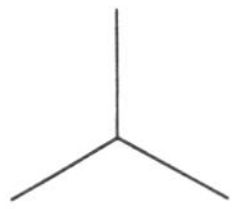

```
      H
      |
    H—C—H
    H |  H
    | |  |
H—C—C—C—H          CH₃
    | |  |          |
    H H  H      CH₃—CH—CH₃
```

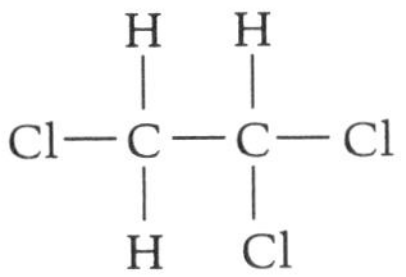

C₂H₃Cl₃

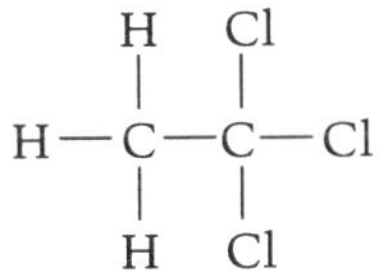

```
    H  Cl
    |  |
H—C—C—Cl           CH₃—CCl₃
    |  |
    H  Cl

    H  H
    |  |
Cl—C—C—Cl          CH₂Cl—CHCl₂
    |  |
    H  Cl
```

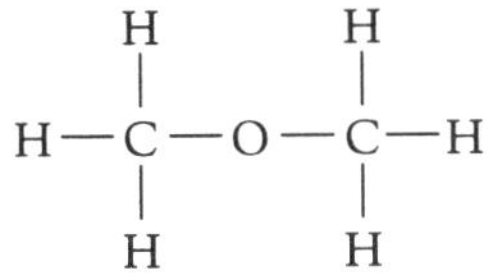

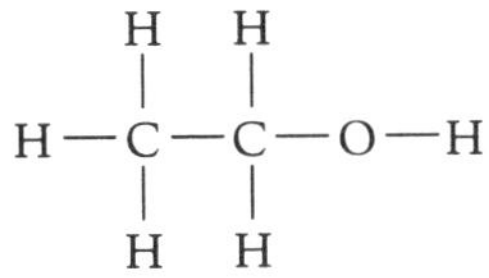

C₂H₆O

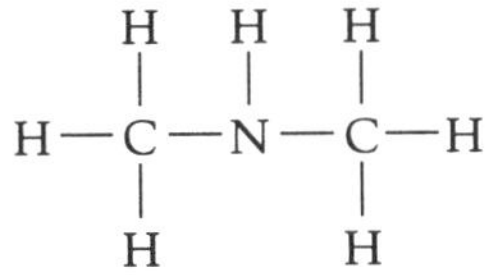

```
    H       H
    |       |
H—C—O—C—H          CH₃—O—CH₃
    |       |
    H       H

    H  H
    |  |
H—C—C—O—H          CH₃—CH₂—OH
    |  |
    H  H
```

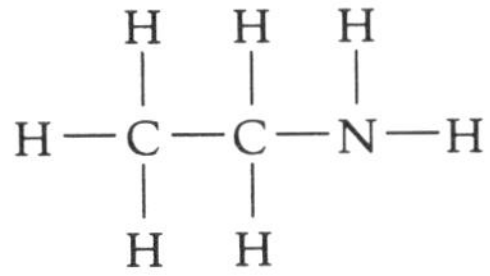

C₂H₇N

```
    H  H  H
    |  |  |
H—C—N—C—H          CH₃—NH—CH₃
    |     |
    H     H

    H  H  H
    |  |  |
H—C—C—N—H          CH₃—CH₂—NH₂
    |  |
    H  H
```

INVESTIGATION 9.2

Classifying Organic Compounds (page 239)

Problem

How can selected organic compounds be classified according to the functional groups in their molecular structures?

ORGANIC CHEMISTRY 101

Evidence

Consumer Product		Selected Organic Compound		
Product Name	Use	Toxicity	IUPAC Name	Structural Diagram
1. Rubbing alcohol (generic)	Massage, antiseptic	3	2-propanol	$CH_3-CH(OH)-CH_3$
2. Cutex® regular	Nail polish remover	3	propanone (acetone)	$CH_3-CO-CH_3$
3. Vinegar (generic)	Pickles, food, cleaner	?	ethanoic acid (acetic acid)	$CH_3-CO-OH$
4. OTO®	Testing chlorine levels in pools	?	4, 4'-diamino-3, 3'-dimethylbiphenyl	(biphenyl with two CH_3 and two NH_2 groups)
5. ASA (generic)	Pain medicine	4	acetylsalicylic acid	(benzene ring with $-COOH$ and $-O-CO-CH_3$)
6. Prestone®	Radiator antifreeze	3	1, 2-ethanediol	$CH_2(OH)-CH_2(OH)$
7. Liquid Paper®	Correction fluid	3	1, 1, 1-trichloroethane	CH_3-CCl_3
8. Plastic repair kit	Repairing holes and tears in plastic	3	butanone	$CH_3-CO-CH_2-CH_3$
9. LP Record	Recording and playing back music	?	polyvinylchloride (PVC)	$-(CHCl-CH_2)_n-$
10. Vitamin C (generic)	Vitamin supplement	1	ascorbic acid	(ascorbic acid structure)

Analysis

According to the evidence from the products labelled 1 to 10, the following functional groups and examples were found.

Functional Group	Examples
—OH	1, 6, 10
—C(=O)—	2, 8
—C(=O)—OH	3, 5
—C(=O)—O—	5, 10
—Cl	7, 9
—NH$_2$	4
—C=C—	10
(benzene ring)	4, 5

Exercise (page 248)

5. (a)

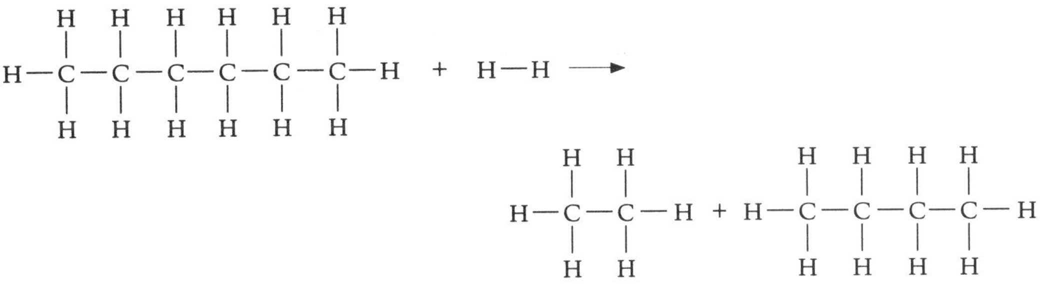

$$CH_3-CH_2-CH_2-CH_2-CH_2-CH_3 + H-H \longrightarrow CH_3-CH_3 + CH_3-CH_2-CH_2-CH_3$$

(b)

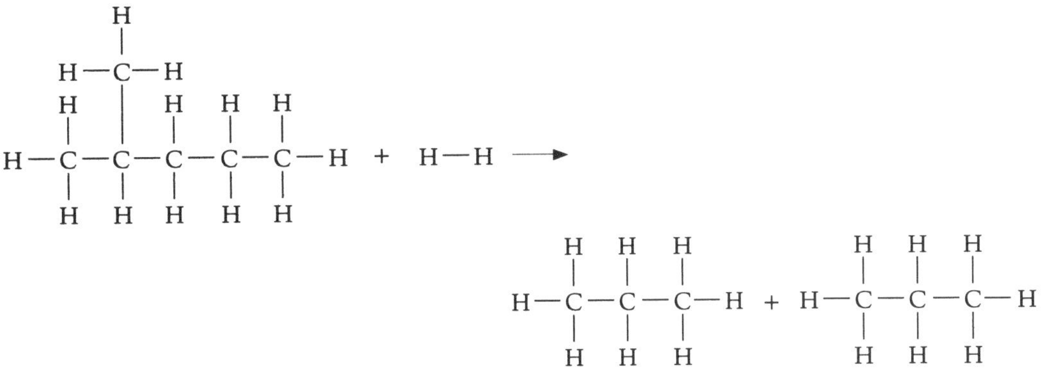

2-methylpentane + H–H →

$$CH_3-CH_2-CH_3 + CH_3-CH_2-CH_3$$

(c)

2,2-dimethylbutane + H–H → CH_3-CH_3 + 2,2-dimethylpropane (neopentane)

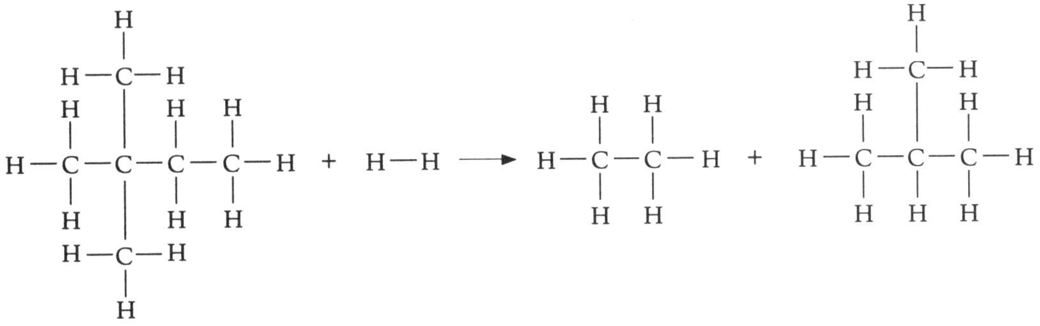

6. (a) propane + pentane → octane + hydrogen

(b) 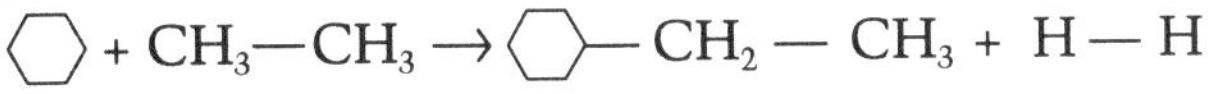 ⬡ + CH_3—CH_3 → ⬡—CH_2—CH_3 + H—H

(c) hexane → 2,2-dimethylbutane

(d)

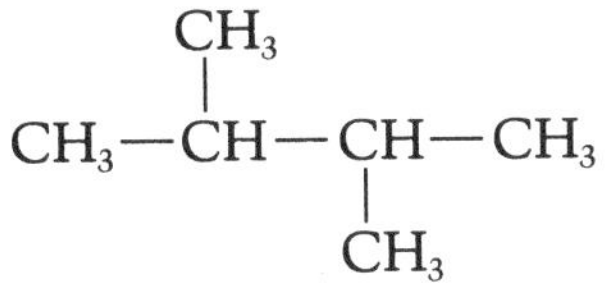

$$CH_3-\underset{\underset{CH_3}{|}}{CH}-\underset{\underset{CH_3}{|}}{CH}-CH_3 \quad \text{2,3-dimethylbutane}$$

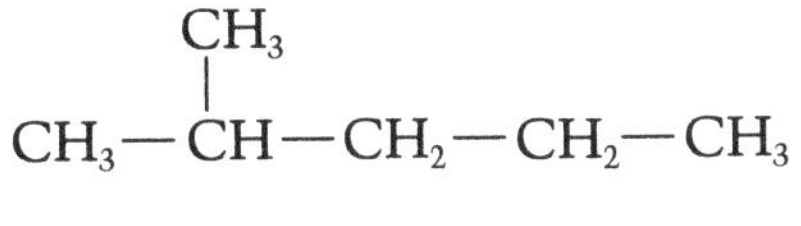

$$CH_3-\underset{\underset{CH_3}{|}}{CH}-CH_2-CH_2-CH_3 \quad \text{2-methylpentane}$$

7. (a) → carbon dioxide + water

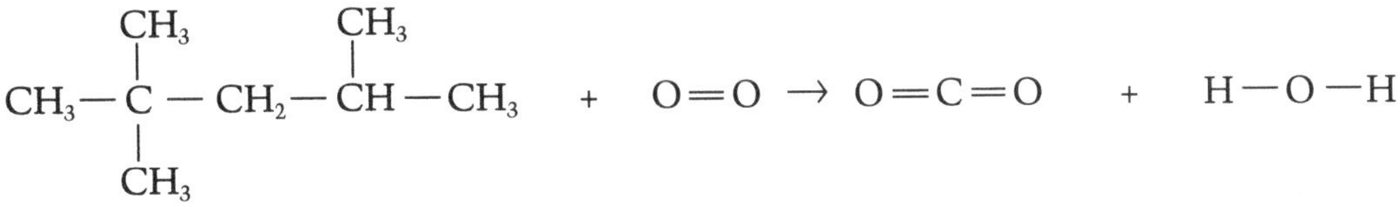

$$CH_3-\underset{\underset{CH_3}{|}}{\overset{\overset{CH_3}{|}}{C}}-CH_2-\underset{\underset{}{|}}{\overset{\overset{CH_3}{|}}{CH}}-CH_3 + O=O \rightarrow O=C=O + H-O-H$$

(b) → $O=C=O$ + H—O—H

2-methylbutane + oxygen → carbon dioxide + water

(c) → $O=C=O$ + H—O—H

cyclopentane + oxygen → carbon dioxide + water

INVESTIGATION 9.3

Structures and Properties of Isomers (page 251)

Problem

What are the structures and physical properties of the isomers of C_4H_8 and C_4H_6?

Evidence/Analysis

C_4H_8

IUPAC Name	Structure	M.P.(°C)	B.P.(°C)
1-butene	$CH_2\!=\!CH-CH_2-CH_3$	−185.3	−6.3
2-butene	$CH_3-CH\!=\!CH-CH_3$	(cis)−138.9 (trans)−105.5	3.7 0.9
methylpropene	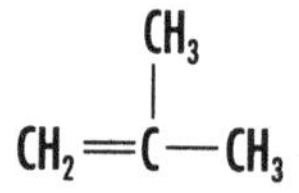 $CH_2\!=\!C(CH_3)-CH_3$	−140.3	−6.9
cyclobutane	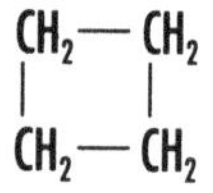 CH_2-CH_2 / CH_2-CH_2 (ring)	−50	12
methylcyclopropane	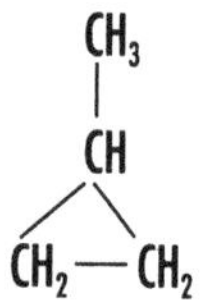 CH_3-CH on cyclopropane ring (CH_2-CH_2)	−117.2	4 to 5

C_4H_6

IUPAC Name	Structure	M.P.(°C)	B.P.(°C)
1-butyne	$CH\equiv C-CH_2-CH_3$	−127.7	8.1
2-butyne	$CH_3-C\equiv C-CH_3$	−32.2	27
cyclobutene	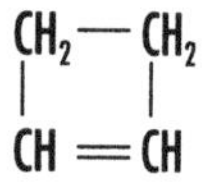 CH_2-CH_2 / $CH\!=\!CH$ (ring)	2	–
methylcyclopropene	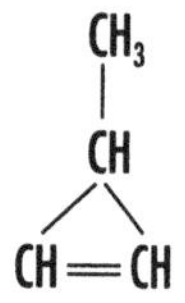 CH_3-CH on cyclopropene ring $(CH\!=\!CH)$	(substance not known)	
1,2-butadiene	$CH_2\!=\!C\!=\!CH-CH_3$	−136.2	10.8
1,3-butadiene	$CH_2\!=\!CH-CH\!=\!CH_2$	−108.9	−4.4

Exercise (page 253)

8. (a) $CH_2=CH-CH_2-CH_3 \rightarrow CH\equiv CH + CH_3-CH_3$

   (b) $CH_3-CH_2-\underset{\underset{CH_3}{|}}{CH}-CH_2-CH_2-CH_2-CH_3 \rightarrow$
   $CH_3-CH=CH-CH_3 + CH_3-CH_2-CH_2-CH_3$

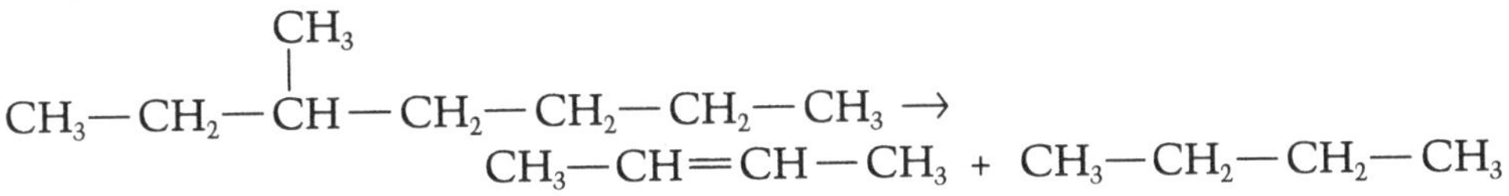

   (c) $CH_3-CH_2-\underset{\underset{CH_3}{|}}{CH}-CH_2-CH_2-CH_2-CH_3 \rightarrow$
   $CH_2=CH-CH_3 + CH_2=\underset{\underset{CH_3}{|}}{C}-CH_2-CH_3 + H-H$

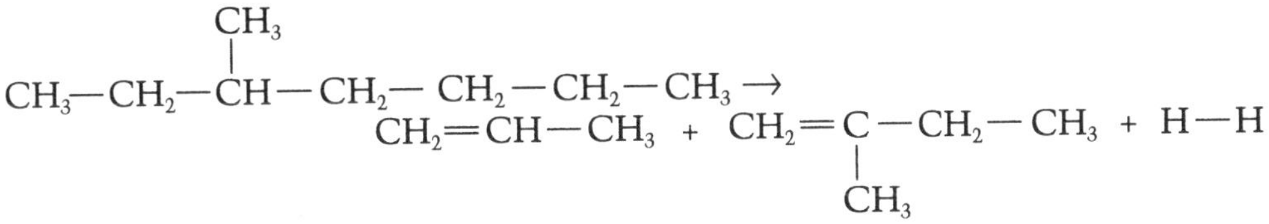

   (d) $\text{C}_6\text{H}_5-C_3H_7 \rightarrow \text{C}_6\text{H}_5-CH_3 + CH_2=CH_2$

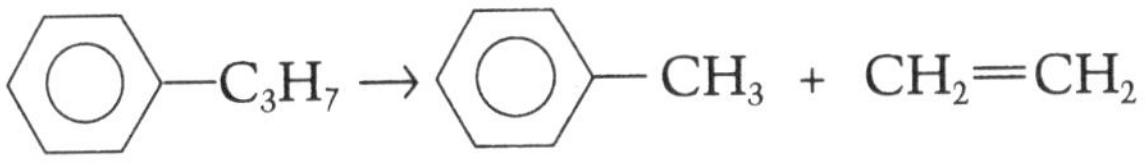

9. (a) $CH_2=\underset{\underset{CH_3}{|}}{C}-CH_2-CH_2-CH_3 \rightarrow CH_2=\underset{\underset{CH_3}{|}}{C}-\overset{\overset{CH_3}{|}}{CH}-CH_3$

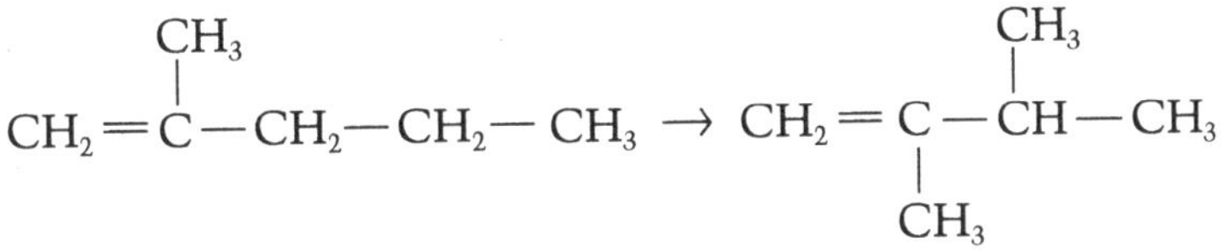

   (b) ethylbenzene → 1,2-dimethylbenzene

   (c) 1,3-dimethylbenzene   1,4-dimethylbenzene

10. (a) ethyne + oxygen → carbon dioxide + water
    $CH\equiv CH + O=O \rightarrow O=C=O + H-O-H$

    (b) methylbenzene + oxygen → carbon dioxide + water
    $\text{C}_6\text{H}_5\text{-CH}_3 + O=O \rightarrow O=C=O + H-O-H$

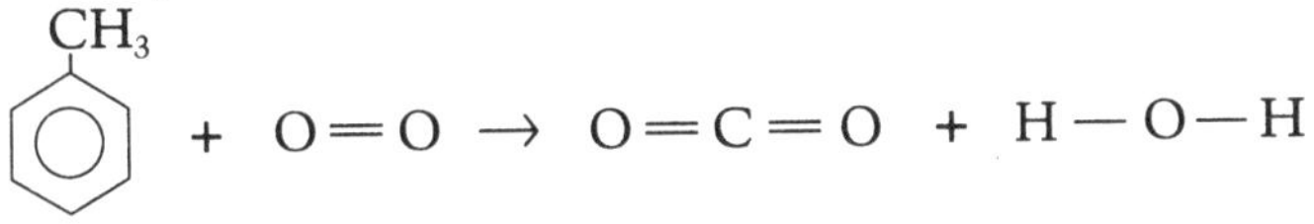

11. (a) addition

    methyl-2-butene   methylbutane
    $CH_3-\underset{\underset{CH_3}{|}}{C}=CH-CH_3 \rightarrow CH_3-\underset{\underset{CH_3}{|}}{CH}-CH_2-CH_3$

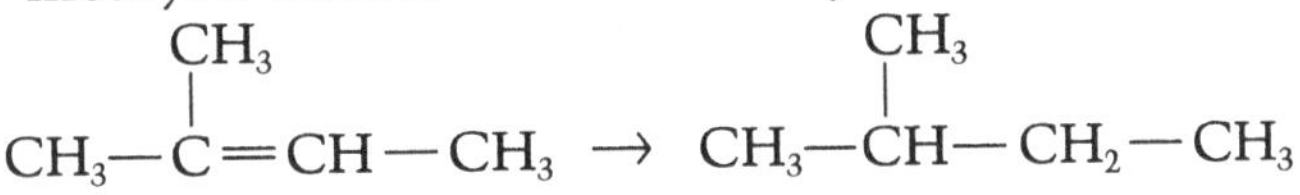

    (b) cracking

    ethylbenzene → phenylethene
    $\text{C}_6\text{H}_5-C_2H_5 \rightarrow \text{C}_6\text{H}_5-CH=CH_2$

    (c) addition

    2-butyne → 2-butene
    $CH_3-C\equiv C-CH_3 \rightarrow CH_3-CH=CH-CH_3$

    (d) addition

    benzene + ethene → ethylbenzene (or phenylethane)
    $\text{C}_6\text{H}_6 + CH_2=CH_2 \rightarrow \text{C}_6\text{H}_5-C_2H_5$

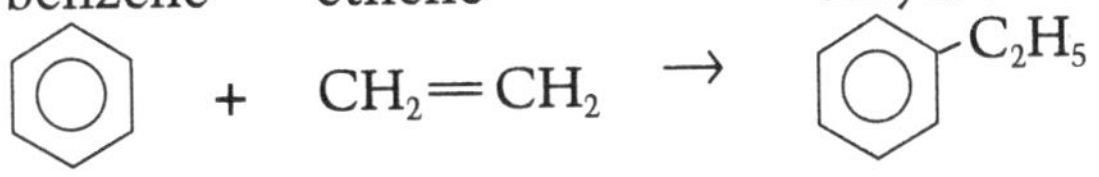

(e) combustion

3-ethyl-4-methyl-2-pentene

$$\text{CH}_3-\text{CH}=\overset{\overset{\text{C}_2\text{H}_5}{|}}{\underset{\underset{\text{CH}_3}{|}}{\text{C}}}-\text{CH}-\text{CH}_3$$

12.

```
                    hydrocarbons
                   /            \
              aliphatic        aromatic
             /    |    \
        alkyne  alkene  alkane
        _____/  _____/
          unsaturated      saturated
```

13. *A brainstorming session (small group or whole class) would be useful as a starting point to identify various perspectives. Minimize discussion particularly of pros and cons, other than noting that opposite positions exist. At this point, students should be directed to do some reading and library research. Either a debate or report could follow. Clear directions are important.*
References: Pages 255–256 of the student book; Alberta Education Senior High Resource Manual *for reports and debates.*

Lab Exercise 9A
Molecular Structure of Unknown Liquid (page 255)

Problem

What is the molecular structure of an unknown liquid that is known to be organic?

Analysis

$$m_C = \frac{91.1}{100} \times 92.2 \text{ g} = 84.0 \text{ g}, \ n_C = 84.0 \text{ g} \times \frac{1 \text{ mol}}{12.01 \text{ g}} = 6.99 \text{ mol}$$

$$m_H = \frac{8.9}{100} \times 92.2 \text{ g} = 8.2 \text{ g}, \ n_H = 8.2 \text{ g} \times \frac{1 \text{ mol}}{1.01 \text{ g}} = 8.1 \text{ mol}$$

According to the evidence, the compound is saturated (alkane or aromatic) and has the molecular formula C_7H_8. Since the formula does not fit the 2n + 2 rule for alkanes, the compound must be aromatic. The structure of the compound is likely,

 methylbenzene

Exercise (page 259)

14. (a) substitution

→ tetrachloromethane + hydrogen chloride

$$\underset{\underset{\text{Cl}}{|}}{\overset{\overset{\text{Cl}}{|}}{\text{Cl}-\text{C}-\text{H}}} + \text{Cl}-\text{Cl} \rightarrow \underset{\underset{\text{Cl}}{|}}{\overset{\overset{\text{Cl}}{|}}{\text{Cl}-\text{C}-\text{Cl}}} + \text{H}-\text{Cl}$$

(b) addition

$$CH_2=CH-CH_3 + Br-Br \rightarrow Br-CH_2-\underset{\underset{Br}{|}}{CH}-CH_3 \rightarrow \text{1,2-dibromopropane}$$

(c) addition

$$CH_2=CH_2 + H-I \rightarrow CH_3-CH_2-I \rightarrow \text{iodoethane}$$

(d) substitution

$$CH_3-CH_3 + Cl-Cl \rightarrow CH_3-CH_2-Cl + H-Cl \rightarrow \text{chloroethane + hydrogen chloride}$$

(e) addition

$$\text{chloroethene + fluorine} \rightarrow \underset{F\quad\;\; F}{\underset{|\quad\;\; |}{CH_2-CH-Cl}} \rightarrow \text{1-chloro-1,2-difluoroethane}$$

(f) addition

$$\rightarrow \underset{\underset{Cl}{|}}{CH_2}-CH_2-CH_2-CH_3 + CH_3-\underset{\underset{Cl}{|}}{CH}-CH_2-CH_3$$

1-butene + hydrogen chloride → 1-chlorobutane + 2-chlorobutane

(g) → HCl + [1,2-dichlorobenzene] + [1,3-dichlorobenzene] + [1,4-dichlorobenzene]

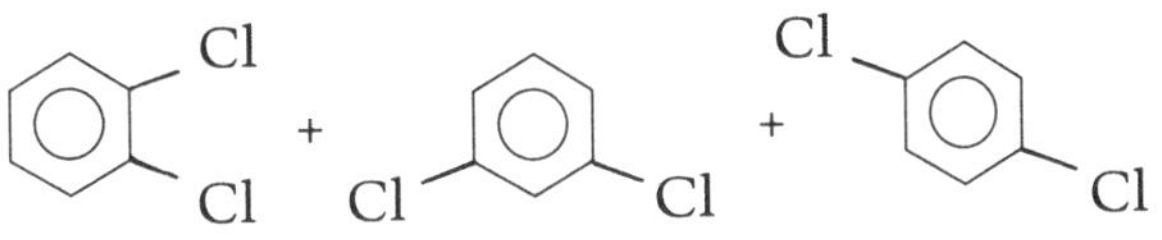

chlorobenzene + chlorine → hydrogen chloride + 1,2-dichlorobenzene + 1,3-dichlorobenzene + 1,4-dichlorobenzene

15. (a) $\rightarrow CH_2=CH_2 + H-O-H + Cl^-$

chloroethane + hydroxide ion → ethene + water + chloride ion

(b) $\rightarrow CH_2=CH-CH_2-CH_3 + CH_3-CH=CH-CH_3 + H-O-H + Cl^-$

2-chlorobutane + hydroxide ion → 1-butene + 2-butene + water + chloride ion

16. (a) Equal amounts of chlorine are reacted with ethyne (acetylene). The product of this reaction is then reacted with hydrogen chloride gas.

(b) (Discussion) Two molecules of chlorine may add to an ethyne molecule to form 1,1,2,2-tetrachloroethane. A significant, but common, experimental complication is that the product contains a mixture of chlorinated hydrocarbons that are difficult to separate from each other.

Exercise (page 261)

17. (a) $CH_3-CH=CH-CH_3 + H-O-H \rightarrow CH_3-\underset{\underset{OH}{|}}{CH}-CH_2-CH_3$

(b) ethene + hydrogen hypochlorite → 2-chloroethanol

18. (a) $CH_3-CH_2-CH_2-OH \rightarrow CH_2=CH-CH_3 + H-O-H$
    (b) 1-butanol $\rightarrow$ 1-butene + water
    $\rightarrow CH_2=CH-CH_2-CH_3 + H-O-H$

19. (a) $C_2H_5OH_{(l)} + 3O_{2(g)} \rightarrow 2CO_{2(g)} + 3H_2O_{(g)}$
    (b) $2C_3H_7OH_{(l)} + 9O_{2(g)} \rightarrow 6CO_{2(g)} + 8H_2O_{(g)}$

INVESTIGATION 9.4

Synthesizing an Ester (page 265)

Problem

What are some physical properties of ethyl ethanoate (ethyl acetate) and methyl salicylate?

Evidence/Analysis

	Solubility	Odor
ethyl acetate	liquids mixed	sweet "gluey" smell
methyl salicylate	two layers formed	wintergreen smell

Exercise (page 268)

20.

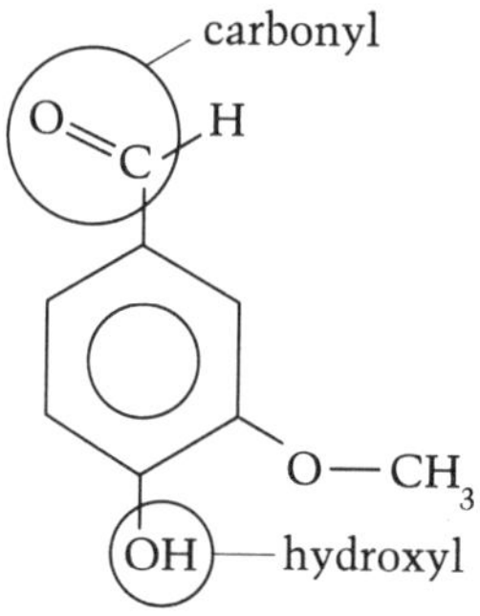

vanillin — carbonyl, hydroxyl

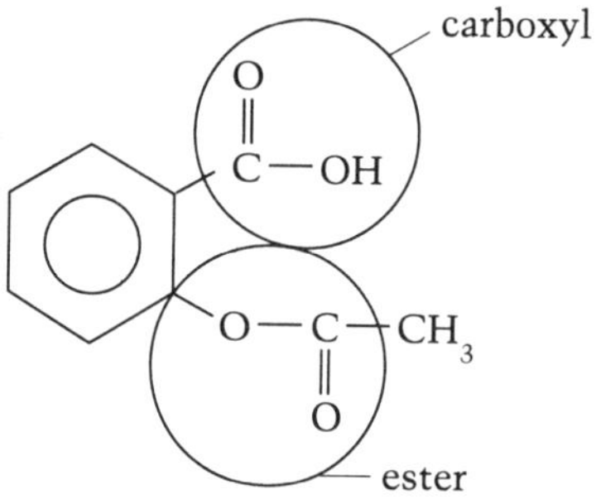

acetylsalicylic acid (ASA) — carboxyl, ester

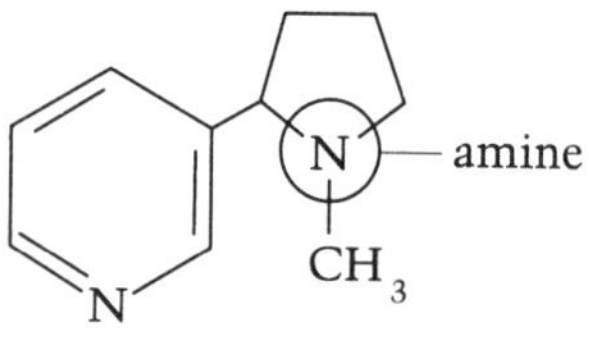

nicotine — amine

urea — amide

(bottom structure) — carboxyl, hydroxyl, carboxyl, carboxyl

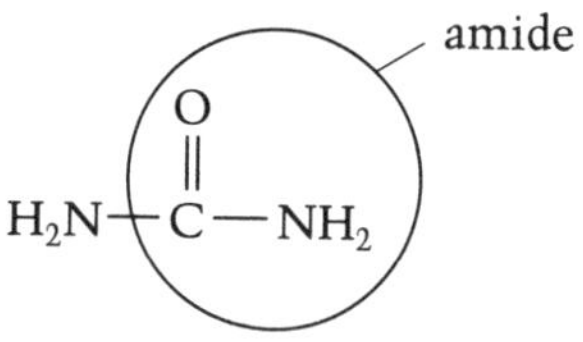

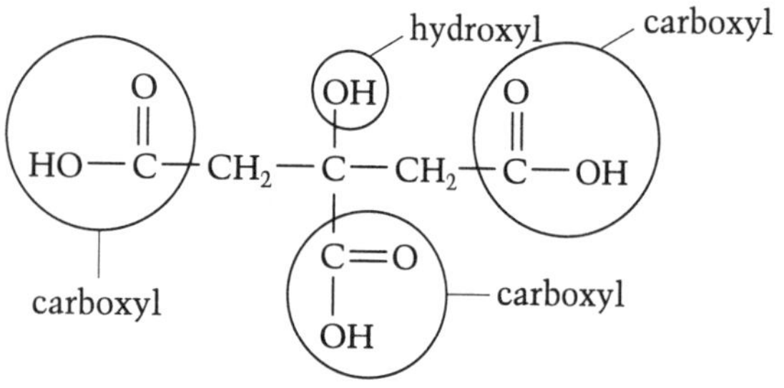

vanillin (flavoring agent), ASA (common drug for headaches), nicotine (addictive agent in cigarettes), urea (found in urine, fertilizer), citric acid (found in citrus fruits)

21. (a) neutralization
$\rightarrow$ NaCH$_3$COO + HOH
sodium acetate   water

(b) esterification
$\rightarrow$ CH$_3$CH$_2$COOCH$_3$ + HOH
methyl propanoate   water

(c) neutralization
$\rightarrow$ KC$_6$H$_5$COO + HOH
potassium benzoate   water

(d) esterification
$\rightarrow$ HCOOC$_2$H$_5$ + HOH
ethyl methanoate   water

22.
$$\begin{array}{l} CH_2-O-\overset{\overset{O}{\|}}{C}-(CH_2)_{16}CH_3 \\ | \\ CH_2-O-\overset{\overset{O}{\|}}{C}-(CH_2)_{16}CH_3 + 3H_2O \\ | \\ CH_2-O-\overset{\overset{O}{\|}}{C}-(CH_2)_{16}CH_3 \end{array}$$

23. (a) substitution
alkane + halogen $\rightarrow$ organic halide + hydrogen halide

CH$_3$—CH$_3$ + Br—Br $\rightarrow$ CH$_3$—CH$_2$—Br + H—Br

(b) addition
alkene + halogen $\rightarrow$ organic halide

CH$_2$=CH—CH$_3$ + Cl—Cl $\rightarrow$ CH$_2$Cl—CHCl—CH$_3$

(c) substitution
aromatic + halogen $\rightarrow$ organic halide + hydrogen halide

C$_6$H$_6$ + I—I $\rightarrow$ C$_6$H$_5$I + H—I

(d) elimination
alkyl halide + hydroxide ion $\rightarrow$ alkene + water + halide ion

CH$_3$—CH$_2$—CH$_2$—CH$_2$Cl + OH$^-$ $\rightarrow$ CH$_3$—CH$_2$—CH=CH$_2$ + HOH + Cl$^-$

(e) esterification
   carboxylic acid + alcohol → ester + water

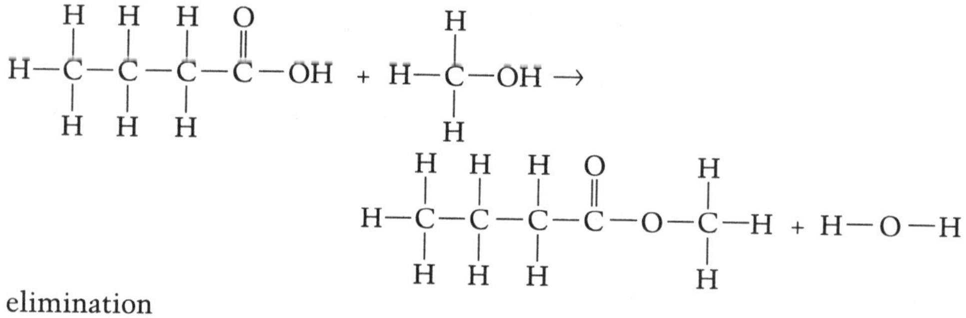

   H H H O   H
   | | | ‖ |
   H−C−C−C−C−OH + H−C−OH →
   | | | |
   H H H   H

                H H H O   H
                | | | ‖ |
                H−C−C−C−C−O−C−H + H−O−H
                | | | |
                H H H   H

(f) elimination
   alcohol → alkene + water

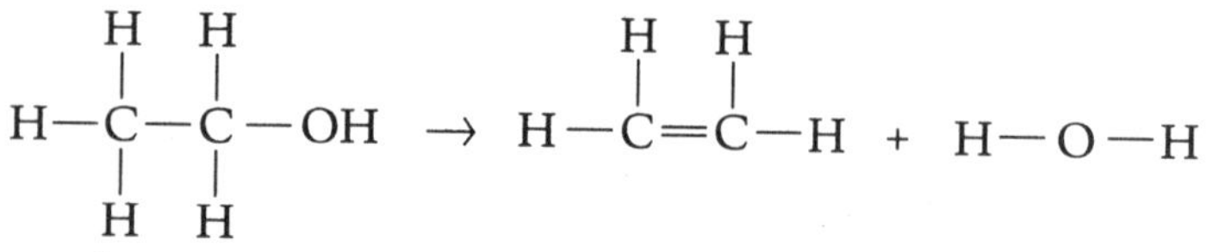

   H H   H H
   | | | |
   H−C−C−OH →   H−C=C−H + H−O−H
   | |
   H H

(g) combustion
   aromatic + oxygen → carbon dioxide + water vapor

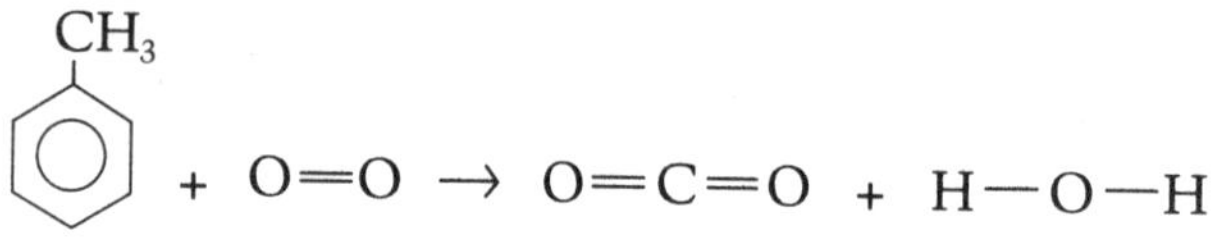

   CH₃
   |
   ⌬ + O=O → O=C=O + H−O−H

(h) other (oxidation)
   alcohol → aldehyde → carboxylic acid

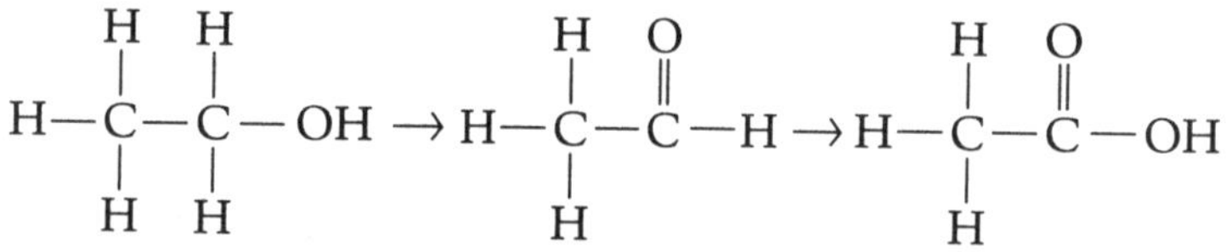

   H H   H O   H O
   | | | ‖ | ‖
   H−C−C−OH → H−C−C−H → H−C−C−OH
   | | | |
   H H   H   H

(i) other (oxidation)
   alcohol → ketone

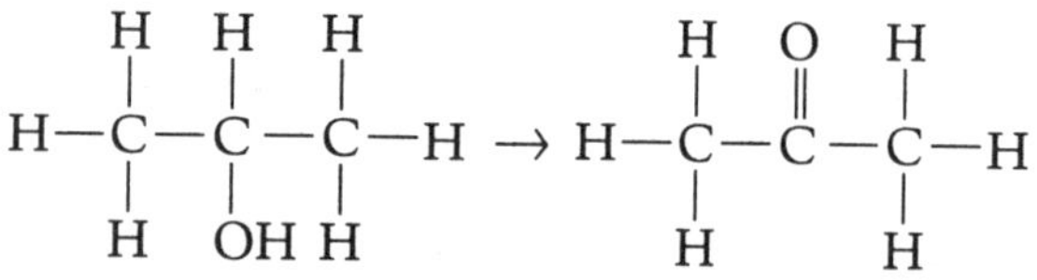

   H H H   H O H
   | | | | ‖ |
   H−C−C−C−H → H−C−C−C−H
   | | | | |
   H OH H   H   H

(j) condensation
   ammonia + carboxylic acid → amide + water

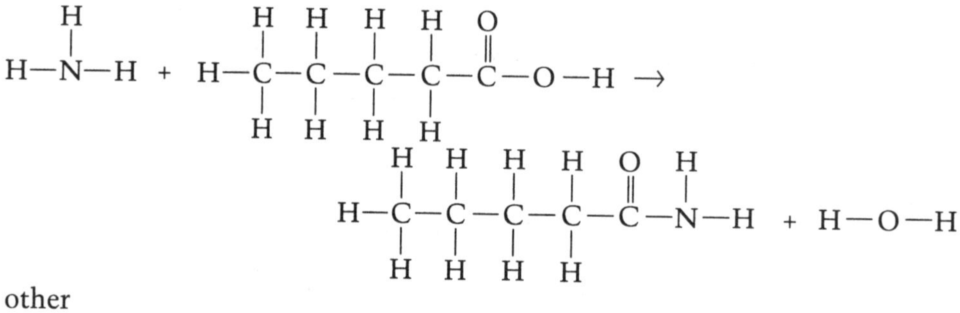

   H   H H H H O
   | | | | | ‖
   H−N−H + H−C−C−C−C−C−O−H →
              | | | |
              H H H H

                H H H H O H
                | | | | ‖ |
                H−C−C−C−C−C−N−H + H−O−H
                | | | |
                H H H H

(k) other
   ammonia + alkane → amine + hydrogen

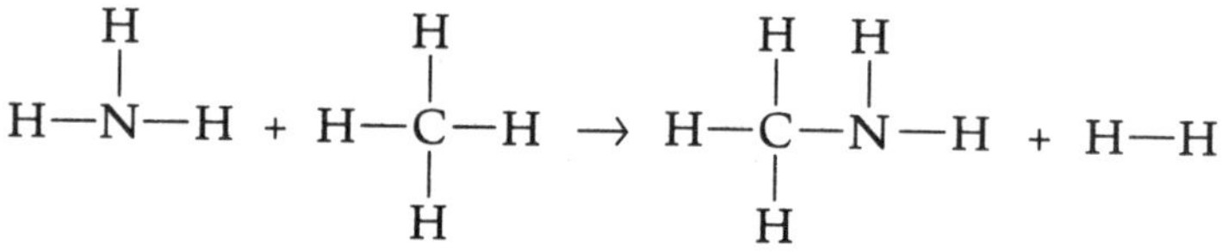

   H   H   H H
   | | | |
   H−N−H + H−C−H → H−C−N−H + H−H
              | |
              H   H

110   CHEMICAL DIVERSITY AND SYSTEMS OF BONDING

Lab Exercise 9B
Chemical Analysis of an Organic Compound (page 269)

Problem
What are the molecular formula and structure of an unknown organic substance?

Analysis

$$n_{gas} = 500 \text{ mL} \times \frac{1 \text{ mol}}{24.8 \text{ L}} = 20.2 \text{ mmol}$$

$$M_{gas} = \frac{0.94 \text{ g}}{0.0202 \text{ mol}} = 47 \text{ g/mol}$$

$$m_C = \frac{53.3}{100} \times 47 \text{ g} = 25 \text{ g}, \quad n_C = 25 \text{ g} \times \frac{1 \text{ mol}}{12.01 \text{ g}} = 2.1 \text{ mol}$$

$$m_H = \frac{15.7}{100} \times 47 \text{ g} = 7.4 \text{ g}, \quad n_H = 7.4 \text{ g} \times \frac{1 \text{ mol}}{1.01 \text{ g}} = 7.2 \text{ mol}$$

$$m_N = \frac{31.0}{100} \times 47 \text{ g} = 15 \text{ g}, \quad n_N = 15 \text{ g} \times \frac{1 \text{ mol}}{14.01 \text{ g}} = 1.0 \text{ mol}$$

According to the evidence, the molecular formula is C_2H_7N, and the likely structural diagram is:

```
   H  H  H                  H     H
   |  |  |                  |     |
H—C—C—N—H      or      H—C—N—C—H
   |  |                     |     |
   H  H                     H     H
```

Evaluation
The experimental design is judged to be adequate to answer the problem. A more precise design would be to use a mass spectrometer to determine the molar mass. However, this would not likely change the final answer. Some diagnostic tests on the unknown substance would improve the certainty of the answer.

Exercise (page 275)

24. propene (also known as propylene)
25. water
26. Each of the monomer units contains six carbon atoms.
27.
```
      F  F                    F  F  F  F
      |  |                    |  |  |  |
   n  C=C     →     · · —C—C—C—C— · ·
      |  |                    |  |  |  |
      F  F                    F  F  F  F
```
28.
```
      H  Cl                   H  Cl H  Cl
      |  |                    |  |  |  |
   n  C=C     →     · · —C—C—C—C— · ·
      |  |                    |  |  |  |
      H  H                    H  H  H  H
```

ORGANIC CHEMISTRY

29. *The equation is not balanced.*

$$CH_2(OH)-CH(OH)-CH_2(OH) + C_6H_4(COOH)_2 \rightarrow$$

[structure showing two glycerol units esterified with phthalate groups, with continuing polymer bonds indicated by dots]

$$+ H-O-H$$

30. (Discussion) A beneficial use of polyethylene is as an inexpensive raw material for manufacturing containers and wrappings for foods and other goods. Polyethylene is not biodegradable and contributes to the garbage disposal problem. Another problem is that polyethylene is made from non-renewable fossil resources. Alternatives to polyethylene include paper and biodegradable cellulose polymers. In actual practice, polyethylene is often recycled to reduce the problems due to its uses.

INVESTIGATION 9.5

Some Properties and Reactions of Organic Compounds (page 276)

Problem

What observations can be made about the properties and reactions of some organic compounds?

Evidence

PART I OBSERVATIONS			
Property	1-butanol	2-butanol	2-methyl-2-butanol
solubility in water (g/100 mL)	7.9	12.5	miscible
melting point (°C)	−89.5	−89	25.5
boiling point (°C)	117.2	99.5	82.3
reaction with $Na_{(s)}$	• gas bubbles	• fewer gas bubbles • some white solid formed later	• fewest gas bubbles • a lot of white solid formed later
reaction with $KMnO_{4(aq)}$	• initial color change to red • two clear layers eventually formed	• some color change to red • two layers formed: top pink, bottom red-brown with some precipitate	• some color change to red • single layer with a slight brown precipitate
reaction with $HCl_{(aq)}$	• no changes observed	• no changes observed	• a cloudy white mixture formed

Part II Observations

Cyclohexane
- Initially, no changes were observed with $KMnO_{4(aq)}$.
- After 5 min, the color changed to grey-purple.

112   *CHEMICAL DIVERSITY AND SYSTEMS OF BONDING*

Cyclohexene
- The color changed to green.
- After 5 min, a brown precipitate formed.

Part III Observations
- Benzoic acid did not completely dissolve in water after shaking.
- Blue litmus paper turned pink when dipped into the mixture.
- The solid disappeared on addition of $NaOH_{(aq)}$.
- A white precipitate formed on addition of $HCl_{(aq)}$.

Analysis

Part I
The three chemical tests performed provided clear evidence to distinguish among primary, secondary, and tertiary alcohols and may be used as diagnostic tests.

Part II
The reaction with basic potassium permanganate solution provided a noticeable difference between cyclohexane (saturated) and cyclohexene (unsaturated) that can be used as a diagnostic test.

Part III
Benzoic acid appeared to have a low solubility, but still dissolved slightly and ionized.
$C_6H_5COOH_{(s)} \rightarrow C_6H_5COO^-_{(aq)} + H^+_{(aq)}$
Sodium hydroxide neutralized benzoic acid.
$C_6H_5COOH_{(aq)} + NaOH_{(aq)} \rightarrow NaC_6H_5COO_{(aq)} + HOH_{(l)}$
Hydrochloric acid reacted with sodium benzoate.
$NaC_6H_5COO_{(aq)} + HCl_{(aq)} \rightarrow C_6H_5COOH_{(s)} + NaCl_{(aq)}$

Evaluation

The experimental design produced sufficient evidence for the development of diagnostic tests for some organic substances. However, more evidence is required to determine the products formed in all of the reactions.

Overview (page 279)

Review

1. fuels, petrochemical feedstock
2. Carbon atoms can form combinations of single, double, or triple covalent bonds with up to four other atoms, and they can bond to other carbon atoms to form very large structures.
3. Organic compounds are the basis of all known life forms.
4. *Answers will vary.*
    (a) C_nH_{2n+2}, CH_4, methane, fuel for heating
    (b) C_nH_{2n}, C_2H_4, ethene, petrochemical feedstock
    (c) C_nH_{2n-2}, C_2H_2, ethyne, fuel for welding
    (d) C_6H_5R, $C_6H_5CH_3$, methylbenzene, solvent in lacquers
    (e) RX, CCl_2F_2, freon (CFC-12), refrigerant
    (f) ROH, C_2H_5OH, ethanol, gasoline additive
    (g) R(H)CHO, CH_3CHO, ethanal, believed to cause alcohol "hangover"
    (h) R_1COR_2, CH_3COCH_3, propanone, solvent in plastic cements
    (i) R(H)COOH, CH_3COOH, ethanoic acid, vinegar
    (j) $R_1(H)COOR_2$, CH_3COOCH_3, methyl ethanoate, manufacture of artificial leather

5. (a) alkane, $CH_3-CH(CH_3)-CH_2-CH_3$

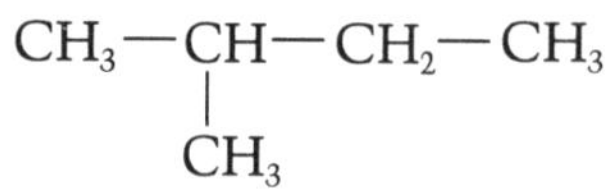

(b) aromatic, $CH_3-CH_2-C_6H_5$

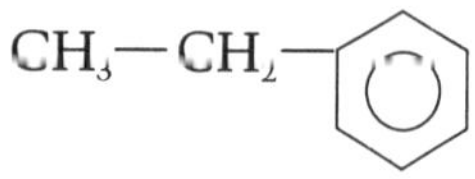

(c) alkyne, $CH_3-CH_2-C{\equiv}C-CH_2-CH_3$

(d) alkene,
$$CH_3-C(CH_3)=C(CH_3)-CH_2-CH_3$$

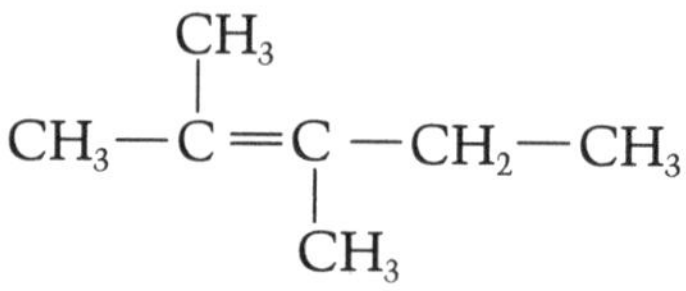

(e) alcohol, $CH_3-CH(OH)-CH_3$

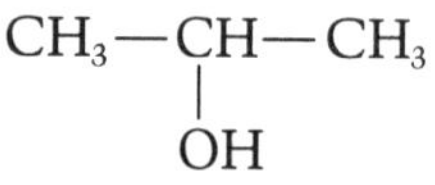

(f) amine, CH_3-NH_2

(g) ester, $CH_3-C(=O)-O-CH_3$

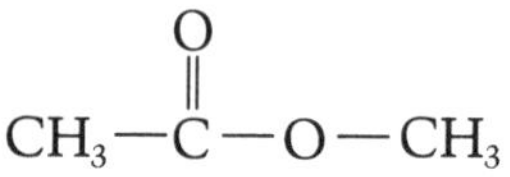

6. (a) 1,2-dichloroethane, organic halide
(b) methylpropene, alkene
(c) butane, alkane
(d) ethanal, aldehyde
(e) butanone, ketone
(f) ethanamide, amide

7. Polymer molecules are made up of many similar small molecules linked together. Cellulose and starch are found in living systems. Polyethylene is a manufactured polymer.

8. (a) reforming

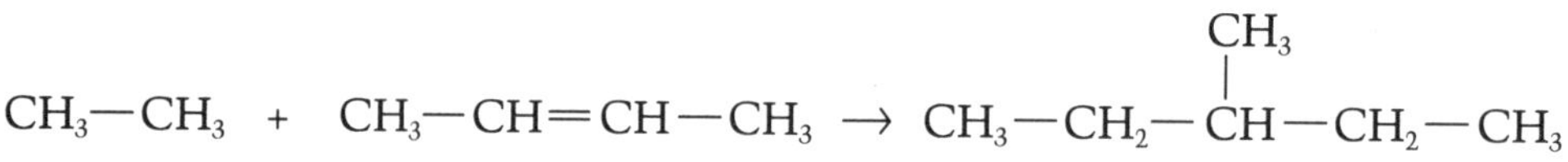

$$CH_3-CH_3 \;+\; CH_3-CH{=}CH-CH_3 \;\rightarrow\; CH_3-CH_2-CH(CH_3)-CH_2-CH_3$$

(b) cracking

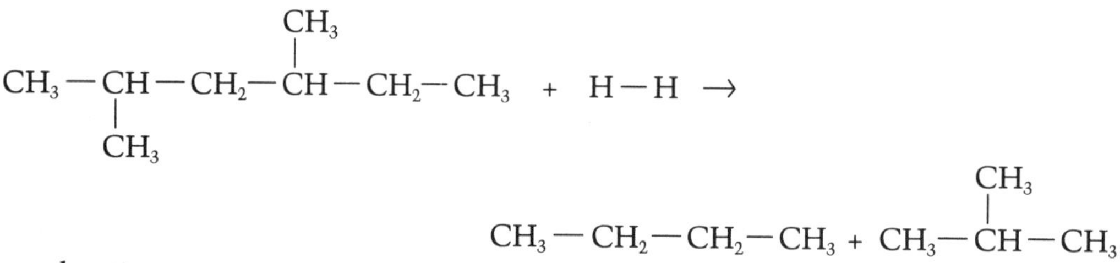

$$CH_3-CH(CH_3)-CH_2-CH(CH_3)-CH_2-CH_3 \;+\; H-H \;\rightarrow$$

$$CH_3-CH_2-CH_2-CH_3 \;+\; CH_3-CH(CH_3)-CH_3$$

(c) combustion

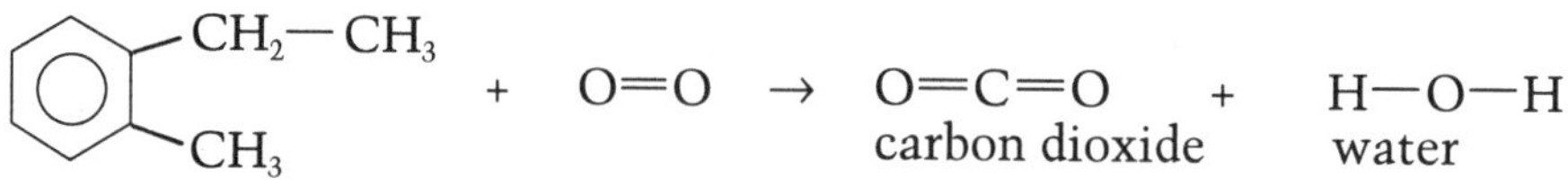

$$C_6H_4(CH_2CH_3)(CH_3) \;+\; O{=}O \;\rightarrow\; O{=}C{=}O \;+\; H-O-H$$
$$\text{carbon dioxide} \quad \text{water}$$

(d) addition

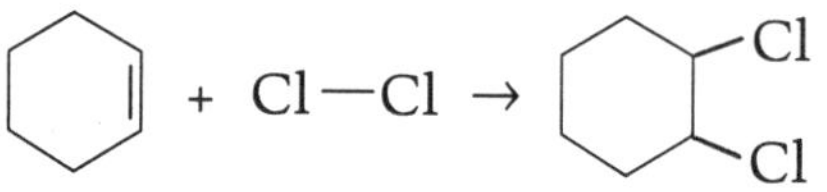

cyclohexene $+ \; Cl-Cl \;\rightarrow$ 1,2-dichlorocyclohexane

114   CHEMICAL DIVERSITY AND SYSTEMS OF BONDING

(e) reforming
propane + pentane → 2-methylheptane + hydrogen

(f) substitution
cyclopentane + bromine → bromocyclopentane + hydrogen bromide

                                 ⌬-Br   H—Br

(g) esterification
butanoic acid + 1-propanol →
    propyl butanoate   +   water

$$CH_3-CH_2-CH_2-\overset{\overset{O}{\|}}{C}-O-CH_2-CH_2-CH_3 \quad H-O-H$$

(h) elimination
2-chloropropane + hydroxide ion →
                        propene   +   water   +   chloride ion
                        $CH_2=CH-CH_3$ H—O—H Cl^-

9. (a) substitution

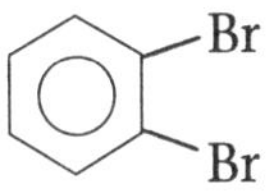

1,2-dibromobenzene + Br—Br →
1,2,3-tribromobenzene + 1,2,4-tribromobenzene + H—Br

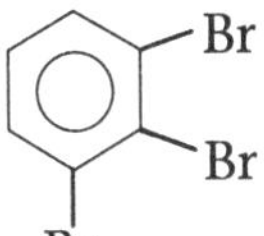

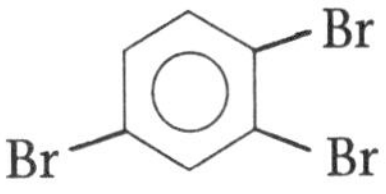

(b) addition
$CH_2=CH-CH_2-CH_3 + H-O-H \rightarrow$

                                                             OH
                                                             |
                 HO—$CH_2-CH_2-CH_2-CH_3$ + $CH_3-CH-CH_2-CH_3$
                 1-butanol   2-butanol

(c) addition
2-pentyne + hydrogen iodide →

$$CH_3-\underset{\underset{I}{|}}{\overset{\overset{I}{|}}{C}}-CH_2-CH_2-CH_3 + CH_3-\overset{\overset{I}{|}}{\underset{\underset{I}{|}}{CH}}-CH-CH_2-CH_3 +$$

2,2-diiodopentane   2,3-diiodopentane

$$CH_3-CH_2-\underset{\underset{I}{|}}{\overset{\overset{I}{|}}{C}}-CH_2-CH_3$$

3,3-diiodopentane

ORGANIC CHEMISTRY   115

(d) elimination

2-chorohexane + hydroxide ion → $CH_2=CH-(CH_2)_3-CH_3$ +

$CH_3-CH=CH-(CH_2)_2-CH_3$ | Cl^- | $H-O-H$
2-hexene   chloride ion   water

10. (a) $CH_3-\overset{O}{\overset{\|}{C}}-OH + CH_3-CH_2-OH \rightarrow$

$CH_3-\overset{O}{\overset{\|}{C}}-O-CH_2-CH_3 + H-O-H$

(b) $CH_2=CH_2 + H-O-H \rightarrow CH_3-CH_2-OH$

(c) $CH_3-\underset{Cl}{\underset{|}{CH}}-CH_3 + OH^- \rightarrow CH_2=CH-CH_3 + H-O-H + Cl^-$

(d) $n\ CF_2=CF_2 \rightarrow (-CF_2-CF_2-)_n$

(e) $Cl-CH_2-CH_2-Cl + OH^- \rightarrow CH_2=CH-Cl + H-O-H + Cl^-$

$CH_2=CH-Cl + OH^- \rightarrow CH\equiv CH + H-O-H + Cl^-$

(f) $CH_4 + Cl-Cl \rightarrow CH_3-Cl + H-Cl$

$CH_3-Cl + Cl-Cl \rightarrow Cl-CH_2-Cl + H-Cl$

$Cl-CH_2-Cl + F-F \rightarrow Cl-\underset{F}{\underset{|}{CH}}-Cl + H-F$

$Cl-\underset{F}{\underset{|}{CH}}-Cl + F-F \rightarrow Cl-\underset{F}{\underset{|}{\overset{F}{\overset{|}{C}}}}-Cl + H-F$

Extensions

11. The benefits of CFCs: CFCs are used as refrigerants in air conditioners and refrigerators, and to make plastic foams for furniture and insulation. CFCs are non-poisonous and non-flammable under normal conditions, and they are easily converted back and forth between the gas and liquid phases.

    The problems with CFCs: Scientific studies indicate that CFCs can break down the ozone molecules in Earth's upper atmosphere. Ozone in the upper atmosphere screens out ultraviolet radiation that is harmful to life on Earth. As CFCs reach the upper atmosphere, ultraviolet rays cause them to break apart and release chlorine atoms, which in turn react with the ozone molecules, converting them to ordinary oxygen molecules.

    The main focus should be on the answer with supporting arguments for the question, "Should governments ban CFCs?" Students should be encouraged to develop a position that could be defended in a class discussion.

12. The labelled peaks in the mass spectrograph of ethanol are likely:
    a (45 g/mol) $CH_3CH_2O^+$
    b (31 g/mol) CH_2OH^+
    c (29 g/mol) $CH_3CH_2^+$
    d (27 g/mol) CH_3C^+
    e (15 g/mol) CH_3^+

Lab Exercise 9C
Determining Structure (page 281)

Problem
What are the structure and the IUPAC name of C_3H_4O?

Analysis
The molar mass of 56 g/mol indicates that C_3H_4O is the molecular formula for the compound. The fast reaction with bromine is consistent with the compound containing a $C=C$ or $C\equiv C$ bond. The red precipitate produced with Fehling's solution suggests that the compound is an aldehyde. According to the evidence, C_3H_4O is propenal, whose structure is:

$$\begin{array}{c} \text{H} \quad\quad\quad \text{O} \\ | \quad\quad\quad\quad \| \\ \text{H}-\text{C}=\text{C}-\text{C}-\text{H} \\ | \\ \text{H} \end{array}$$

Lab Exercise 9D
Determining Percent Yield (page 281)

Problem
What is the percent yield in the initial substitution reaction between methane and chlorine?

Analysis

$$CH_4 \;+\; Cl_2 \;\rightarrow\; CH_3Cl + HCl$$
1.00 kg   2.46 kg
16.05 g/mol   50.49 g/mol

If a complete reaction of 1.00 kg of methane occurs,

$$n_{CH_4} = 1.00 \text{ kg} \times \frac{1 \text{ mol}}{16.05 \text{ g}} = 0.0623 \text{ kmol}$$

$$n_{CH_3Cl} = 0.0623 \text{ kmol} \times \frac{1}{1} = 0.0623 \text{ kmol}$$

$$m_{CH_3Cl} = 0.0623 \text{ kmol} \times \frac{50.49 \text{ g}}{1 \text{ mol}} = 3.15 \text{ kg}$$

Yield of chloromethane:

$$\% \text{ yield} = \frac{2.46 \text{ kg}}{3.15 \text{ kg}} \times 100 = 78.2\%$$

According to the evidence, the yield of chloromethane is 78.2% by mass.

UNIT V
ENERGY AND CHANGE

10 Energy Changes

Exercise (page 286)

Answers to questions 1 to 4 will vary. The intent is to generate discussion and to increase student awareness about the significance of energy in their lives. Student values will be reflected in their answers.

1. furnace, refrigerator (essential); calculator (practical), watch (efficient), car (convenient); CD player, electric toothbrush (non-essential)
2. For the examples given in question 1, the technologically useful forms of energy are: heat (furnace and refrigerator); electrical energy (calculator and watch); mechanical (watch, car, and electric toothbrush); sound (CD player)
3. Furnaces and cars usually burn fossil fuels directly (chemical energy); any electrical device uses energy which may be generated from the combustion of fossil fuels (chemical energy), from the fission of uranium (nuclear energy), or from hydroelectric power plants (solar energy).
4. Turn the thermostat down and lights out when not needed. Drive in car pools, use public transportation, ride a bike, or walk whenever possible. Use a sealant to close cracks in the house to reduce air infiltration.
5. The main advantages are that they are renewable and generally more environmentally friendly.

Exercise (page 288)

6. c_{ice} = 2.01 J/(g•°C), c_{water} = 4.19 J/(g•°C), c_{steam} = 2.01 J/(g•°C)
   The constant c_{ice} is used to calculate the quantity of heat, q, flowing into or out of a sample of ice when it changes in temperature. The constant c_{water} is used to calculate the quantity of heat, q, flowing into or out of a sample of liquid water when it changes in temperature. The constant c_{steam} is used to calculate the quantity of heat, q, flowing into or out of a sample of steam when it changes in temperature.
   The temperature change in a calculation of a quantity of heat is always a positive value. It is obtained by subtracting the initial temperature from the final temperature when there is a rise of temperature, or by taking the absolute value of the temperature difference, $|t_f - t_i|$, when there is a drop of temperature. Emphasize the cancellation of units in discussion.

7. $q = vc\Delta t$
   $= 1.50 \text{ L} \times \dfrac{4.19 \text{ kJ}}{\text{L} \cdot °\text{C}} \times (98.7 - 18.0)°\text{C}$
   $= 507 \text{ kJ}$

8. $q = mc\Delta t$
   $= 100 \text{ kg} \times \dfrac{2.01 \text{ J}}{\text{g} \cdot °\text{C}} \times (210 - 100)°\text{C}$
   $= 22.1 \text{ MJ}$

9. $q = mc\Delta t$

$= 2.5 \text{ kg} \times \dfrac{0.86 \text{ J}}{\text{g} \cdot °\text{C}} \times (350 - 15)°\text{C}$

$= 0.72 \text{ MJ}$

10. $q = vc\Delta t$

$v = \dfrac{q}{c\Delta t}$

$= \dfrac{250 \text{ kJ}}{3.7 \text{ kJ/(L} \cdot °\text{C)} \times 10°\text{C}}$

$= 6.8 \text{ L}$

11. (a) $q = vc\Delta t$

$= 100 \text{ L} \times \dfrac{4.19 \text{ kJ}}{\text{L} \cdot °\text{C}} \times (45 - 10)°\text{C}$

$= 15 \text{ MJ}$

(b) *Using the unrounded quantity of solar energy from (a),*
money saved = $1500 \times 15 \text{ MJ} \times 0.351¢/\text{MJ} = \$77$

12. (a) $q = vc\Delta t$

$= 100 \text{ L} \times \dfrac{4.19 \text{ kJ}}{\text{L} \cdot °\text{C}} \times (70 - 45)°\text{C}$

$= 10 \text{ MJ}$

(b) *Using the unrounded quantity of heat from (a),*
cost = $1500 \times 10 \text{ MJ} \times 0.351¢/\text{MJ} = \$55$

(c) (Discussion) *Have the students answer from at least one perspective other than the obvious economic perspective. Ethical and ecological perspectives fit in well here.*

13. In western Canada, natural gas water heaters are most common. Natural gas is a non-renewable resource.

INVESTIGATION 10.1

Designing and Evaluating a Water Heater (page 289)

Problem

What is the best design for a simple water heater?

Prediction

According to energy concepts and my experience, the best design is a metal can insulated with fibreglass insulation held in place by an outer layer of aluminum foil (shiny side on the inside). The top of the can is sealed with a circular piece of polystyrene.

Materials

lab apron	tape
safety glasses	thermometer
small soup can	centigram balance
3 – 4 cm thick strip of fibreglass insulation	stirring rod
	ring clamp and stand
2 – 3 cm thick piece of polystyrene	laboratory burner and striker
aluminum foil	watch or clock

Cross-Section of Water Heater

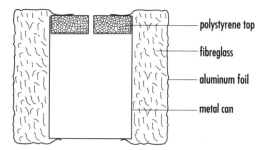

- polystyrene top
- fibreglass
- aluminum foil
- metal can

Evidence

	Trial 1	Trial 2*
mass of empty water heater (g)	67.11	84.10
mass of heater and water (g)	280.74	297.73
initial temperature of water (°C)	19.8	19.5
temperature of heated water (°C)	60.2	61.0
final temperature of water (°C) (after 10 min)	56.8	58.0

- In both trials, the bottom of the can was heated with a laboratory burner.
* In trial 2, a circular piece of fibreglass insulation was used on the bottom of the can after removing it from the heat source.

Analysis

	Trial 1 (no insulation on bottom)	Trial 2 (insulation on bottom)
mass of water (g)	213.63	213.63
Δt for heating (°C)	40.4	41.5
q during heating (kJ)	36.2	37.1
total mass of heater (g)	280.74	297.73
specific energy (kJ/kg)	129	125
Δt for cooling (°C)	3.4	3.0
q during cooling (kJ)	3.0	2.7
cooling rate (kJ/min)	0.30	0.27

Sample Calculation (Trial 1):

$$q_{(heating)} = mc\Delta t = 213.63 \text{ g} \times \frac{4.19 \text{ J}}{\text{g} \cdot °C} \times (60.2 - 19.8)°C = 36.2 \text{ kJ}$$

$$\text{specific energy} = \frac{36.2 \text{ kJ}}{0.280\,74 \text{ kg}} = 129 \text{ kJ/kg}$$

$$q_{(cooling)} = mc\Delta t = 213.63 \text{ g} \times \frac{4.19 \text{ J}}{\text{g} \cdot °C} \times (60.2 - 56.8)°C = 3.0 \text{ kJ}$$

$$\text{cooling rate} = \frac{3.0 \text{ kJ}}{10 \text{ min}} = 0.30 \text{ kJ/min}$$

The modification of adding insulation to the bottom of the can after removing it from the heat source slightly lowered both the specific energy and the cooling rate.

Evaluation

The general design of the water heater is very simple and involves inexpensive materials. To get more reliable results after many uses, the aluminum foil should be secured to the bottom edges of the can. The modification used in trial 2 produced a slightly poorer specific energy and a slightly better cooling rate. The significance of this modification is difficult to evaluate since the experimental uncertainties are not known. However, the results suggest an improvement to reduce the cooling rate as expected.

An evaluation of the prediction would require a comparison of several water heaters of different designs.

Exercise (page 295)

14. exothermic: l→s (solidification), g→l (condensation), g→s (sublimation)
    endothermic: s→l (fusion), l→g (vaporization), s→g (sublimation)

15. $\Delta H_{vap} = nH_{vap}$
    $= 100 \text{ g} \times \dfrac{1 \text{ mol}}{18.02 \text{ g}} \times \dfrac{40.8 \text{ kJ}}{1 \text{ mol}}$
    $= 226 \text{ kJ}$

16. (a) The heating curve for ethylene glycol from –50°C to +250°C is shown below.

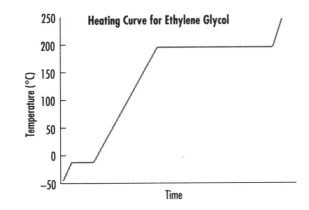

(b) $\Delta H_{vap} = nH_{vap}$
$= 500 \text{ g} \times \dfrac{1 \text{ mol}}{62.08 \text{ g}} \times \dfrac{58.8 \text{ kJ}}{1 \text{ mol}}$
$= 474 \text{ kJ}$

17. $\Delta H_{cond} = nH_{cond}$
$= 1.00 \text{ Mg} \times \dfrac{1 \text{ mol}}{18.02 \text{ g}} \times \dfrac{40.8 \text{ kJ}}{1 \text{ mol}}$
$= 2.26 \text{ GJ}$

$\Delta H_{fr} = nH_{fr}$
$= 1.00 \text{ Mg} \times \dfrac{1 \text{ mol}}{18.02 \text{ g}} \times \dfrac{6.03 \text{ kJ}}{1 \text{ mol}}$
$= 0.335 \text{ GJ}$

A total of 2.60 GJ would be required to change the vapor to a solid.

18. $\Delta H_{cond} = nH_{cond}$

$= 100 \text{ kg} \times \dfrac{1 \text{ mol}}{18.02 \text{ g}} \times \dfrac{40.8 \text{ kJ}}{1 \text{ mol}}$

$= 226 \text{ MJ}$

19. $\Delta H_s = nH_s$

$= 1.00 \text{ kg} \times \dfrac{1 \text{ mol}}{322.24 \text{ g}} \times \dfrac{78.0 \text{ kJ}}{1 \text{ mol}}$

$= 242 \text{ kJ}$

20. The released energy is transferred to the surroundings, usually causing an increase in temperature of the surroundings.

Exercise (page 297)

21. (a) The heating curve for snow at −15°C to liquid water at +37°C is shown below.

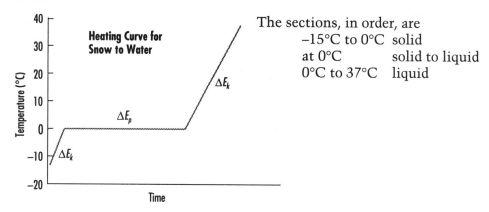

The sections, in order, are
−15°C to 0°C   solid
at 0°C   solid to liquid
0°C to 37°C   liquid

(b) (on graph)

(c) −15°C to 0°C $q = mc\Delta t$
    s → l at 0°C $\Delta H_{fus} = nH_{fus}$
    0°C to 37°C $q = mc\Delta t$

(d) $\Delta E_{total} = \underset{\text{(solid warming)}}{q} + \underset{\text{(melting)}}{\Delta H_{fus}} + \underset{\text{(liquid warming)}}{q}$

$= mc\Delta t + nH_{fus} + mc\Delta t$

$= 750 \text{ g} \times \dfrac{2.01 \text{ J}}{\text{g} \cdot °C} \times (0 - (-15))°C + 750 \text{ g} \times \dfrac{1 \text{ mol}}{18.02 \text{ g}} \times \dfrac{6.03 \text{ kJ}}{1 \text{ mol}}$

$+ 750 \text{ g} \times \dfrac{4.19 \text{ J}}{\text{g} \cdot °C} \times (37 - 0)°C$

$= 390 \text{ kJ}$

(e) If 750 g of snow at −15°C is eaten, then 390 kJ of energy would have to be used to change it into water at body temperature (37°C). The body would have 390 kJ less energy with which to fight hypothermia.

22. (a) $\Delta E_{total} = \underset{\text{(liquid warming)}}{q} + \underset{\text{(liquid vaporizing)}}{\Delta H_{vap}}$

$= mc\Delta t + nH_{vap}$

$= 100 \text{ kg} \times \dfrac{4.19 \text{ J}}{\text{g} \cdot °C} \times (100 - 10)°C + 100 \text{ kg} \times \dfrac{1 \text{ mol}}{18.02 \text{ g}} \times \dfrac{40.8 \text{ kJ}}{1 \text{ mol}}$

$= 38 \text{ MJ} + 226 \text{ MJ}$

$= 264 \text{ MJ}$

(b) $\Delta E_{total} = \underset{\text{(liquid warming)}}{q} + \underset{\text{(liquid vaporizing)}}{\Delta H_{vap}} + \underset{\text{(steam warming)}}{q}$

$= mc\Delta t + nH_{vap} + mc\Delta t$

$= 27 \text{ Mg} \times \dfrac{4.19 \text{ J}}{\text{g} \cdot {}^\circ\text{C}} \times (100 - 70){}^\circ\text{C} + 27 \text{ Mg} \times \dfrac{1 \text{ mol}}{18.02 \text{ g}} \times \dfrac{40.8 \text{ kJ}}{1 \text{ mol}}$

$+ 27 \text{ Mg} \times \dfrac{2.01 \text{ J}}{\text{g} \cdot {}^\circ\text{C}} \times (260 - 100){}^\circ\text{C}$

$= 3.4 \text{ GJ} + 61.1 \text{ GJ} + 8.7 \text{ GJ}$

$= 73.2 \text{ GJ}$

(c) (Discussion) Some possible advantages and disadvantages are listed in the table below.

Technology	Advantages	Disadvantages
chemical	• technology already developed • economically viable	• greenhouse/polluting gases • depletion of fossil fuels
nuclear	• high energy output for small quantity of fuel	• waste extremely hazardous • societal fears
hydro	• little pollution • fossil fuels conserved	• ecological/social disruptions from dam building
geothermal	• little pollution • fossil fuels conserved	• not easy to develop in all areas
solar	• little pollution • fossil fuels conserved • "free"	• source not continuous (due to darkness and cloudiness)
tidal	• little pollution • fossil fuels conserved • "free"	• technology developed only in coastal areas
ocean thermal	• little pollution • fossil fuels conserved	• technology developed only in coastal areas

Exercise (page 300)

23. (a) $H_{fr} = -6.03$ kJ/mol
    (b) $H_{vap} = +40.8$ kJ/mol

24. **Freezing of One Mole of Water**

E_p (kJ)

$H_2O_{(l)}$

$\Delta H_{fr} = -6.03$ kJ

$H_2O_{(s)}$

Vaporization of One Mole of Water

E_p (kJ)

$H_2O_{(g)}$

$\Delta H_{vap} = +40.8$ kJ

$H_2O_{(l)}$

25. (a) $H_{sub} = H_{melting} + H_{vap}$
$= (+6.03 \text{ kJ/mol}) + (+40.8 \text{ kJ/mol})$
$= +46.8 \text{ kJ/mol}$

(b) **Sublimation of One Mole of Ice**

E_p (kJ) axis: $H_2O_{(s)}$ lower level → $H_2O_{(g)}$ upper level; $\Delta H_{sub} = +46.8 \text{ kJ}$

26. (a) **Vaporization of 1.00 mol of Liquid Ammonia**

E_p (kJ) axis: $NH_{3(l)}$ lower level → $NH_{3(g)}$ upper level; $\Delta H_{vap} = +1.37 \text{ kJ}$

(b) **Condensation of 1.00 mol of Freon-12 Gas**

E_p (kJ) axis: $CCl_2F_{2(g)}$ upper level → $CCl_2F_{2(l)}$ lower level; $\Delta H_{cond} = -35.0 \text{ kJ}$

(c) **Solidification of 1.00 mol of Glauber's Salt**

E_p (kJ) axis: $Na_2SO_4 \cdot 10H_2O_{(aq)}$ upper level → $Na_2SO_4 \cdot 10H_2O_{(s)}$ lower level; $\Delta H_{solidification} = -78.0 \text{ kJ}$

INVESTIGATION 10.2

Molar Enthalpy of Solution (page 303)

Problem

What is the molar enthalpy of solution of an ionic compound?

Materials

lab apron
safety glasses
$NaNO_{3(s)}$
weighing boat
centigram balance
laboratory scoop
100 mL graduated cylinder
calorimeter apparatus
wash bottle of distilled water

Evidence

mass of sodium nitrate (each trial) = 8.50 g

TEMPERATURE CHANGES — DISSOLVING OF SODIUM NITRATE			
Trial	t_i (°C)	t_f (°C)	Δt (°C)
1	21.5	16.3	5.2
2	21.5	16.5	5.0
3	21.5	16.2	5.3

Analysis

$$n_{NaNO_3} = 8.50 \text{ g} \times \frac{1 \text{ mol}}{85.00 \text{ g}} = 0.100 \text{ mol}$$

$$\Delta H_{s\,NaNO_3} = q_{(calorimeter)}$$

$$nH_s = vc\Delta t$$

$$0.100 \text{ mol} \times H_s = 0.100 \text{ L} \times \frac{4.19 \text{ kJ}}{\text{L} \cdot °C} \times 5.2°C$$

$$H_s = 22 \text{ kJ/mol}$$

According to the evidence collected in this experiment, the molar enthalpy of solution for sodium nitrate is +22 kJ/mol.

Exercise (page 303)

27. Three assumptions made in student investigations involving simple calorimeters are:
    - No heat is transferred between the calorimeter and the outside environment.
    - Any heat absorbed or released by the calorimeter materials is negligible.
    - A dilute aqueous solution has the same density and specific heat capacity as pure water.

28. The measurements that limit the certainty in calorimetry experiments such as Investigation 10.2 are:
    - the measurement of the temperature with a thermometer, which limits the certainty to two significant digits if the temperature change is less than 10°C;
    - the specific heat capacity, which limits the certainty to three significant digits even if the temperature change, volume, and mass measurements are more certain.

29.
$$\Delta H_s = q$$
$$\text{(urea)} \quad \text{(calorimeter)}$$
$$nH_s = vc\Delta t$$
$$10 \text{ g} \times \frac{1 \text{ mol}}{60.07 \text{ g}} \times H_s = 0.150 \text{ L} \times \frac{4.19 \text{ kJ}}{\text{L}\cdot°\text{C}} \times 3.7°\text{C}$$
$$H_s = 14 \text{ kJ/mol}$$

The molar enthalpy of solution for urea is reported as +14 kJ/mol.

30.
$$\Delta H_{dil} = q$$
$$\text{HCl} \quad \text{(calorimeter)}$$
$$nH_{dil} = vc\Delta t$$
$$43.1 \text{ mL} \times \frac{11.6 \text{ mol}}{1 \text{ L}} \times H_{dil} = 500 \text{ mL} \times \frac{4.19 \text{ kJ}}{\text{L}\cdot°\text{C}} \times 2.6°\text{C}$$
$$H_{dil} = 11 \text{ kJ/mol}$$

The molar enthalpy of dilution for hydrochloric acid is reported as –11 kJ/mol.

31.
$$\Delta H_{fr} = q$$
$$\text{Ga} \quad \text{(calorimeter)}$$
$$nH_{fr} = mc\Delta t$$
$$10.0 \text{ g} \times \frac{1 \text{ mol}}{69.72 \text{ g}} \times H_{fr} = 50 \text{ g} \times \frac{4.19 \text{ J}}{\text{g}\cdot°\text{C}} \times (27.8 - 24.0)°\text{C}$$
$$H_{fr} = 5.55 \text{ kJ/mol}$$

The molar enthalpy of solidification of gallium is reported as –5.55 kJ/mol. In addition to the usual assumptions for simple calorimeters (question 27), it is assumed in this case that any heat transferred by the gallium, once it has solidified, is negligible.

Lab Exercise 10A
Designing a Calorimetry Lab (page 304)

Problem
What mass of ammonium nitrate should be dissolved to produce a temperature increase of 5.0°C?

Prediction
According to the law of conservation of energy, 17 g of ammonium nitrate should be dissolved as shown by the following calculations.
$$\Delta H_s = q$$
$$\text{NH}_4\text{NO}_4 \quad \text{(calorimeter)}$$
$$nH_s = mc\Delta t$$
$$n_{\text{NH}_4\text{NO}_4} \times \frac{25 \text{ kJ}}{1 \text{ mol}} = 0.250 \text{ L} \times \frac{4.19 \text{ kJ}}{\text{L}\cdot°\text{C}} \times 5.0°\text{C}$$
$$n_{\text{NH}_4\text{NO}_4} = 0.21 \text{ mol}$$
$$m_{\text{NH}_4\text{NO}_4} = 0.21 \text{ mol} \times \frac{80.06 \text{ g}}{1 \text{ mol}}$$
$$= 17 \text{ g}$$

Lab Exercise 10B
Molar Enthalpy of a Phase Change (page 304)

Problem
What is the molar enthalpy of fusion of ice?

Prediction
According to Table 10.3, page 293, the molar enthalpy of fusion of ice is +6.03 kJ/mol.

Analysis
mass of calorimeter water = 103.26 g – 3.76 g = 99.50 g
mass of ice used = 120.59 g – 103.26 g = 17.33 g

$$\Delta E_{total} \text{ (ice and ice water)} = q \text{ (calorimeter water)}$$

$$\Delta H_{fus} + q = q$$
$$nH_{fus} + mc\Delta t = mc\Delta t$$

$$17.33 \text{ g} \times \frac{1 \text{ mol}}{18.02 \text{ g}} \times H_{fus} + 17.33 \text{ g} \times \frac{4.19 \text{ J}}{\text{g} \cdot °C} \times (15.6 - 0.0)°C$$

$$= 99.50 \text{ g} \times \frac{4.19}{\text{g} \cdot °C} \times (32.4 - 15.6)°C$$

$$H_{fus} = 6.11 \text{ kJ/mol}$$

According to the evidence collected, the molar enthalpy of fusion of ice is +6.11 kJ/mol.

Evaluation
The experimental design is judged to be adequate to answer the problem with no obvious flaws or better alternatives. The major sources of uncertainty include the mass and temperature measurements and the drying of the ice before adding it to the calorimeter. These uncertainties might account for 5–10% as a rough estimate.

The percent difference of the experimental result is 1% as shown by the following calculation.

$$\% \text{ difference} = \frac{|6.11 \text{ kJ/mol} - 6.03 \text{ kJ/mol}|}{6.03 \text{ kJ/mol}} \times 100 = 1\%$$

Because the percent difference is very low, the prediction is judged to be verified and the value given in the reference table is acceptable.

INVESTIGATION 10.3

Molar Enthalpy of Reaction (page 305)

Problem
What is the molar enthalpy of neutralization for sodium hydroxide when 50 mL of aqueous 1.0 mol/L sodium hydroxide reacts with an excess quantity of 1.0 mol/L sulfuric acid?

Experimental Design

Excess sulfuric acid was mixed with sodium hydroxide in a simple laboratory calorimeter and the maximum temperature change recorded. According to the balanced chemical equation,

$H_2SO_{4(aq)} + 2NaOH_{(aq)} \rightarrow Na_2SO_{4(aq)} + 2H_2O_{(l)}$

and the information given in the problem, 50 mL of sodium hydroxide will require a minimum of 25 mL of sulfuric acid for neutralization.

Materials

lab apron
safety glasses
calorimeter apparatus
(2) medicine droppers
(2) 250 mL beakers

150 mL of 1.0 mol/L $NaOH_{(aq)}$
150 mL of 1.0 mol/L $H_2SO_{4(aq)}$
(2) 50 mL graduated cylinders
wash bottle of distilled water

Procedure

1. Measure 50 mL of 1.0 mol/L $H_2SO_{4(aq)}$ in a graduated cylinder and transfer into the calorimeter.
2. Measure 50 mL of 1.0 mol/L of $NaOH_{(aq)}$ in a graduated cylinder.
3. Record the initial temperature of each solution.
4. Add the $NaOH_{(aq)}$ to the calorimeter and quickly cover the calorimeter.
5. Stir and record the highest temperature obtained.
6. Dispose of the final solution into the sink.
7. Rinse and dry the calorimeter.
8. Repeat steps 1 to 7 two more times.

Evidence

		t_f of mixture (°C)		
Reactants	t_i (°C)	Trial 1	Trial 2	Trial 3
50.0 mL of 1.0 mol/L $H_2SO_{4(aq)}$	22.1	29.2	29.0	29.0
50.0 mL of 1.0 mol/L $NaOH_{(aq)}$	21.8			

Analysis

average temperature change = $((29.2 + 29.0 + 29.0)/3 - (22.1 + 21.8)/2)°C = 7.1°C$

$$\Delta H_{neutralization \; NaOH} = q_{(calorimeter)}$$

$$nH_{neutralization} = vc\Delta t$$

$$50.0 \; mL \times \frac{1.0 \; mol}{1 \; L} \times H_{neutralization} = 100.0 \; mL \times \frac{4.19 \; kJ}{L \cdot °C} \times 7.1°C$$

$$H_{neutralization} = 60 \; kJ/mol$$

According to the evidence, the molar enthalpy of neutralization for sodium hydroxide is reported as –60 kJ/mol.

Evaluation

The percent difference of this experiment is 5% as shown below.

$$\% \; difference = \frac{|60 \; kJ/mol - 57 \; kJ/mol|}{57 \; kJ/mol} \times 100 = 5\%$$

Since the percent difference is within the acceptable value and may be accounted for by normal experimental uncertainties, the simple calorimeter design and assumptions used are judged to be acceptable.

Exercise (page 305)

32. (a) $q = mc\Delta t$

$= 100.00 \text{ g} \times \dfrac{4.19 \text{ J}}{\text{g} \cdot {}^\circ\text{C}} \times 5.0{}^\circ\text{C}$

$= 2.1 \text{ kJ}$

(b) $\Delta E_{total} = \underset{\text{(water)}}{q} + \underset{\text{(cups)}}{q} + \underset{\text{(stirrer)}}{q} + \underset{\text{(thermometer)}}{q}$

$= 100.00 \text{ g} \times \dfrac{4.19 \text{ J}}{\text{g} \cdot {}^\circ\text{C}} \times 5.0{}^\circ\text{C} + 3.58 \text{ g} \times \dfrac{0.30 \text{ J}}{\text{g} \cdot {}^\circ\text{C}} \times 5.0{}^\circ\text{C}$

$+ 9.45 \text{ g} \times \dfrac{0.84 \text{ J}}{\text{g} \cdot {}^\circ\text{C}} \times 5.0{}^\circ\text{C} + 7.67 \text{ g} \times \dfrac{0.87 \text{ J}}{\text{g} \cdot {}^\circ\text{C}} \times 5.0{}^\circ\text{C}$

$= 2.2 \text{ kJ}$

(c) Using the unrounded values from parts (a) and (b),

% error $= \dfrac{|2.2 \text{ kJ} - 2.1 \text{ kJ}|}{2.2 \text{ kJ}} \times 100 = 4\%$

(d) The assumption of negligible heat transfer to the solid calorimeter materials is acceptable since the percent error introduced is very low and is likely less than the total experimental uncertainties.

INVESTIGATION 10.4

Designing a Calorimeter for Combustion Reactions (page 306)

Problem

Which design for a metal can calorimeter gives the largest molar enthalpy of combustion for paraffin wax?

Prediction

According to energy concepts and experience gained in Investigation 10.1, an insulated metal can with an aluminum foil skirt around the bottom would be a good design for a simple calorimeter.

Materials

lab apron
safety glasses
metal can calorimeter from
 previous investigation
aluminum foil
tape
thermometer
centigram balance
stirring rod
ring clamp and stand
matches
medicine dropper
paraffin candle
watch glass
100 mL graduated cylinder

Procedure

1. Light the candle and drip some molten wax onto a watch glass. Immediately press the bottom of the candle into the wax.
2. Measure the total mass of the candle and the watch glass.
3. Assemble the calorimeter and set it on a ring clamp attached to a stand.
4. Adjust the height so that the top of the candle is level with the bottom of the aluminum foil skirt.

5. Measure 100.0 mL of water in a graduated cylinder and pour into the calorimeter can.
6. Record the initial temperature of the water.
7. Cover the calorimeter and insert the thermometer so that it is in the middle of the water.
8. Light the candle and let it burn for 5 to 10 min.
9. Blow out the candle. Stir the water and record the maximum temperature reached by the water.
10. Measure the final mass of the candle and the watch glass.
11. Empty the calorimeter and dry the inside.
12. Repeat steps 3 to 11 twice.

Calorimeter for Combustion Reactions

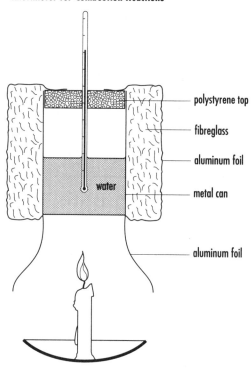

Evidence

	Trial 1	Trial 2	Trial 3
initial mass of candle (g)	95.49	94.93	94.17
final mass of candle (g)	94.93	94.17	93.42
initial temperature of water (°C)	22.5	23.0	22.6
final temperature of water (°C)	47.0	60.2	52.8

- Each trial used 100.0 mL of water.
- The candle went out twice in trial 2; it was immediately relit, and the calorimeter was moved up slightly.
- A considerable amount of black soot covered the bottom of the can by trial 3; the candle was a little below the aluminum skirt in this trial.

- Air currents affected the candle flame throughout, especially when the wick became longer.
- The size of the flame increased as the experiment progressed; the wick was trimmed before trial 3 to give a smaller flame.

Analysis

For Trial 1:
$$\Delta H_c \text{ (paraffin)} = q \text{ (water)}$$
$$nH_c = vc\Delta t$$
$$(95.49 - 94.93)\text{ g} \times \frac{1\text{ mol}}{352.77\text{ g}} \times H_c = 0.1000\text{ L} \times \frac{4.19\text{ kJ}}{\text{L}\cdot°\text{C}} \times (47.0 - 22.5)°\text{C}$$
$$H_c = 6.5\text{ MJ/mol}$$

For Trial 2: $H_c = 7.2$ MJ/mol
For Trial 3: $H_c = 6.0$ MJ/mol

According to the evidence collected using an insulated metal can calorimeter, the average molar enthalpy of combustion for paraffin wax is –6.6 MJ/mol.

Evaluation

The design of the metal can calorimeter was adequate to determine the molar enthalpy of combustion for paraffin wax. However, there were some uncertainties which could be due to the flaws in the design. The metal ring supporting the can was also exposed to the flame and did get quite hot by the end of a trial. The position of the tip of the flame and the size of the flame varied. Air currents also affected the flame. The design used was not the best available compared with other class results. A candle inside a can (open at the bottom) with an inner can for the water gave an average result of approximately –10 MJ/mol. This other design is just as simple, inexpensive, and probably more reliable than the design used in this experiment.

Exercise (page 308)

33.
$$\Delta H_c \text{ (acetylene)} = q \text{ (calorimeter)}$$
$$nH_c = C\Delta t$$
$$1.12\text{ g} \times \frac{1\text{ mol}}{26.04\text{ g}} \times H_c = \frac{6.49\text{ kJ}}{°\text{C}} \times 8.55°\text{C}$$
$$H_c = 1.29\text{ MJ/mol}$$

The molar enthalpy of combustion for acetylene is reported as –1.29 MJ/mol.

34. $\Delta H_h \text{ (zeolite)} = q \text{ (calorimeter)}$
$$= C\Delta t$$
$$= \frac{157\text{ kJ}}{°\text{C}} \times (73 - 27)°\text{C}$$
$$= 7.2\text{ MJ}$$

The enthalpy change of hydration for zeolite is –7.2 MJ.

35. Other factors include the environmental impact, cost, and availability of such fuels, and their technological application to existing cars.

Lab Exercise 10C
Calibrating a Bomb Calorimeter (page 308)

Problem

What is the heat capacity of a newly assembled oxygen bomb calorimeter?

Analysis

Trial 1:
$$\Delta H_{c \text{ (benzoic acid)}} = q_{\text{(calorimeter)}}$$
$$nH_c = C\Delta t$$
$$1.024 \text{ g} \times \frac{1 \text{ mol}}{122.13 \text{ g}} \times \frac{3231 \text{ kJ}}{1 \text{ mol}} = C \times (27.99 - 24.96)°C$$
$$C = 8.94 \text{ kJ/°C}$$

Trials 2 and 3 give results of 8.96 kJ/°C and 8.95 kJ/°C, respectively. According to the evidence, the average heat capacity of the oxygen bomb calorimeter is 8.95 kJ/°C.

Lab Exercise 10D
Energy Content of Foods (page 309)

Problem

Which substance, fat or sugar, has the higher energy content in kilojoules per mole?

Analysis

$$\Delta H_{c \text{ (fat)}} = q_{\text{(calorimeter)}}$$
$$nH_c = C\Delta t$$
$$1.14 \text{ g} \times \frac{1 \text{ mol}}{284.54 \text{ g}} \times H_c = \frac{8.57 \text{ kJ}}{°C} \times (30.28 - 25.00)°C$$
$$H_c = 11.3 \text{ MJ/mol}$$

According to the evidence, fat (stearic acid) has the higher energy content since it has a greater molar enthalpy of combustion (−11.3 MJ/mol) than sucrose (−5.63 MJ/mol) has.

Evaluation

Assuming that molar enthalpy of combustion is a suitable measure of energy content, the experimental design is adequate. The problem was answered and there do not appear to be any better designs. A major uncertainty is due to the difference in conditions and reactions between a combustion in a bomb calorimeter and biological reactions inside a human body. I am not very certain that the question has been answered by this experiment.

Overview (page 310)

Review

1. combustion of natural gas to heat buildings, cook meals, heat water, etc.; combustion of gasoline and diesel fuel to power cars and trucks; reaction of carbohydrates in the body to maintain body temperature

2. geothermal energy (hot springs), solar energy (water cycle), nuclear energy (uranium)
3. When heat is transferred between substances, a change in temperature occurs.
$q = mc\Delta t$, $q = vc\Delta t$, $q = C\Delta t$
4. $q = vc\Delta t$

$= 2.57 \text{ L} \times \dfrac{4.19 \text{ kJ}}{\text{L} \cdot °\text{C}} \times (95.0 - 3.0)°\text{C}$

$= 991 \text{ kJ}$
5. phase changes (different state is formed), chemical changes (new substances are formed), and nuclear changes (new elements or subatomic particles are formed)
6. endothermic: fusion (melting) < vaporization < sublimation (s → g)
exothermic: solidification < condensation < sublimation (g → s)
7. Phase, chemical, and nuclear changes all involve changes in potential energy. None of these are believed to involve a kinetic energy change.
8. The molar enthalpy of a chemical change (10^2 to 10^4 kJ/mol) is approximately ten to one hundred times the molar enthalpy of a phase change (10^0 to 10^2 kJ/mol).
9. Nuclear changes produce the largest quantities of energy because they involve the strongest bonds.
10. $\Delta H_{fus} = nH_{fus}$

$= 9.53 \text{ g} \times \dfrac{1 \text{ mol}}{18.02 \text{ g}} \times \dfrac{6.03 \text{ kJ}}{1 \text{ mol}}$

$= 3.19 \text{ kJ}$
11. A positive sign is used to report an endothermic molar enthalpy and a negative sign is used to report an exothermic molar enthalpy. During endothermic reactions, the potential energy of the chemical system increases as energy is gained from the surroundings. During exothermic reactions, the potential energy of the chemical system decreases as energy is lost to the surroundings.
12. (a) insulated container, thermometer, known quantity of water
    (b) law of conservation of energy
    (c) The calorimeter is isolated from the surroundings — no heat is transferred between the calorimeter and the outside environment. A dilute aqueous solution has the same density and specific heat capacity as pure water. Any heat absorbed or released by the calorimeter is negligible.
13. specific heat capacity: c, J/(g·°C)   change in potential energy: ΔE_p, J
    heat capacity: C, J/°C   amount of a substance: n, mol
    temperature change: Δt, °C   molar enthalpy: H, kJ/mol
    heat: q, J   enthalpy change: ΔH, kJ
14.
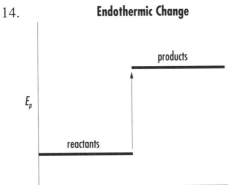
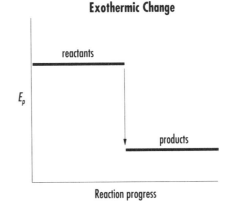

Applications

15. $\quad q = mc\Delta t$
$\quad\quad 16 \text{ kJ} = 938 \text{ g} \times c \times (35.0 - 19.5)°C$
$\quad\quad c_{brick} = 1.1 \text{ J/(g•°C)}$

16. $q = vc\Delta t$
$\quad = (8.0 \times 4.0 \times 0.10) \text{ m}^3 \times \dfrac{2.1 \text{ MJ}}{\text{m}^3 \text{•°C}} \times (30.0 - 18.0)°C$
$\quad = 81 \text{ MJ}$

17. (a) $q = vc\Delta t$
$\quad = (10.0 \times 11.0 \times 2.40) \text{ m}^3 \times \dfrac{0.0012 \text{ MJ}}{\text{m}^3 \text{•°C}} \times (20.5 - (-25.0))°C$
$\quad = 14 \text{ MJ}$

(b) Caulk any cracks such as those around windows and doors and make sure fireplace and furnace dampers are completely closed.
Other suggestions include closing drapes at night, minimizing times the outside doors are open, and taping plastic over windows (on the inside frame).

18. Steam at 100°C will release the energy of condensation to the surroundings (the person) in addition to the heat transferred when the water at 100°C cools to 35°C.

$\Delta E_{total} = \underset{\text{(steam)}}{\Delta H_{cond}} + \underset{\text{(water)}}{q}$

$\quad = nH_{cond} + mc\Delta t$

$\quad = 2.50 \text{ g} \times \dfrac{1 \text{ mol}}{18.02 \text{ g}} \times \dfrac{40.8 \text{ kJ}}{1 \text{ mol}} + 2.50 \text{ g} \times \dfrac{4.19 \text{ J}}{\text{g•°C}} \times (100 - 35)°C$

$\quad = 5.66 \text{ kJ} + 0.68 \text{ kJ}$
$\quad = 6.34 \text{ kJ}$

Water at 100°C will transfer only 0.68 kJ of heat as it cools from 100°C to 35°C.

19. (a) The heating curve for ice at –25°C to steam at 115°C is shown below.

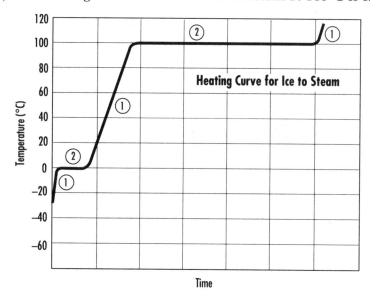

(b) Regions labelled 1 correspond to temperature changes and those labelled 2 correspond to phase changes.

20. (a) A sketch of the cooling curve for chlorine from 25°C to –150°C is shown below. (boiling/condensation point = –35°C, freezing point = –101°C)

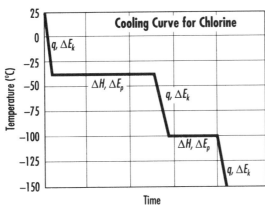

21. $\Delta E_{total} = \underset{\text{(ice)}}{q} + \underset{\text{(ice)}}{\Delta H_{fus}} + \underset{\text{(water)}}{q}$

$= mc\Delta t + nH_{fus} + mc\Delta t$

$= 2.39 \text{ kg} \times \dfrac{2.01 \text{ J}}{\text{g} \cdot °\text{C}} \times (0 - (-12.4))°\text{C} + 2.39 \text{ kg} \times \dfrac{1 \text{ mol}}{18.02 \text{ g}} \times \dfrac{6.03 \text{ kJ}}{1 \text{ mol}}$

$+ 2.39 \text{ kg} \times \dfrac{4.19 \text{ J}}{\text{g} \cdot °\text{C}} \times (97.8 - 0)°\text{C}$

$= 59.6 \text{ kJ} + 800 \text{ kJ} + 979 \text{ kJ}$

$= 1.839 \text{ MJ}$

22. $\Delta E_{total} = \underset{\text{(water)}}{q} + \underset{\text{(water)}}{\Delta H_{vap}} + \underset{\text{(steam)}}{q}$

$= mc\Delta t + nH_{vap} + mc\Delta t$

$= 1.00 \text{ Mg} \times \dfrac{4.19 \text{ J}}{\text{g} \cdot °\text{C}} \times (100 - 85)°\text{C} + 1.00 \text{ Mg} \times \dfrac{1 \text{ mol}}{18.02 \text{ g}} \times \dfrac{40.8 \text{ kJ}}{1 \text{ mol}}$

$+ 1.00 \text{ Mg} \times \dfrac{2.01 \text{ J}}{\text{g} \cdot °\text{C}} \times (250 - 100)°\text{C}$

$= 63 \text{ MJ} + 2.26 \text{ GJ} + 302 \text{ MJ}$

$= 2.63 \text{ GJ}$

23. (a) The heating curve for pressurized water heated from 85°C to 120°C as water and to 150°C as steam is shown below.

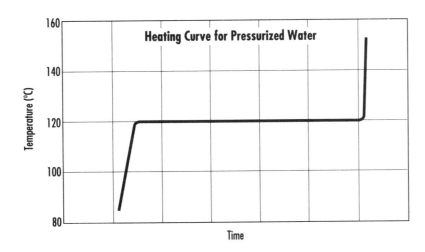

(b) $\Delta E_{total} = \underset{\text{(water)}}{q} + \underset{\text{(water)}}{\Delta H_{vap}} + \underset{\text{(steam)}}{q}$

$= mc\Delta t + nH_{vap} + mc\Delta t$

$= 120 \text{ kg} \times \dfrac{4.19 \text{ J}}{\text{g}\cdot°\text{C}} \times (120 - 85)°\text{C} + 120 \text{ kg} \times \dfrac{1 \text{ mol}}{18.02 \text{ g}} \times \dfrac{40.8 \text{ kJ}}{1 \text{ mol}}$

$+ 120 \text{ kg} \times \dfrac{2.01 \text{ J}}{\text{g}\cdot°\text{C}} \times (150 - 120)°\text{C}$

$= 18 \text{ MJ} + 272 \text{ MJ} + 7.2 \text{ MJ}$

$= 297 \text{ MJ}$

24. $\Delta H_{vap} = nH_{vap}$

$= 1.00 \text{ kg} \times \dfrac{1 \text{ mol}}{17.04 \text{ g}} \times \dfrac{1.37 \text{ kJ}}{1 \text{ mol}}$

$= 80.4 \text{ kJ}$

25. (a) $\Delta H_{cond} = nH_{cond}$

$= 100 \text{ kg} \times \dfrac{1 \text{ mol}}{30.08 \text{ g}} \times \dfrac{15.65 \text{ kJ}}{1 \text{ mol}}$

$= 52.0 \text{ MJ}$

(b) high pressure and low temperature

(c) $\Delta H_{vap} = nH_{vap}$

$1.00 \text{ MJ} = n \times \dfrac{15.65 \text{ kJ}}{1 \text{ mol}}$

$n_{C_2H_6} = 63.9 \text{ mol}$

(d) $\underset{\text{(ethane)}}{\Delta H_{vap}} = \underset{\text{(air)}}{q}$

$nH_{vap} = vc\Delta t$

$1.00 \text{ kg} \times \dfrac{1 \text{ mol}}{30.08 \text{ g}} \times \dfrac{15.65 \text{ kJ}}{1 \text{ mol}} = v \times \dfrac{0.0012 \text{ MJ}}{\text{m}^3\cdot°\text{C}} \times (29 - 19)°\text{C}$

$V_{air} = 43 \text{ m}^3$

26. $\underset{\text{(methane)}}{\Delta H_c} = \underset{\text{(water)}}{q}$

$nH_c = vc\Delta t$

$n \times \dfrac{802 \text{ kJ}}{1 \text{ mol}} = 3.77 \text{ L} \times \dfrac{4.19 \text{ kJ}}{\text{L}\cdot°\text{C}} \times (98.6 - 16.8)°\text{C}$

$n_{CH_4} = 1.61 \text{ mol}$

$m_{CH_4} = 1.61 \text{ mol} \times \dfrac{16.05 \text{ g}}{1 \text{ mol}}$

$= 25.9 \text{ g}$

27. $\underset{\text{(octane)}}{\Delta H_c} = \underset{\text{(engine)}}{q}$

$nH_c = C\Delta t$

$n \times \dfrac{1.3 \text{ MJ}}{1 \text{ mol}} = \dfrac{105 \text{ kJ}}{°\text{C}} \times (120 - 18)°\text{C}$

$n_{C_8H_{18}} = 8.2 \text{ mol}$

$m_{C_8H_{18}} = 8.2 \text{ mol} \times \dfrac{114.26 \text{ g}}{1 \text{ mol}}$

$= 0.94 \text{ kg}$

28. $$q_{\text{(brass)}} = q_{\text{(water)}}$$
$$mc\Delta t = mc\Delta t$$
$$77.5 \text{ g} \times c \times (98.7 - 23.5)°C = 102.76 \text{ g} \times \frac{4.19 \text{ J}}{\text{g} \cdot °C} \times (23.5 - 18.5)°C$$
$$c_{\text{brass}} = 0.37 \text{ J/(g} \cdot °C) \text{ or } 0.369 \text{ J/(g} \cdot °C)$$

Due to the temperature change, the answer has a certainty of two significant digits.

29. $$\Delta H_{c \text{ (butter)}} = q_{\text{(calorimeter)}}$$
$$mh_c = C\Delta t \qquad \text{Let } h_c \text{ be the required quantity.}$$
$$3.00 \text{ g} \times h_c = \frac{9.22 \text{ kJ}}{°C} \times (31.89 - 19.62)°C$$
$$h_c = 37.7 \text{ kJ/g}$$

Following the usual convention this value would be reported as –37.7 kJ/g.

Extensions

30. (a) Determine the boiling point, freezing point, specific or volumetric heat capacity, or molar enthalpies of solidification, vaporization, and combustion. Compare one or more of these experimental values to known values from a chemical reference.
    (b) Perform a combustion and mass spectrometer chemical analysis to determine the molecular formula.
    Perform esterification reactions with several carboxylic acids and identify the esters produced.

31. *Survey of use and costs could be done to match the billing period (e.g., monthly) of the utility. Some energy examples are as follows:*
    electrical — lights, kitchen appliances such as stoves and toasters, stereo, television, clothes and hair dryers, iron, computer, power tools
    chemical — furnace, hot water heater, barbecue, fireplace, candles
    A plan to reduce energy consumption would include more efficient use of energy and closer monitoring of the actual consumption. Ideally, a plan is implemented after use and costs have been surveyed for a period of time. The effect of the plan could then be evaluated by comparing before and after costs.

32. (a) *Problem*
    How energy efficient is my kettle?
    Experimental Design
    A measured volume of water at a known temperature is heated for a specific length of time and the final temperature determined. The efficiency is determined by the ratio of heat transferred to the water (output) and electrical energy used by the kettle (input).
    Most kettles have a 90% or higher efficiency.
    (b) *Students will likely suggest insulation for the kettle to reduce heat transferred to the kettle itself and the surroundings. It is debatable whether the energy efficiency gained is worth the additional cost and energy consumed in the manufacture of the insulation.*

Lab Exercise 10E
Solidification of Wax (page 313)

Problem
What is the molar enthalpy of solidification of paraffin wax?

Analysis
The average temperature change is 6.6°C.

$$\Delta H_{solid \atop (paraffin)} = q_{(calorimeter)}$$

$$nH_{solid} = vc\Delta t$$

$$25.00 \text{ g} \times \frac{1 \text{ mol}}{352.77 \text{ g}} \times H_{solid} = 0.150 \text{ L} \times \frac{4.19 \text{ kJ}}{\text{L} \cdot {}^\circ\text{C}} \times 6.6{}^\circ\text{C}$$

$$H_{solid} = 59 \text{ kJ/mol}$$

According to the evidence, the molar enthalpy of solidification for paraffin wax is reported as –59 kJ/mol.

Evaluation
The experimental design is judged to be inadequate even though the problem was answered. This design is flawed since it is not possible to determine the moment when all of the wax has solidified. It is likely that some liquid wax may exist on the inside of a solid piece. It is also likely that some of the solid has cooled significantly below its freezing point before the final temperature of

11 Reaction Enthalpies

Exercise (page 319)

1. (a) $CH_3OH_{(l)} + \frac{3}{2}O_{2(g)} \rightarrow CO_{2(g)} + 2H_2O_{(g)}$ $\Delta H_c° = -638.0$ kJ

   (b) $C_{(s)} + \frac{1}{4}S_{8(s)} \rightarrow CS_{2(l)}$ $\Delta H_f° = +89.0$ kJ

   (c) $ZnS_{(s)} + \frac{3}{2}O_{2(g)} \rightarrow ZnO_{(s)} + SO_{2(g)}$ $\Delta H_c° = -441.3$ kJ

   (d) $Fe_2O_{3(s)} \rightarrow 2Fe_{(s)} + O_{2(g)}$ $\Delta H_{sd}° = +824.2$ kJ

2. (a) $H_c° = -241.8$ kJ/mol for $H_{2(g)}$
   (b) $H_c° = -283.6$ kJ/mol for $NH_{3(g)}$
   (c) $H_c° = +81.6$ kJ/mol for $N_{2(g)}$
   (d) $H_c° = -372.8$ kJ/mol for $Fe_{(s)}$

3. (a) $C_3H_{8(g)} + 5O_{2(g)} \rightarrow 3CO_{2(g)} + 4H_2O_{(g)}$ $\Delta H_c° = -2.04$ MJ
   $C_3H_{8(g)} + 5O_{2(g)} \rightarrow 3CO_{2(g)} + 4H_2O_{(g)} + 2.04$ MJ

   (b) $\frac{1}{2}N_{2(g)} + \frac{1}{2}O_{2(g)} \rightarrow NO_{(g)}$ $\Delta H_f° = +90.2$ kJ

   $\frac{1}{2}N_{2(g)} + \frac{1}{2}O_{2(g)} + 90.2$ kJ $\rightarrow NO_{(g)}$

   (c) $C_2H_5OH_{(l)} + 3O_{2(g)} \rightarrow 2CO_{2(g)} + 3H_2O_{(g)}$ $\Delta H_c° = -1.28$ MJ
   $C_2H_5OH_{(l)} + 3O_{2(g)} \rightarrow 2CO_{2(g)} + 3H_2O_{(g)} + 1.28$ MJ

4. (a) $H_2SO_{4(aq)} + 2NaOH_{(aq)} \rightarrow Na_2SO_{4(aq)} + 2H_2O_{(l)}$ $\Delta H_r° = -114$ kJ

   (b) $H_n°_{H_2SO_4} = \dfrac{-114 \text{ kJ}}{1 \text{ mol}} = -114$ kJ/mol

   (c) $H_n°_{NaOH} = \dfrac{-114 \text{ kJ}}{2 \text{ mol}} = -57$ kJ/mol

5. (a) $H_{2(g)} + \frac{1}{2}O_{2(g)} \rightarrow H_2O_{(g)}$ $\Delta H_c° = -241.8$ kJ

   $H_2O_{(g)} \rightarrow H_{2(g)} + \frac{1}{2}O_{2(g)}$ $\Delta H_d° = +241.8$ kJ

   (b) The values are the same for both, but the signs are opposite.
   Generalization: $\Delta H_{forward} = -\Delta H_{reverse}$

6. (a) $2C_{(s)} + H_{2(g)} + 228.2$ kJ $\rightarrow C_2H_{2(g)}$

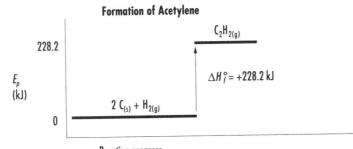

Formation of Acetylene

(b) $Al_2O_{3(s)} + 1675.7 \text{ kJ} \rightarrow 2Al_{(s)} + \frac{3}{2}O_{2(g)}$

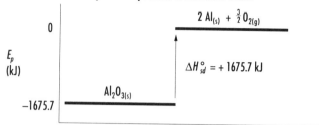

(c) $C_{(s)} + O_{2(g)} \rightarrow CO_{2(g)} + 393.5 \text{ kJ}$

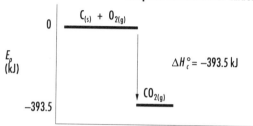

Exercise (page 323)

7.
$2Al_{(s)} + \frac{3}{2}O_{2(g)} \rightarrow Al_2O_{3(s)}$ $\hspace{2cm} \Delta H° = -1675.7 \text{ kJ}$

$Fe_2O_{3(s)} \rightarrow 2Fe_{(s)} + \frac{3}{2}O_{2(g)}$ $\hspace{2cm} \Delta H° = +824.2 \text{ kJ}$

$Fe_2O_{3(s)} + 2Al_{(s)} \rightarrow Al_2O_{3(s)} + 2Fe_{(s)}$ $\hspace{2cm} \Delta H_r° = -851.5 \text{ kJ}$

8.
$C_{(s)} + \frac{1}{2}O_{2(g)} \rightarrow CO_{(g)}$ $\hspace{2cm} \Delta H° = -110.5 \text{ kJ}$

$H_2O_{(g)} \rightarrow H_{2(g)} + \frac{1}{2}O_{2(g)}$ $\hspace{2cm} \Delta H° = +241.8 \text{ kJ}$

$H_2O_{(g)} + C_{(s)} \rightarrow CO_{(g)} + H_{2(g)}$ $\hspace{2cm} \Delta H_r° = +131.3 \text{ kJ}$

9.
$CO_{(g)} \rightarrow C_{(s)} + \frac{1}{2}O_{2(g)}$ $\hspace{2cm} \Delta H° = +110.5 \text{ kJ}$

$C_{(s)} + O_{2(g)} \rightarrow CO_{2(g)}$ $\hspace{2cm} \Delta H° = -393.5 \text{ kJ}$

$H_{2(g)} + \frac{1}{2}O_{2(g)} \rightarrow H_2O_{(g)}$ $\hspace{2cm} \Delta H° = -241.8 \text{ kJ}$

$CO_{(g)} + H_{2(g)} + O_{2(g)} \rightarrow CO_{2(g)} + H_2O_{(g)}$ $\hspace{2cm} \Delta H_r° = -524.8 \text{ kJ}$

10.
$CO_{(g)} \rightarrow C_{(s)} + \frac{1}{2}O_{2(g)}$ $\hspace{2cm} \Delta H° = +110.5 \text{ kJ}$

$CO_{2(g)} + 2H_2O_{(g)} \rightarrow CH_{4(g)} + 2O_{2(g)}$ $\hspace{2cm} \Delta H° = +802.7 \text{ kJ}$

$C_{(s)} + O_{2(g)} \rightarrow CO_{2(g)}$ $\hspace{2cm} \Delta H° = -393.5 \text{ kJ}$

$3H_{2(g)} + \frac{3}{2}O_{2(g)} \rightarrow 3H_2O_{(g)}$ $\hspace{2cm} \Delta H° = -725.4 \text{ kJ}$

$3H_{2(g)} + CO_{(g)} \rightarrow CH_{4(g)} + H_2O_{(g)}$ $\hspace{2cm} \Delta H_r° = -205.7 \text{ kJ}$

Lab Exercise 11A
Analysis Using Hess's Law (page 324)

Problem

What is the standard enthalpy change for the production of hydrogen from methane and steam?

$$CH_{4(g)} + H_2O_{(g)} \rightarrow CO_{(g)} + 3H_{2(g)} \qquad \Delta H_r^\circ = ?$$

Analysis

Many solutions are possible. Two of them are given below. Encourage students to find other solutions to emphasize this point in Hess's law: the net enthalpy change is independent of the route.

(2)	$CH_{4(g)} + 2O_{2(g)} \rightarrow CO_{2(g)} + 2H_2O_{(g)}$	$\Delta H^\circ = -802.7$ kJ
(from 3)	$CO_{2(g)} + H_{2(g)} \rightarrow CO_{(g)} + H_2O_{(g)}$	$\Delta H^\circ = +41.2$ kJ
(from 4)	$4H_2O_{(g)} \rightarrow 4H_{2(g)} + 2O_{2(g)}$	$\Delta H^\circ = +967.2$ kJ

$$CH_{4(g)} + H_2O_{(g)} \rightarrow CO_{(g)} + 3H_{2(g)} \qquad \Delta H_r^\circ = +205.7 \text{ kJ}$$

According to the evidence from equations 2, 3, and 4, the standard enthalpy change is +205.7 kJ.

(from 1)	$C_{(s)} + \frac{1}{2}O_{2(g)} \rightarrow CO_{(g)}$	$\Delta H^\circ = -110.5$ kJ
(from 4)	$H_2O_{(g)} \rightarrow H_{2(g)} + \frac{1}{2}O_{2(g)}$	$\Delta H^\circ = +241.8$ kJ
(from 5)	$CH_{4(g)} \rightarrow C_{(s)} + 2H_{2(g)}$	$\Delta H^\circ = +74.4$ kJ

$$CH_{4(g)} + H_2O_{(g)} \rightarrow CO_{(g)} + 3H_{2(g)} \qquad \Delta H_r^\circ = +205.7 \text{ kJ}$$

According to the evidence from equations 1, 4, and 5, the standard enthalpy change is +205.7 kJ.

Lab Exercise 11B
Testing Hess's Law (page 325)

Problem

What is the standard molar enthalpy of combustion of pentane?

Prediction

According to Hess's law and the data given, the standard molar enthalpy of combustion of pentane is −3508.8 kJ/mol.

$C_5H_{12(l)} \rightarrow 5C_{(s)} + 6H_{2(g)}$	$\Delta H^\circ = +173.5$ kJ
$5C_{(s)} + 5O_{2(g)} \rightarrow 5CO_{2(g)}$	$\Delta H^\circ = -1967.5$ kJ
$6H_{2(g)} + 3O_{2(g)} \rightarrow 6H_2O_{(g)}$	$\Delta H^\circ = -1450.8$ kJ
$6H_2O_{(g)} \rightarrow 6H_2O_{(l)}$	$\Delta H^\circ = -264.0$ kJ

$$C_5H_{12(l)} + 8O_{2(g)} \rightarrow 5CO_{2(g)} + 6H_2O_{(l)} \qquad \Delta H_c^\circ = -3508.8 \text{ kJ}$$

$$H_{c\,C_5H_{12}}^\circ = \frac{\Delta H_c^\circ}{n} = \frac{-3508.8 \text{ kJ}}{1 \text{ mol}} = -3508.8 \text{ kJ/mol}$$

Analysis

$$\Delta H_{c\,(\text{pentane})} = q_{(\text{calorimeter})}$$

$$nH_c = vc\Delta t$$

$$2.15 \text{ g} \times \frac{1 \text{ mol}}{72.17 \text{ g}} \times H_c = 1.24 \text{ L} \times \frac{4.19 \text{ kJ}}{\text{L}\cdot°\text{C}} \times (37.6 - 18.4)°\text{C}$$

$$H_c = 3.35 \text{ MJ/mol}$$

According to the evidence collected in this experiment, the molar enthalpy of combustion of pentane is reported as −3.35 MJ/mol.

Evaluation

The major uncertainties are due to measurement and the difference in conditions for obtaining the predicted and experimental answers. The predicted value corresponds to SATP conditions, whereas the experimental value does not.

The percent difference of this experiment is 4.6% as shown by the following calculation.

$$\% \text{ difference} = \frac{|3.35 \text{ MJ/mol} - 3.5088 \text{ MJ/mol}|}{3.5088 \text{ MJ/mol}} \times 100 = 4.6\%$$

The prediction is judged to be verified. The percent difference is reasonably low and is likely due to the uncertainties mentioned. Hess's law appears to be acceptable since the prediction was verified.

INVESTIGATION 11.1

Applying Hess's Law (page 325)

Problem

What is the molar enthalpy of combustion for magnesium?

Procedure

1. Measure precisely about 1.00 g of magnesium oxide in a clean, dry calorimeter.
2. Obtain 50.0 mL of 1.00 mol/L $HCl_{(aq)}$ in a 50 mL graduated cylinder.
3. Measure the initial temperature of the $HCl_{(aq)}$.
4. Add the acid to the magnesium oxide and quickly cover the calorimeter.
5. Stir and record the maximum temperature of the mixture.
6. Dispose of the contents in the sink, rinse and dry the calorimeter cup.
7. Obtain about 15 cm of magnesium ribbon and clean it with steel wool.
8. Measure the mass of the clean magnesium ribbon.
9. Obtain 50.0 mL of $HCl_{(aq)}$ in a 50 mL graduated cylinder.
10. Place the acid in the calorimeter and record the initial temperature.
11. Add the magnesium ribbon to the calorimeter, making sure all of the metal is in the acid.

12. Cover, stir, and record the maximum temperature.
13. Dispose of the contents in the sink.

Evidence

Reactant	Mass (g)	Initial Temperature of $HCl_{(aq)}$ (°C)	Final Temperature of Mixture (°C)
magnesium oxide	0.91	23.6	36.0
magnesium metal	0.16	23.6	37.7

- 50.0 mL of 1.00 mol/L $HCl_{(aq)}$ were used for each reaction.

Analysis

$$\Delta H_{r_{MgO}} = q_{(calorimeter)}$$

$$nH_r = vc\Delta t$$

$$0.91 \text{ g} \times \frac{1 \text{ mol}}{40.31 \text{ g}} \times H_r = 0.0500 \text{ L} \times \frac{4.19 \text{ kJ}}{\text{L} \cdot °\text{C}} \times 12.4°\text{C}$$

$$H_r = 0.12 \text{ MJ/mol}$$

According to the evidence, the molar enthalpy of reaction for magnesium oxide is reported as -0.12 MJ/mol.

$$\Delta H_{r_{Mg}} = q_{(calorimeter)}$$

$$nH_r = vc\Delta t$$

$$0.16 \text{ g} \times \frac{1 \text{ mol}}{24.31 \text{ g}} \times H_r = 0.0500 \text{ L} \times \frac{4.19 \text{ kJ}}{\text{L} \cdot °\text{C}} \times 14.1°\text{C}$$

$$H_r = 0.45 \text{ MJ/mol}$$

According to the evidence, the molar enthalpy of reaction for magnesium is reported as -0.45 MJ/mol.

According to Hess's law, the three equations can be combined as follows.

$MgO_{(s)} + 2HCl_{(aq)} \rightarrow MgCl_{2(aq)} + H_2O_{(l)}$ $\Delta H = 1 \text{ mol} \times -0.12 \text{ MJ/mol} = -0.12 \text{ MJ}$

$Mg_{(s)} + 2HCl_{(aq)} \rightarrow MgCl_{2(aq)} + H_{2(g)}$ $\Delta H = 1 \text{ mol} \times -0.45 \text{ MJ/mol} = -0.45 \text{ MJ}$

$H_{2(g)} + \frac{1}{2}O_{2(g)} \rightarrow H_2O_{(l)}$ $\Delta H = -0.2858 \text{ MJ}$

$Mg_{(s)} + \frac{1}{2}O_{2(g)} \rightarrow MgO_{(s)}$ $\Delta H_c = -0.62 \text{ MJ}$

Therefore, $H_{c_{Mg}} = -0.62$ MJ/mol.

According to the evidence collected and Hess's law, the molar enthalpy of combustion for magnesium is -0.62 MJ/mol.

Evaluation

The experimental design is judged to be adequate because the problem was answered. A bomb calorimeter could be used as an alternative. However, the design used in this experiment is simpler and less expensive. I am reasonably confident in the results obtained. The procedure was adequate, although additional trials would improve the certainty of the answer. The technological skills were simple and adequate.

> Overall, I am moderately certain of the results. Sources of uncertainty include limitations of the measurements made and the non-SATP conditions under which the results were obtained. I would estimate a 5% total for these uncertainties.
>
> The accuracy of the results obtained is reflected by the percent difference shown.
>
> $$\% \text{ difference} = \frac{|0.62 \text{ MJ/mol} - 0.6016 \text{ MJ/mol}|}{0.6016 \text{ MJ/mol}} \times 100 = 2\%$$
>
> The prediction based on the table of standard molar enthalpies of formation is judged to be verified since the experimental result agrees well with the predicted value. The percent difference is likely due to the uncertainties discussed. The table of standard molar enthalpies appears to be acceptable for predicting molar enthalpies since the prediction was verified. I am quite confident in this judgment.

Exercise (page 329)

11. (a) $\Delta H_r^\circ = \Sigma n H_{fp}^\circ - \Sigma n H_{fr}^\circ$
    $= (1 \text{ mol} \times -110.5 \text{ kJ/mol} + 3 \text{ mol} \times 0 \text{ kJ/mol})$
    $- (1 \text{ mol} \times -74.4 \text{ kJ/mol} + 1 \text{ mol} \times -241.8 \text{ kJ/mol})$
    $= +205.7 \text{ kJ}$

    (b) $\Delta H_r^\circ = \Sigma n H_{fp}^\circ - \Sigma n H_{fr}^\circ$
    $= (1 \text{ mol} \times -393.5 \text{ kJ/mol} + 1 \text{ mol} \times 0 \text{ kJ/mol})$
    $- (1 \text{ mol} \times -110.5 \text{ kJ/mol} + 1 \text{ mol} \times -241.8 \text{ kJ/mol})$
    $= -41.2 \text{ kJ}$

    (c) $3H_{2(g)} + N_{2(g)} \rightarrow 2NH_{3(g)}$ $\Delta H_f^\circ = n H_f^\circ = 2 \text{ mol} \times -45.9 \text{ kJ/mol} = -91.8 \text{ kJ}$

12. (a) $\Delta H_c^\circ = \Sigma n H_{fp}^\circ - \Sigma n H_{fr}^\circ$
    $= (4 \text{ mol} \times +90.2 \text{ kJ/mol} + 6 \text{ mol} \times -241.8 \text{ kJ/mol})$
    $- (4 \text{ mol} \times -45.9 \text{ kJ/mol} + 5 \text{ mol} \times 0 \text{ kJ/mol})$
    $= -906.4 \text{ kJ}$

    $H_{c_{NH_3}}^\circ = \dfrac{-906.4 \text{ kJ}}{4 \text{ mol}} = -226.6 \text{ kJ/mol}$

    (b) $\Delta H_c^\circ = \Sigma n H_{fp}^\circ - \Sigma n H_{fr}^\circ$
    $= (2 \text{ mol} \times +33.2 \text{ kJ/mol}) - (2 \text{ mol} \times +90.2 \text{ kJ/mol} + 1 \text{ mol} \times 0 \text{ kJ/mol})$
    $= -114.0 \text{ kJ}$

    $H_{c_{NO}}^\circ = \dfrac{-114.0 \text{ kJ}}{2 \text{ mol}} = -57.0 \text{ kJ/mol}$

    (c) $\Delta H_r^\circ = \Sigma n H_{fp}^\circ - \Sigma n H_{fr}^\circ$
    $= (2 \text{ mol} \times -174.1 \text{ kJ/mol} + 1 \text{ mol} \times +90.2 \text{ kJ/mol})$
    $- (3 \text{ mol} \times +33.2 \text{ kJ/mol} + 1 \text{ mol} \times -285.8 \text{ kJ/mol})$
    $= -71.8 \text{ kJ}$

    $H_{r_{NO_2}}^\circ = \dfrac{-71.8 \text{ kJ}}{3 \text{ mol}} = -23.9 \text{ kJ/mol}$

13. (a) $\Delta H_r^\circ = \Sigma n H_{fp}^\circ - \Sigma n H_{fr}^\circ$
    $= (1 \text{ mol} \times -365.6 \text{ kJ/mol})$
    $- (1 \text{ mol} \times -45.9 \text{ kJ/mol} + 1 \text{ mol} \times -174.1 \text{ kJ/mol})$
    $= (-365.6 \text{ kJ}) - (-220.0 \text{ kJ})$
    $= -145.6 \text{ kJ}$

(b)

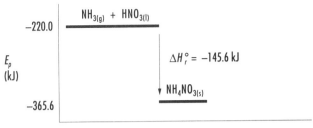

Reaction of Ammonia and Nitric Acid

14. (Discussion) *The following statements are examples of what can be presented from a variety of perspectives. The concept of a multi-perspective approach to an STS topic is important and students should be able to generate examples from a number of perspectives. Note that a perspective can be either positive or negative and that an evaluation requires a judgment of the relative importance of a point of view. (See Appendix D, page 538, for a list of perspectives.)*

- From a scientific perspective, each step in the fertilizer manufacturing process can be described with a chemical equation.
- From a technological perspective, the producer uses the most efficient and reliable processes available in order to facilitate continuous production.
- From an ecological perspective, the fertilizers are mostly beneficial, although run-off from fields into lakes and rivers can cause some problems with weed growth, etc.
- From an economic perspective, the technology must be the most economic possible, in order for the producer to be competitive.
- From a political perspective, the Canadian producers of fertilizers seek markets in the United States under the provisions of the Canada–U.S. free trade agreement.
- From a legal perspective, the packaging must display the appropriate WHMIS symbols and the consumer must be provided with the appropriate MSDS sheets.
- From an ethical perspective, the technology is not benefiting the starving peoples of the world to the extent needed.
- From a social perspective, communities have been built around fertilizer plants. Fertilizers have allowed people to farm less than ideal agricultural land with increased crop yield that would feed large numbers of people in cities.
- From a militaristic perspective, fertilizers were made possible because explosives were developed for military purposes. (The Haber process, page 443, was originally needed for the production of ammonia-based explosives.)
- From a mystical perspective, some people believe that natural, "organic" fertilizers are more acceptable than technologically-produced fertilizers.

Lab Exercise 11C
Testing $\Delta H_r°$ from Formation Data (page 330)

Problem
What is the molar enthalpy of combustion of methanol?

Prediction

According to the table of standard molar enthalpies of formation (Appendix F, page 551) and the chemical equation, the molar enthalpy of combustion of methanol is –726.0 kJ/mol.

$2CH_3OH_{(l)} + 3O_{2(g)} \rightarrow 2CO_{2(g)} + 4H_2O_{(l)}$

$\Delta H_r° = \Sigma nH_{fp}° - \Sigma nH_{fr}°$
$= (2 \text{ mol} \times -393.5 \text{ kJ/mol} + 4 \text{ mol} \times -285.8 \text{ kJ/mol})$
$- (2 \text{ mol} \times -239.1 \text{ kJ/mol} + 3 \text{ mol} \times 0 \text{ kJ/mol})$
$= -1452.0 \text{ kJ}$

$H_{c_{CH_3OH}}° = \dfrac{-1452.0 \text{ kJ}}{2 \text{ mol}} = -726.0 \text{ kJ/mol}$

Analysis

$\Delta H_{c \text{ (methanol)}} = q_{\text{(calorimeter)}}$

$nH_c = C\Delta t$

$4.38 \text{ g} \times \dfrac{1 \text{ mol}}{32.05 \text{ g}} \times H_c = \dfrac{10.9 \text{ kJ}}{°C} \times (27.9 - 20.4)°C$

$H_c = 0.60 \text{ MJ/mol}$

According to the evidence, the molar enthalpy of combustion of methanol is reported as –0.60 MJ/mol.

Evaluation

The experimental design using a bomb calorimeter is a common design and is adequate to answer the problem. There are no better alternatives for the study of combustion reactions. Other than the measurements, the non-SATP temperatures in the experiment also contribute to uncertainties in the results. The small differences between the experimental results and the predictions in previous experiments show that this is not a major factor. I estimate that a 5% difference would be reasonable for this experiment.

The experimental result was not very accurate as shown by the 18% difference.

% difference $= \dfrac{|0.60 \text{ MJ/mol} - 0.7260 \text{ MJ/mol}|}{0.7260 \text{ MJ/mol}} \times 100 = 18\%$

Because of the relatively large percent difference, the prediction is falsified. It is not clear how this difference can be accounted for by the uncertainties mentioned. However, since the evidence of only one trial has been reported, additional experimentation is required before judging the table of standard molar enthalpies of formation as unacceptable.

Lab Exercise 11D
Determining Standard Molar Enthalpies of Formation (page 330)

Problem

What is the standard molar enthalpy of formation for hexane, $C_6H_{14(l)}$?

Analysis

$$C_6H_{14(l)} + \tfrac{19}{2} O_{2(g)} \rightarrow 6CO_{2(g)} + 7H_2O_{(l)}$$

$\Delta H_c^\circ = 1 \text{ mol} \times -4162.9 \text{ kJ/mol} = -4169.2 \text{ kJ}$
$\Delta H_c^\circ = \Sigma nH_{fp}^\circ - \Sigma nH_{fr}^\circ$
$-4162.9 \text{ kJ} = (6 \text{ mol} \times -393.5 \text{ kJ/mol} + 7 \text{ mol} \times -285.8 \text{ kJ/mol})$
$\quad\quad\quad - (1 \text{ mol} \times H_f^\circ + 19/2 \text{ mol} \times 0 \text{ kJ/mol})$
$H_f^\circ = -198.7 \text{ kJ/mol}$

According to the evidence provided, the standard molar enthalpy of formation for hexane is reported as –198.7 kJ/mol.

Lab Exercise 11E
Calorimetry of a Nuclear Reaction (page 332)

Problem

What is the molar enthalpy of nuclear decay for a radioactive isotope?

Experimental Design

A measured quantity of a particular isotope is placed into a calorimeter, and the temperature change of the water is determined after a specific period of time. The rate of decay of the isotope, determined from a reference book, is used to calculate the amount of the isotope that decayed during the calorimetry experiment.

$\Delta H_{decay} = q$
(isotope) (calorimeter)

$nH_{decay} = C\Delta t$

Exercise (page 334)

Answers to questions 15 to 18 will vary depending upon the locale and personal opinion.

15. (a) In Alberta, the two most common energy sources for heating are gas and coal. In eastern Canada, oil, coal, solar energy (hydroelectric), and nuclear energy are commonly used.
    (b) By the year 2030, it is likely that solar energy (direct and indirect hydroelectricity) will be very common. Nuclear energy may be very common if nuclear fusion reactors can be developed economically.
    (c) Solar energy and geothermal energy are the two most environmentally friendly choices.

There are several alternatives in handling questions 16, 17, and 18. Any of these questions could be used as a class project or assigned as an individual project.

16. This question could be answered using the format of either group presentations followed by questions from the class or a debate. There is a possibility of six different groups — coal, natural gas, hydro, solar, and nuclear power plants, and the alternative of energy conservation practices eliminating the need for the power plant.

Each group could be required to present and defend:
- brief description of the plant (or practices) including resources used (or saved)
- the main technological component or design
- suggested location(s) with reasons
- why their choice is the best — using technological, ecological, and economic perspectives, and at least one other perspective
- what the trade-offs of their choice are

17. Cold fusion refers to the claimed discovery by Stanley Pons and Martin Fleischmann of the fusion of hydrogen during the electrolysis of heavy water under ordinary laboratory conditions. Pons and Fleischmann claimed an output (as heat) of up to fifty times the input electrical energy. Their experiment failed to meet a crucial criterion of scientific work — the ability to be replicated by independent scientists. This "breakthrough" was controversial primarily in the way it was released and promoted in the media without the usual scientific accountability.

    The story of cold fusion is an excellent study of the nature of science and scientific knowledge. The December 1989 issue of the Discover *magazine has a good review of cold fusion along with a comparison of the more widely accepted "hot" fusion.*

18. Some possibilities of heating alternatives include:
    - insulating curtains
    - automatic set-back thermostats
    - waste heat from local industries
    - smaller houses
    - earth-sheltered buildings
    - street designed so that houses face south

    Some possibilities of transportation alternatives include:
    - car pools
    - natural gas powered vehicles
    - computer monitoring and adjusting due to driving conditions
    - more people working at home with computer links to an office
    - decentralization of offices and stores in large urban centres
    - improved public transportation
    - solar assisted electric cars
    - super conductor levitated monorails
    - bicycle lane and roadways

    Some possibilities of industrial alternatives include:
    - recycling materials
    - recycling energy as heat
    - natural lighting
    - clustering of different industries to share resources and reduce costs

    Students should select their home or a local business to report on present energy use, and on products and processes presently used or possible in the future to reduce energy use.

Exercise (page 337)

19. $H°_{c_{C_{52}H_{16}O}} = \dfrac{\Delta H°_c}{n} = \dfrac{-44.0 \text{ MJ}}{2 \text{ mol}} = -22.0 \text{ MJ/mol}$

    $\Delta H°_c = nH°_c$

    $= 100 \text{ kg} \times \dfrac{1 \text{ mol}}{656.68 \text{ g}} \times \dfrac{22.0 \text{ MJ}}{1 \text{ mol}}$

    $= 3.35 \text{ GJ}$

20. Some alternative energy sources are nuclear, hydro, direct solar, geothermal, tidal, and ocean thermal.
    See the answer to question 22 (c) in Chapter 10 for some advantages and disadvantages. This question can be used as an extension to previous discussions. Students could pursue one alternative in more detail.

21. $H^\circ_{c_{C_8H_{18}}} = \dfrac{\Delta H^\circ_c}{n} = \dfrac{-10148.2 \text{ kJ}}{2 \text{ mol}} = -5074.1 \text{ kJ/mol}$

    $\Delta H^\circ_c = nH^\circ_c$

    $= 1.00 \text{ kg} \times \dfrac{1 \text{ mol}}{114.26 \text{ g}} \times \dfrac{5074.1 \text{ kJ}}{1 \text{ mol}}$

    $= 44.4 \text{ MJ}$

22. (a) $2CH_3OH_{(l)} + 3O_{2(g)} \rightarrow 2CO_{2(g)} + 4H_2O_{(g)}$

    $\Delta H^\circ_r = \Sigma nH^\circ_{fp} - \Sigma nH^\circ_{fr}$
    $= (2 \text{ mol} \times -393.5 \text{ kJ/mol} + 4 \text{ mol} \times -241.8 \text{ kJ/mol})$
    $\quad - (2 \text{ mol} \times -239.1 \text{ kJ/mol} + 3 \text{ mol} \times 0 \text{ kJ/mol})$
    $= -1276.0 \text{ kJ}$

    $H^\circ_{c_{CH_3OH}} = \dfrac{-1276.0 \text{ kJ}}{2 \text{ mol}} = -638.0 \text{ kJ/mol}$

    $\Delta H^\circ_c = 1.00 \text{ kg} \times \dfrac{1 \text{ mol}}{32.05 \text{ g}} \times \dfrac{638.0 \text{ kJ}}{1 \text{ mol}}$

    $= 19.9 \text{ MJ}$

    (b) $2H_{2(g)} + O_{2(g)} \rightarrow 2H_2O_{(g)}$

    $\Delta H^\circ_r = \Sigma nH^\circ_{fp} - \Sigma nH^\circ_{fr}$
    $= (2 \text{ mol} \times -241.8 \text{ kJ/mol})$
    $\quad - (2 \text{ mol} \times 0 \text{ kJ/mol} + 1 \text{ mol} \times 0 \text{ kJ/mol})$
    $= -483.6 \text{ kJ}$

    $H^\circ_{c_{H_2}} = \dfrac{-483.6 \text{ kJ}}{2 \text{ mol}} = -241.8 \text{ kJ/mol}$

    $\Delta H^\circ_c = 1.00 \text{ kg} \times \dfrac{1 \text{ mol}}{2.02 \text{ g}} \times \dfrac{241.8 \text{ kJ}}{1 \text{ mol}}$

    $= 120 \text{ MJ}$

    (c) Methanol has a lower heat content per kilogram than octane, while hydrogen has a higher heat content per kilogram than octane.

    (d) Examples include cost, environmental impact, compatibility with existing vehicles, and safety of storing and handling.

23. (a) $CH_{4(g)} + 2O_{2(g)} \rightarrow CO_{2(g)} + 2H_2O_{(g)}$

    $\Delta H^\circ_r = \Sigma nH^\circ_{fp} - \Sigma nH^\circ_{fr}$
    $= (1 \text{ mol} \times -393.5 \text{ kJ/mol} + 2 \text{ mol} \times -241.8 \text{ kJ/mol})$
    $\quad - (1 \text{ mol} \times -74.4 \text{ kJ/mol} + 2 \text{ mol} \times 0 \text{ kJ/mol})$
    $= -802.7 \text{ kJ}$

    $H^\circ_{c_{CH_4}} = \dfrac{-802.7 \text{ kJ}}{1 \text{ mol}}$

$$\Delta H_c^\circ \underset{\text{(methane)}}{} = q \underset{\text{(water)}}{}$$

$$nH_c^\circ = vc\Delta t$$

$$n \times \frac{802.7 \text{ kJ}}{1 \text{ mol}} = 100 \text{ L} \times \frac{4.19 \text{ kJ}}{\text{L} \cdot {}^\circ\text{C}} \times (70 - 5){}^\circ\text{C}$$

$$n_{CH_4} = 34 \text{ mol}$$

(b) Improve the insulation of the storage tank and water lines; make sure that the gas burner is cleaned and adjusted for maximum efficiency; heat already partially-heated water.

(c) Solar energy may be used to preheat water.
See questions 11 and 12 in Chapter 10.

24. $\Delta H_r^\circ = nH_r^\circ$

$$= 4.26 \text{ mol} \times \frac{1.9 \times 10^{10} \text{ kJ}}{1 \text{ mol}}$$

$$= 8.1 \times 10^{10} \text{ kJ}$$

Lab Exercise 11F
Molar Enthalpy of Formation (page 339)

Problem

What is the molar enthalpy of formation for cyclohexane?

Analysis

$$\Delta H_c \underset{\text{(cyclohexane)}}{} = q \underset{\text{(calorimeter)}}{}$$

$$nH_c = C\Delta t$$

$$1.43 \text{ g} \times \frac{1 \text{ mol}}{84.18 \text{ g}} \times H_c = \frac{10.5 \text{ kJ}}{{}^\circ\text{C}} \times (26.67 - 20.32){}^\circ\text{C}$$

$$H_c = 3.92 \text{ MJ/mol}$$

Therefore, the molar enthalpy of combustion for $C_6H_{12(l)}$ is reported as -3.92 MJ/mol.

$C_6H_{12(l)} + 9O_{2(g)} \rightarrow 6CO_{2(g)} + 6H_2O_{(l)}$
$\Delta H_c = 1 \text{ mol} \times -3.92 \text{ MJ/mol} = -3.92 \text{ MJ}$
$\Delta H_c = \Sigma nH_{fp} - \Sigma nH_{fr}$
$-3.92 \text{ MJ} = (6 \text{ mol} \times -393.5 \text{ kJ/mol} + 6 \text{ mol} \times -285.8 \text{ kJ/mol})$
$\qquad - (1 \text{ mol} \times H_f + 9 \text{ mol} \times 0 \text{ kJ/mol})$
$H_f = -151 \text{ kJ/mol}$

According to the evidence collected, the molar enthalpy of formation for cyclohexane is -151 kJ/mol.

Overview (page 340)

Review

1. molar enthalpy of a substance in a specific reaction; enthalpy change, ΔH, of a reaction; enthalpy change as a term in an equation; potential energy diagram

2. If ΔH is negative, then the energy term is on the product side. If ΔH is positive, then the energy term is on the reactant side.

3. (a) When one mole of butane is burned to produce carbon dioxide and water vapor, 2657 kJ of energy is released when the initial and final conditions are at SATP.

(b) $C_4H_{10(g)} + \frac{13}{2} O_{2(g)} \rightarrow 4CO_{2(g)} + 5H_2O_{(g)}$ $\quad\quad \Delta H_c° = -2657$ kJ

(c) $C_4H_{10(g)} + \frac{13}{2} O_{2(g)} \rightarrow 4CO_{2(g)} + 5H_2O_{(g)} + 2657$ kJ

(d) According to the table of molar enthalpies in Appendix F, page 551, the value of $H_f°$ for butane is –125.6 kJ/mol. This is not the molar enthalpy of combustion of butane; it represents the quantity of heat released per mole of butane formed from its elements, carbon and hydrogen, as illustrated in the equation below.
$4C_{(s)} + 5H_{2(g)} \rightarrow C_4H_{10(g)}$ $\quad\quad \Delta H_f° = -125.6$ kJ

4. The reference zero point of potential energy for chemical reactions is the potential energy of elements in their most stable form at SATP. The standard molar enthalpy of formation of any element in its most stable form at SATP is therefore assigned as zero.

5. During an endothermic reaction, potential energy increases. During an exothermic reaction, potential energy decreases.

6. List any two of the following: Hess's law, enthalpies of formation method, bond energies.

7. Enthalpy changes are independent of the way the system changes from its initial state to its final state.

8. If a chemical equation is reversed, then its ΔH changes its sign. If the coefficients of a chemical equation are altered by multiplying or dividing by a constant factor, then the ΔH is altered in the same way.

9. (a) $\Delta H_r° = \Sigma nH_{fp}° - \Sigma nH_{fr}°$

(b) The enthalpy change for a chemical reaction equals (the sum of the amount times the molar enthalpy of formation for each product) minus (the sum of the amount times the molar enthalpy of formation for each reactant).

(c) The summation term for the products represents the total potential energy of the products, while the summation term for the reactants represents the total potential energy of the reactants.

10. (a) Nuclear reactions typically have much larger energy changes than chemical reactions.

(b) Any endothermic reaction has a positive ΔH, and any exothermic reaction has a negative ΔH.

11. (a) Heating: Choose among solar energy, geothermal energy, biomass gas, electricity from nuclear reactions, or others.
Transportation: Choose among alcohol/gasohol and hydrogen fuels, batteries and fuel cells, or others.
Industry: Choose among solar energy, nuclear energy, hydroelectricity, or others.

(b) Heating: improved insulation
Transportation: car pools and mass transit
Industry: recovery of waste heat

Applications

12. $C_3H_{8(g)} + 5O_{2(g)} \rightarrow 3CO_{2(g)} + 4H_2O_{(g)} + 2.25$ MJ
or
$C_3H_{8(g)} + 5O_{2(g)} \rightarrow 3CO_{2(g)} + 4H_2O_{(g)}$ $\quad\quad \Delta H_c° = -2.25$ MJ

13. (a) $H_{r_{CO_2}}° = \dfrac{+129 \text{ kJ}}{1 \text{ mol}} = +129$ kJ/mol

(b) $2NaHCO_{3(s)} \rightarrow Na_2CO_{3(s)} + H_2O_{(g)} + CO_{2(g)}$ $\Delta H_r° = +129\text{ kJ}$

(c)

Decomposition of Baking Soda

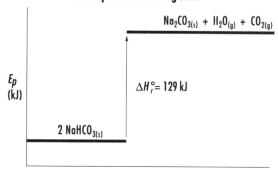

14. (a) $2CO_{(g)} + 2NO_{(g)} \rightarrow N_{2(g)} + 2CO_{2(g)}$ $\Delta H_r° = -746\text{ kJ}$

(b) $\underset{NO}{H_r°} = \dfrac{-746\text{ kJ}}{2\text{ mol}} = -373\text{ kJ/mol}$

(c) $\Delta H_r° = 500\text{ g} \times \dfrac{1\text{ mol}}{30.01\text{ g}} \times \dfrac{373\text{ kJ}}{1\text{ mol}}$

$= 6.21\text{ MJ}$

15. (a) $\underset{N_2H_4}{H_c°} = \dfrac{-400\text{ kJ}}{1\text{ mol}} = -400\text{ kJ/mol}$

(b) $\underset{O_2}{H_c°} = \dfrac{-400\text{ kJ}}{3\text{ mol}} = -133\text{ kJ/mol}$

(c) $\Delta H_c° = 8.00\text{ g} \times \dfrac{1\text{ mol}}{32.00\text{ g}} \times \dfrac{133\text{ kJ}}{1\text{ mol}}$

$= 33.3\text{ kJ}$

16. (a) $CuS_{(s)} + \frac{3}{2}O_{2(g)} \rightarrow CuO_{(s)} + SO_{2(g)}$

$\Delta H_c° = \Sigma nH_{fp}° - \Sigma nH_{fr}°$

$= (1\text{ mol} \times -157.3\text{ kJ/mol} + 1\text{ mol} \times -296.8\text{ kJ/mol})$

$- (1\text{ mol} \times -53.1\text{ kJ/mol} + \frac{3}{2}\text{ mol} \times 0\text{ kJ/mol})$

$= -401.0\text{ kJ}$

(b)

Roasting of Copper(II) Sulfide

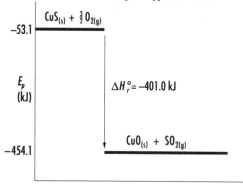

17. (a) $C_6H_{12}O_{6(s)} + 6O_{2(g)} \rightarrow 6CO_{2(g)} + 6H_2O_{(l)}$

$\Delta H_r° = \Sigma nH_{fp}° - \Sigma nH_{fr}°$

$$= (6 \text{ mol} \times -393.5 \text{ kJ/mol} + 6 \text{ mol} \times -285.8 \text{ kJ/mol})$$
$$- (1 \text{ mol} \times -1273.1 \text{ kJ/mol} + 6 \text{ mol} \times 0 \text{ kJ/mol})$$
$$= -2802.7 \text{ kJ}$$

(b) The equations are the same, although in respiration, liquid water is produced while in combustion, water vapor is produced. Also, respiration occurs more slowly than combustion; therefore, it releases its energy more slowly.

18. (a) $2N_{2(g)} + 6H_2O_{(l)} \rightarrow 4NH_{3(g)} + 3O_{2(g)}$

$$\Delta H_r^\circ = \Sigma n H_{fp}^\circ - \Sigma n H_{fr}^\circ$$
$$= (4 \text{ mol} \times -45.9 \text{ kJ/mol} + 3 \text{ mol} \times 0 \text{ kJ/mol})$$
$$- (2 \text{ mol} \times 0 \text{ kJ/mol} + 6 \text{ mol} \times -285.8 \text{ kJ/mol})$$
$$= +1531.2 \text{ kJ}$$

(b) $H_{r\ NH_3}^\circ = \dfrac{+1531.2 \text{ kJ}}{4 \text{ mol}} = +382.8 \text{ kJ/mol}$

$$\Delta H_r^\circ = 1.00 \text{ kg} \times \dfrac{1 \text{ mol}}{17.04 \text{ g}} \times \dfrac{382.8 \text{ kJ/mol}}{1 \text{ mol}}$$
$$= 22.5 \text{ MJ}$$

(c) Area $= 22.5 \text{ MJ} \times \dfrac{1 \text{ m}^2}{3.60 \text{ MJ}} = 6.24 \text{ m}^2$

(d) The major assumption is that all of the collected energy is transferred to the reaction.

19. (a) $\Delta H_{cond} = 1.00 \text{ mol} \times -40.8 \text{ kJ/mol} = -40.8 \text{ kJ}$
    (b) $\Delta H_f^\circ = 1.00 \text{ mol} \times -285.8 \text{ kJ/mol} = -286 \text{ kJ}$
    (c) $\Delta H_r = 1.00 \text{ mol} \times -1.7 \times 10^9 \text{ kJ/mol} = -1.7 \times 10^9 \text{ kJ}$
    (d) Distance for $\Delta H_{cond \ (steam)} = 40.8 \text{ kJ} \times \dfrac{1 \text{ cm}}{100 \text{ kJ}} = 0.408 \text{ cm}$

    Distance for $\Delta H_{f \ (water)}^\circ = 286 \text{ kJ} \times \dfrac{1 \text{ cm}}{100 \text{ kJ}} = 2.86 \text{ cm}$

    Distance for $\Delta H_{r \ (helium-4)} = 1.7 \times 10^9 \text{ kJ} \times \dfrac{1 \text{ cm}}{100 \text{ kJ}} = 1.7 \times 10^7 \text{ cm}$

    Number of pages $= 1.70 \times 10^7 \text{ cm} \times \dfrac{1 \text{ page}}{28 \text{ cm}} = 6.1 \times 10^5 \text{ pages}$

20. (a) $\Delta H_r^\circ = \Sigma n H_{fp}^\circ - \Sigma n H_{fr}^\circ$
$$= (1 \text{ mol} \times +52.5 \text{ kJ/mol} + 1 \text{ mol} \times 0 \text{ kJ/mol})$$
$$- (1 \text{ mol} \times -83.8 \text{ kJ/mol})$$
$$= +136.3 \text{ kJ}$$

(b)

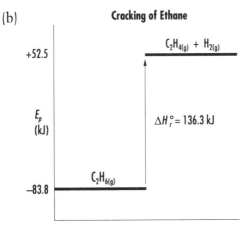

21. $2C_2H_{2(g)} + 5O_{2(g)} \rightarrow 4CO_{2(g)} + 2H_2O_{(g)}$

$\Delta H_c° = \Sigma nH_{fp}° - \Sigma nH_{fr}°$
$= (4 \text{ mol} \times -393.5 \text{ kJ/mol} + 2 \text{ mol} \times -241.8 \text{ kJ/mol})$
$- (2 \text{ mol} \times +228.2 \text{ kJ/mol} + 5 \text{ mol} \times 0 \text{ kJ/mol})$
$= -2514.0 \text{ kJ}$

$H_{c_{C_2H_2}}° = \dfrac{-2514.0 \text{ kJ}}{2 \text{ mol}} = -1257.0 \text{ kJ/mol}$

22. Based on the assumption that the most stable compound will have the most negative standard molar enthalpy of formation (alternatively, the most positive molar enthalpy of simple decomposition), the compounds are ranked below in order of increasing stability.

    acetylene, $C_2H_{2(g)}$ $H_f° = +228.2$ kJ/mol $H_{sd}° = -228.2$ kJ/mol
    ethylene, $C_2H_{4(g)}$ $H_f° = +52.5$ kJ/mol $H_{sd}° = -52.5$ kJ/mol
    ethane, $C_2H_{6(g)}$ $H_f° = -83.8$ kJ/mol $H_{sd}° = +83.8$ kJ/mol

23.
$H_{2(g)} + Cl_{2(g)} \rightarrow 2HCl_{(g)}$ $\qquad \Delta H° = -184.6$ kJ
$C_2H_{4(g)} + HCl_{(g)} \rightarrow C_2H_5Cl_{(l)}$ $\qquad \Delta H° = -65.0$ kJ
$C_2H_5Cl_{(l)} \rightarrow C_2H_3Cl_{(g)} + H_{2(g)}$ $\qquad \Delta H° = +138.9$ kJ

$C_2H_{4(g)} + Cl_{2(g)} \rightarrow C_2H_3Cl_{(g)} + HCl_{(g)}$ $\qquad \Delta H_r° = -110.7$ kJ

24. (a) $C_2H_{4(g)} + \frac{1}{2}O_{2(g)} + H_2O_{(l)} \rightarrow C_2H_4(OH)_{2(l)}$

(b) $\Delta H_r° = \Sigma nH_{fp}° - \Sigma nH_{fr}°$
$= (1 \text{ mol} \times -454.8 \text{ kJ/mol})$
$- (1 \text{ mol} \times +52.5 \text{ kJ/mol} + \frac{1}{2} \text{ mol} \times 0 \text{ kJ/mol} + 1 \text{ mol} \times -285.8 \text{ kJ/mol})$
$= -221.5 \text{ kJ}$

(c) **Production of Ethylene Glycol**

A potential energy diagram showing E_p (kJ) with reactants $C_2H_{4(g)} + \frac{1}{2}O_{2(g)} + H_2O_{(l)}$ at −233.3 kJ and product $C_2H_4(OH)_{2(l)}$ at −454.8 kJ, with $\Delta H_r° = -221.5$ kJ.

(d) $H_{r_{C_2H_4(OH)_2}}° = \dfrac{-221.5 \text{ kJ}}{1 \text{ mol}} = -221.5$ kJ/mol

$\Delta H_r° = 1.00 \text{ Mg} \times \dfrac{1 \text{ mol}}{62.08 \text{ g}} \times \dfrac{221.5 \text{ kJ}}{1 \text{ mol}} = 3.57$ GJ

25. (a) $$\Delta H_{r\,(\text{chloroethane})} = q_{(\text{calorimeter})}$$
$$nH_r = C\Delta t$$
$$15.78 \text{ g} \times \frac{1 \text{ mol}}{64.52 \text{ g}} \times H_r = \frac{4.06 \text{ kJ}}{°C} \times (22.92 - 19.03)°C$$
$$H_r = 64.6 \text{ kJ/mol}$$

According to the evidence provided, the molar enthalpy of reaction for chloroethane is reported as –64.6 kJ/mol.

(b) $C_2H_{4(g)} + HCl_{(g)} \rightarrow C_2H_5Cl_{(l)} + 64.6$ kJ

(c) $\Delta H_r° = \Sigma nH_{fp}° - \Sigma nH_{fr}°$
-64.6 kJ $= 1$ mol $\times H_f$
$\qquad - (1$ mol $\times +52.5$ kJ/mol $+ 1$ mol $\times -92.3$ kJ/mol$)$
$H_f = -104.4$ kJ/mol

26. $C_3H_{8(g)} + 5O_{2(g)} \rightarrow 3CO_{2(g)} + 4H_2O_{(g)}$
$\Delta H_c° = \Sigma nH_{fp}° - \Sigma nH_{fr}°$
$= (3$ mol $\times -393.5$ kJ/mol $+ 4$ mol $\times -241.8$ kJ/mol$)$
$\quad - (1$ mol $\times -104.7$ kJ/mol $+ 5$ mol $\times 0$ kJ/mol$)$
$= -2043.0$ kJ

$$H_{c\,C_3H_8} = \frac{-2043.0 \text{ kJ}}{1 \text{ mol}} = -2043.0 \text{ kJ/mol}$$

$$\Delta H_{c\,(\text{propane})} = q_{(\text{water})}$$
$$nH_c = vc\Delta t$$
$$n \times 2043.0 \text{ kJ/mol} = 2.50 \text{ L} \times \frac{4.19 \text{ kJ}}{\text{L}\cdot°C} \times (80 - 10)°C$$
$$n_{C_3H_8} = 0.36 \text{ mol}$$
$$m_{C_3H_8} = 0.36 \text{ mol} \times \frac{44.11 \text{ g}}{1 \text{ mol}} = 16 \text{ g}$$

27. Three possible designs, in order of increasing certainty, are the determination of the boiling point, the determination of the molar enthalpy of combustion, and the determination of the molecular formula from combustion and mass spectrometer analyses.
Some other lower certainty possibilities are the determination of the specific heat capacity and the reaction with an organic acid to produce an ester with a characteristic odor.

Extensions

28. *Possible alternatives include propane, gasoline-alcohol mixtures, alcohols, natural gas, and hydrogen. Students could choose or be assigned one of these alternatives. The question states some specific aspects to include in a report and/or presentation.*

29. *See your Physical Education department for a sample of a cold pack.*
Most cold packs contain ammonium nitrate and a separate pouch of water, which is broken to initiate an endothermic dissolving. Using this design, a new version could include solid citric acid and a separate solution of baking soda.
In the student research, the molar enthalpies of ammonium nitrate dissolving and citric acid–baking soda reacting is determined calorimetrically and compared. Hopefully, students will realize that although the citric acid–baking soda reaction is quite endothermic, it produces a considerable amount of gas. This creates a technical problem in constructing a simple, reliable, and inexpensive product.

30. A bomb calorimeter is altered to start and end the calorimeter water and reaction contents at SATP conditions. Reactants and calorimeter water are initially set at SATP conditions. Once the reaction has been completed, the calorimeter is returned to 25°C by pumping water through a heat exchanger into a separate insulated container of water. The initial and final temperatures of the water in the heat exchanger are measured. The gaseous products of the reaction are also allowed to expand to standard pressure.

Lab Exercise 11G
H_f for Calcium Oxide (page 343)

Problem

What is the enthalpy change for the reaction?

$$Ca_{(s)} + \tfrac{1}{2}O_{2(g)} \rightarrow CaO_{(s)}$$

Prediction

According to the table of standard molar enthalpies of formation, the enthalpy change for the reaction is

$$Ca_{(s)} + \tfrac{1}{2}O_{2(g)} \rightarrow CaO_{(s)} \qquad \Delta H_f^\circ = -634.9 \text{ kJ}$$

Analysis

Reaction of calcium metal with hydrochloric acid:

$$\Delta H_{r\,(calcium)} = q_{(calorimeter)}$$

$$nH_r = vc\Delta t$$

$$0.52 \text{ g} \times \frac{1 \text{ mol}}{40.08 \text{ g}} \times H_r = 0.100 \text{ L} \times \frac{4.19 \text{ kJ}}{L\cdot{}^\circ C} \times (34.5 - 21.3)^\circ C$$

$$H_r = 0.43 \text{ MJ/mol}$$

The molar enthalpy of reaction for calcium is reported as -0.43 MJ/mol.
Reaction of calcium oxide with hydrochloric acid:

$$\Delta H_{r\,(calcium\ oxide)} = q_{(calorimeter)}$$

$$nH_r = vc\Delta t$$

$$1.47 \text{ g} \times \frac{1 \text{ mol}}{56.08 \text{ g}} \times H_r = 0.100 \text{ L} \times \frac{4.19 \text{ kJ}}{L\cdot{}^\circ C} \times (28.0 - 21.1)^\circ C$$

$$H_r = 0.11 \text{ MJ/mol}$$

The molar enthalpy of reaction for calcium oxide is reported as -0.11 MJ/mol. According to Hess's law, the two experimental reaction equations and the equation for the formation of liquid water can be combined to generate the desired equation.

$$Ca_{(s)} + 2HCl_{(aq)} \rightarrow CaCl_{2(aq)} + H_{2(g)} \qquad \Delta H = -0.43 \text{ MJ}$$
$$CaCl_{2(aq)} + H_2O_{(l)} \rightarrow CaO_{(s)} + 2HCl_{(aq)} \qquad \Delta H = +0.11 \text{ MJ}$$
$$H_{2(g)} + \tfrac{1}{2}O_{2(g)} \rightarrow H_2O_{(l)} \qquad \Delta H = -0.2858 \text{ MJ}$$

$$Ca_{(s)} + \tfrac{1}{2}O_{2(g)} \rightarrow CaO_{(s)} \qquad \Delta H_f = -0.60 \text{ MJ}$$

Based on the evidence given, the enthalpy change of the formation of one mole of calcium oxide is reported as -0.60 MJ *(accurate to a percent difference of 5%)*.

Lab Exercise 11H
H_c by Four Methods (page 343)

Problem
What is the molar enthalpy of combustion of methylpropane?

Prediction
In a bomb calorimeter, liquid water is usually produced. Therefore, all predictions are made assuming liquid water.

(1) According to the standard molar enthalpies of formation method, the molar enthalpy of combustion of methylpropane is –2868.8 kJ/mol as shown below.

$$C_4H_{10(g)} + \tfrac{13}{2} O_{2(g)} \rightarrow 4CO_{2(g)} + 5H_2O_{(l)}$$

$$\begin{aligned}\Delta H_c^\circ &= \Sigma n H_{fp}^\circ - \Sigma n H_{fr}^\circ \\ &= (4 \text{ mol} \times -393.5 \text{ kJ/mol} + 5 \text{ mol} \times -285.8 \text{ kJ/mol}) \\ &\quad - (1 \text{ mol} \times -134.2 \text{ kJ/mol} + \tfrac{13}{2} \text{ mol} \times 0 \text{ kJ/mol}) \\ &= -2868.8 \text{ kJ}\end{aligned}$$

$$H_{c \; C_4H_{10}} = \frac{-2868.8 \text{ kJ}}{1 \text{ mol}} = -2868.8 \text{ kJ/mol}$$

(2) According to Hess's law using appropriate formation reaction equations, the molar enthalpy of combustion of methylpropane is –2868.8 kJ/mol as shown by the following reasoning.

$4C_{(s)} + 5H_{2(g)} \rightarrow C_4H_{10(g)}$ $\quad\quad \Delta H_f^\circ = -134.2$ kJ

$C_{(s)} + O_{2(g)} \rightarrow CO_{2(g)}$ $\quad\quad \Delta H_f^\circ = -393.5$ kJ

$H_{2(g)} + \tfrac{1}{2} O_{2(g)} \rightarrow H_2O_{(l)}$ $\quad\quad \Delta H_f^\circ = -285.8$ kJ

These formation equations may be rearranged as follows:

$C_4H_{10(g)} \rightarrow 4C_{(s)} + 5H_{2(g)}$ $\quad\quad \Delta H^\circ = +134.2$ kJ

$4C_{(s)} + 4O_{2(g)} \rightarrow 4CO_{2(g)}$ $\quad\quad \Delta H^\circ = -1574.0$ kJ

$5H_{2(g)} + \tfrac{5}{2} O_{2(g)} \rightarrow 5H_2O_{(l)}$ $\quad\quad \Delta H^\circ = -1429.0$ kJ

$C_4H_{10(g)} + \tfrac{13}{2} O_{2(g)} \rightarrow 4CO_{2(g)} + 5H_2O$ $\quad\quad \Delta H_c^\circ = -2868.8$ kJ

$$H_{c \; C_4H_{10}} = \frac{-2868.8 \text{ kJ}}{1 \text{ mol}} = -2868.8 \text{ kJ/mol}$$

(3) According to the method of bond energies, the molar enthalpy of combustion of methylpropane is –2.57 MJ/mol as shown by the following reasoning.

Most college general chemistry texts contain tables of average bond energies. The values may vary widely. The following bond energies were referenced from Chemistry, Fourth Edition, *Raymond Chang.*

Bond	Bond Energy, E_b (kJ/mol)
C—C	347
C—H	414
C=O	799 (in CO_2)
O—H	460
O=O	499

REACTION ENTHALPIES 157

$$C_4H_{10(g)} + \tfrac{13}{2}O_{2(g)} \rightarrow 4CO_{2(g)} + 5H_2O_{(l)}$$

10 mol C—H $\tfrac{13}{2}$ mol O=O   8 mol C=O   10 mol O—H

3 mol C—C

$\Delta H_c^\circ = \Sigma nE_{br} - \Sigma nE_{bp}$

$\quad = (10 \text{ mol} \times 414 \text{ kJ/mol} + 3 \text{ mol} \times 347 \text{ kJ/mol} + \tfrac{13}{2} \text{ mol} \times 499 \text{ kJ/mol})$

$\quad - (8 \text{ mol} \times 799 \text{ kJ/mol} + 10 \text{ mol} \times 460 \text{ kJ/mol})$

$\quad = -2.57 \text{ MJ}$

$H_{c_{C_4H_{10}}} = \dfrac{-2.57 \text{ MJ}}{1 \text{ mol}} = -2.57 \text{ MJ/mol}$

Analysis

$\Delta H_{c \text{ (methylpropane)}} = q_{\text{(calorimeter)}}$

$nH_c = C\Delta t$

$1.52 \text{ g} \times \dfrac{1 \text{ mol}}{58.14 \text{ g}} \times H_c = \dfrac{9.35 \text{ kJ}}{°C} \times (28.25 - 20.21)°C$

$H_c = 2.88 \text{ MJ/mol}$

According to the evidence gathered by calorimetry and the analysis using the law of conservation of energy, the molar enthalpy of combustion of methylpropane is reported as –2.88 MJ/mol.

Evaluation

The experimental design of calorimetry is a common design and was adequate to answer the problem. Based upon previous experience, I have a great deal of confidence in this design. The usual uncertainties of calorimetry experiments should give a percent difference of 5% or less.

Using the standard molar enthalpies of formation and the Hess's law methods, the predicted molar enthalpy of combustion for methylpropane is accurate to within 0.2%.

$$\% \text{ difference} = \dfrac{|2.88 \text{ MJ/mol} - 2.8688 \text{ MJ/mol}|}{2.8688 \text{ MJ/mol}} \times 100 = 0.2\%$$

Using the bond energy method, the predicted molar enthalpy of combustion for methylpropane is accurate to within 12%.

$$\% \text{ difference} = \dfrac{|2.88 \text{ MJ/mol} - 2.57 \text{ MJ/mol}|}{2.57 \text{ MJ/mol}} \times 100 = 12\%$$

The Hess's law method and the standard molar enthalpies of formation method are equivalent methods as they both use the same data, namely, the standard molar enthalpies of formation. Each method predicts the same molar enthalpy of combustion of methylpropane, which leads to a percent difference of 0.2%. This is an unusually close agreement between the prediction and the empirically determined value. These two predictions are clearly verified and the two methods are considered acceptable. The bond energy method is the most theoretical and provides only an approximate value for the molar enthalpy of combustion. The prediction based on bond energies is inconclusive. The method of bond energies should be restricted to situations where the standard molar enthalpies of formation are not known.

UNIT VI
CHANGE AND SYSTEMS

12 Electrochemistry

Exercise (page 348)

1. (a) Reduction is the production of metals from their compounds.
   (b) Oxidation is the reaction of substances, usually with oxygen or another active nonmetal, to produce a compound.
   (c) An oxidizing agent is a substance that causes or promotes the oxidation of a metal to produce a metal compound.
   (d) A reducing agent is a substance that causes or promotes the reduction of a metal compound to an elemental metal.
   (e) Metallurgy is the science and technology of extracting metals from their naturally occurring compounds and adapting these metals for useful purposes. *Extracting metals from ores generally involves four steps — concentrating, roasting, reducing, and refining.*
   (f) Corrosion is the adverse reaction of metals with oxygen in the environment.

2. (a) oxidation of iron; the oxidizing agent is oxygen
   (b) reduction of lead(II) oxide; the reducing agent is carbon
   (c) reduction of nickel(II) oxide; the reducing agent is hydrogen
   (d) oxidation of tin; the oxidizing agent is bromine
   (e) reduction of iron(III) oxide; the reducing agent is carbon monoxide
   (f) oxidation of copper; the oxidizing agent is nitric acid

3. Three common reducing agents used in metallurgy are $C_{(s)}$, $CO_{(g)}$, and $H_{2(g)}$.

4. Nonmetals serve as oxidizing agents for metals.

5. Technological applications of fire came well before scientific understanding. Although humans have been using fire for tens of thousands of years, the importance of oxygen in burning was not proposed until the late eighteenth century.

INVESTIGATION 12.1

Single Replacement Reactions (page 349)

Problem

What are the products of the single replacement reactions for the following sets of reactants?

Prediction

According to the single replacement reaction generalization, the products are as follows:

1. $Cu_{(s)} + 2AgNO_{3(aq)} \rightarrow 2Ag_{(s)} + Cu(NO_3)_{2(aq)}$
2. $Cl_{2(aq)} + 2NaBr_{(aq)} \rightarrow Br_{2(l)} + 2NaCl_{(aq)}$

3. $Mg_{(s)} + 2HCl_{(aq)} \rightarrow H_{2(g)} + MgCl_{2(aq)}$
4. $Zn_{(s)} + CuSO_{4(aq)} \rightarrow Cu_{(s)} + ZnSO_{4(aq)}$
5. $Cl_{2(aq)} + 2KI_{(aq)} \rightarrow I_{2(s)} + 2KCl_{(aq)}$

Experimental Design

The reactants are mixed and evidence is collected to identify at least one of the predicted products. For reaction 1, silver metal is identified by its silvery appearance and copper(II) nitrate is identified by its blue color in solution. For reaction 2, bromine is identified using the halogen diagnostic test. For reaction 3, hydrogen gas is identified by the hydrogen diagnostic test. For reaction 4, copper metal is identified by its appearance. For reaction 5, iodine is identified by the halogen diagnostic test.

Evidence/Analysis

#	Test Procedure	Observations	Analysis
1	mixture observed	• silvery crystals on copper strip • blue colored solution	$Ag_{(s)}$ $Cu(NO_3)_{2(aq)}$
2	mixture observed before and after adding trichlorotrifluoroethane	• initial yellow-brown color • orange color in organic layer	Br_2 confirmed
3	gas collected and a flame inserted in the test tube	• a pop sound heard	$H_{2(g)}$
4	mixture observed	• red-brown solid formed on zinc strip	$Cu_{(s)}$
5	mixture observed before and after adding trichlorotrifluoroethane	• initial yellow-brown color • violet color in organic layer	I_2 confirmed

Evaluation

The experimental design was adequate to answer the problem with the materials provided. In most cases, only one product was tested. The design could be improved by including additional tests. The procedure and technological skills were adequate for collecting the evidence indicated in the design. I am moderately confident in this design.

All predictions are judged to be verified because they agree with the results obtained. No contradictory evidence was obtained. As a result, the single replacement reaction generalization remains acceptable. Although the certainty is not high due to the limited number of tests, I am reasonably confident in my judgments.

Exercise (page 349)

6. (a) Copper(II) ions are converted to copper atoms.
   (b) $Cu^{2+}_{(aq)} + 2e^- \rightarrow Cu_{(s)}$
   (c) Zinc atoms are converted to zinc ions.
   (d) $Zn_{(s)} \rightarrow Zn^{2+}_{(aq)} + 2e^-$
   (e) The copper(II) ion gains electrons. The zinc atom loses electrons.
   (f) The sulfate ion is unaffected by the reaction. Such substances are called spectator ions.

7. (a) $Ag^+_{(aq)} + e^- \rightarrow Ag_{(s)}$
       $Cu_{(s)} \rightarrow Cu^{2+}_{(aq)} + 2e^-$
   (b) $2H^+_{(aq)} + 2e^- \rightarrow H_{2(g)}$
       $Mg_{(s)} \rightarrow Mg^{2+}_{(aq)} + 2e^-$

8. (a) Bromide ions are converted to bromine molecules.
   (b) $2Br^-_{(aq)} \rightarrow Br_{2(l)} + 2e^-$
   (c) $Cl_{2(g)} + 2e^- \rightarrow 2Cl^-_{(aq)}$
       $2I^-_{(aq)} \rightarrow I_{2(s)} + 2e^-$
   (d) Sodium ions and potassium ions are spectator ions in the reactions involving chlorine.

Exercise (page 352)

9. (a) A redox reaction is a chemical reaction in which electrons are transferred between entities.
   (b) Reduction is a chemical process in which electrons are gained (GER).
   (c) Oxidation is a chemical process in which electrons are lost (LEO).
   (d) An oxidizing agent is a chemical species that removes (gains) electrons from another species causing the oxidation of that species in a redox reaction.
   (e) A reducing agent is a chemical species that donates (loses) electrons to another species causing the reduction of that species in a redox reaction.

10. (a) reduction $2[Cu^{2+}_{(aq)} + 2e^- \rightarrow Cu_{(s)}]$
        oxidation $Sn_{(s)} \rightarrow Sn^{4+}_{(aq)} + 4e^-$
        net $2Cu^{2+}_{(aq)} + Sn_{(s)} \rightarrow 2Cu_{(s)} + Sn^{4+}_{(aq)}$

    (b) reduction $Br_{2(l)} + 2e^- \rightarrow 2Br^-_{(aq)}$
        oxidation $2I^-_{(aq)} \rightarrow I_{2(s)} + 2e^-$
        net $Br_{2(l)} + 2I^-_{(aq)} \rightarrow 2Br^-_{(aq)} + I_{2(s)}$

    (c) reduction $2H^+_{(aq)} + 2e^- \rightarrow H_{2(g)}$
        oxidation $Ca_{(s)} \rightarrow Ca^{2+}_{(aq)} + 2e^-$
        net $2H^+_{(aq)} + Ca_{(s)} \rightarrow H_{2(g)} + Ca^{2+}_{(aq)}$

    (d) reduction $Fe^{3+}_{(s)} + 3e^- \rightarrow Fe_{(l)}$
        oxidation $Al_{(s)} \rightarrow Al^{3+}_{(s)} + 3e^-$
        net $Fe^{3+}_{(s)} + Al_{(s)} \rightarrow Fe_{(l)} + Al^{3+}_{(s)}$

11. (a) *See Question 10 above.*
    (b) In 10 (a), the oxidizing agent is $Cu^{2+}_{(aq)}$ and the reducing agent is $Sn_{(s)}$.
        In 10 (b), the oxidizing agent is $Br_{2(l)}$ and the reducing agent is $I^-_{(aq)}$.
        In 10 (c), the oxidizing agent is $H^+_{(aq)}$ and the reducing agent is $Ca_{(s)}$.
        In 10 (d), the oxidizing agent is $Fe^{3+}_{(s)}$ and the reducing agent is $Al_{(s)}$.

12. Because of atomic theory's prediction that all atoms have attraction for electrons, it is more consistent to think of oxidizing agents "pulling electrons away from reducing agents." No atoms should be thought of as "giving" away electrons; they may lose electrons because of other atoms' stronger attraction for electrons.

13. (a) $FeCl_{3(aq)} + 3NaOH_{(aq)} \rightarrow Fe(OH)_{3(s)} + 3NaCl_{(aq)}$
        or
        $Fe^{3+}_{(aq)} + 3OH^-_{(aq)} \rightarrow Fe(OH)_{3(s)}$
    (b) No, a redox reaction has not taken place because there has not been a gain or loss of electrons. The charge on any entity in the chemical equation remains the same.

INVESTIGATION 12.2

Spontaneity of Redox Reactions (page 353)

Problem
Which combinations of copper, lead, silver, and zinc metals and their aqueous metal ion solutions produce spontaneous reactions?

Prediction
Based on previous assumptions for stoichiometry (page 174), all combinations that result in products different from reactants are spontaneous. Therefore, all metals should react with compounds of other metal ions.

Experimental Design
Each of the four metal strips is cleaned and placed in different metal ion solutions and qualitative evidence for a chemical reaction is observed.

Procedure
1. Obtain approximately 25 mL of each aqueous solution in four labelled beakers.
2. Clean the four metal strips using steel wool.
3. Label and rinse four test tubes with distilled water.
4. Add 2–3 cm of $Cu(NO_3)_{2(aq)}$ to each of the four test tubes.
5. Hang a different metal strip into each test tube and observe for several minutes.
6. Repeat steps 2 to 5 using $Pb(NO_3)_{2(aq)}$, $AgNO_{3(aq)}$, and $Zn(NO_3)_{2(aq)}$ in place of $Cu(NO_3)_{2(aq)}$.
7. Dispose of the lead and silver solutions into a labelled waste container and clean the metal strips for reuse.

Evidence

REACTIONS OF METALS AND METAL IONS				
	$Cu_{(s)}$	$Pb_{(s)}$	$Ag_{(s)}$	$Zn_{(s)}$
$Cu^{2+}_{(aq)}$	slightly cleaner metal surface	red-brown precipitate	no change	red-brown precipitate
$Pb^{2+}_{(aq)}$	no change	no change	no change	black precipitate
$Ag^{+}_{(aq)}$	silver crystals	silver crystals	no change	silver crystals
$Zn^{2+}_{(aq)}$	no change	no change	no change	no change

Analysis
The evidence obtained is consistent with spontaneous, single replacement reactions for the combinations of copper and silver ions, lead and copper(II) ions, lead and silver ions, and zinc and each of copper(II), silver, and lead(II) ions.

Evaluation
The experimental design was adequate to answer the problem since only evidence for a reaction (not the identity of the product) was required. The procedure and technological skills were adequate, although the short observation time did create a little uncertainty for those combinations that did not appear

to react. I would be more certain of the results if those mixtures were left for at least one day.

Overall, the prediction is judged to be falsified since six out of the twelve predicted spontaneous reactions did not give any evidence of a chemical change. The mixture of a metal and a solution of its own ion was predicted to be non-spontaneous and this was verified with the possible exception of the copper system, which would require further testing. The assumption of spontaneous single replacement reactions is judged to be unacceptable since the prediction was clearly falsified. The assumption will need to be restricted, revised, or discarded.

Exercise (page 355)

14. (a) zinc and lead
    (b) copper and silver
    (c) The metals that react spontaneously with $Cu^{2+}_{(aq)}$ are below $Cu_{(s)}$ in the table.
    (d) The other predicted spontaneous reactions are silver ions with copper metal, lead metal, and zinc metal; and lead(II) ions with zinc metal. The hypothesis is verified.

Lab Exercise 12A
Spontaneity of Reactions (page 355)

Problem

What is the relative strength of oxidizing agents among beryllium, cadmium, radium, and vanadium aqueous ions?

Analysis

	Number of Reactions			
	3	2	1	0
ions	$Cd^{2+}_{(aq)}$	$V^{2+}_{(aq)}$	$Be^{2+}_{(aq)}$	$Ra^{2+}_{(aq)}$
metals	$Ra_{(s)}$	$Be_{(s)}$	$V_{(s)}$	$Cd_{(s)}$

A TABLE OF REDOX HALF-REACTIONS

SOA $\quad Cd^{2+}_{(aq)} + 2e^- \rightleftharpoons Cd_{(s)}$
$\qquad V^{2+}_{(aq)} + 2e^- \rightleftharpoons V_{(s)}$
$\qquad Be^{2+}_{(aq)} + 2e^- \rightleftharpoons Be_{(s)}$
$\qquad Ra^{2+}_{(aq)} + 2e^- \rightleftharpoons Ra_{(s)} \quad$ **SRA**

Evaluation

The spontaneity rule is acceptable for the metals and metal ions used in this investigation because the pattern obtained is consistent with Investigation 12.2.

Lab Exercise 12B
Redox Tables — Design 1 (page 356)

Problem

What is the redox half-reaction table for zinc, vanadium, cadmium, and lead?

Analysis

A TABLE OF REDOX HALF-REACTIONS
SOA $\quad Pb^{2+}_{(aq)} + 2e^- \rightleftharpoons Pb_{(s)}$
$\quad\quad\quad Cd^{2+}_{(aq)} + 2e^- \rightleftharpoons Cd_{(s)}$
$\quad\quad\quad Zn^{2+}_{(aq)} + 2e^- \rightleftharpoons Zn_{(s)}$
$\quad\quad\quad V^{2+}_{(aq)} + 2e^- \rightleftharpoons V_{(s)}\quad$ SRA

Synthesis

The three tables are merged by recognizing that silver ions and copper(II) ions are both stronger oxidizing agents than lead(II) ions; beryllium ion and radium ion are both weaker oxidizing agents than vanadium(II) ion.

AN EXTENDED TABLE OF REDOX HALF-REACTIONS
SOA $\quad Ag^+_{(aq)} + e^- \rightleftharpoons Ag_{(s)}$
$\quad\quad\quad Cu^{2+}_{(aq)} + 2e^- \rightleftharpoons Cu_{(s)}$
$\quad\quad\quad Pb^{2+}_{(aq)} + 2e^- \rightleftharpoons Pb_{(s)}$
$\quad\quad\quad Cd^{2+}_{(aq)} + 2e^- \rightleftharpoons Cd_{(s)}$
$\quad\quad\quad Zn^{2+}_{(aq)} + 2e^- \rightleftharpoons Zn_{(s)}$
$\quad\quad\quad V^{2+}_{(aq)} + 2e^- \rightleftharpoons V_{(s)}$
$\quad\quad\quad Be^{2+}_{(aq)} + 2e^- \rightleftharpoons Be_{(s)}$
$\quad\quad\quad Ra^{2+}_{(aq)} + 2e^- \rightleftharpoons Ra_{(s)}\quad$ SRA

Exercise (page 357)

15. **SOA** $\quad Co^{2+}_{(aq)} + 2e^- \rightleftharpoons Co_{(s)}$
    $\quad\quad\quad\quad Zn^{2+}_{(aq)} + 2e^- \rightleftharpoons Zn_{(s)}$
    $\quad\quad\quad\quad Mg^{2+}_{(aq)} + 2e^- \rightleftharpoons Mg_{(s)}\quad$ **SRA**

16. **SOA** $\quad Cu^{2+}_{(aq)} + 2e^- \rightleftharpoons Cu_{(s)}$
    $\quad\quad\quad\quad 2H^+_{(aq)} + 2e^- \rightleftharpoons H_{2(g)}$
    $\quad\quad\quad\quad Cd^{2+}_{(aq)} + 2e^- \rightleftharpoons Cd_{(s)}$
    $\quad\quad\quad\quad Be^{2+}_{(aq)} + 2e^- \rightleftharpoons Be_{(s)}$
    $\quad\quad\quad\quad Ca^{2+}_{(aq)} + 2e^- \rightleftharpoons Ca_{(s)}\quad$ **SRA**

17. The spontaneity rule is empirical because the table is generated from experimental evidence.

    It should be pointed out, however, that reference made to electrons gained or lost during a reaction is theoretical, as is any reference to "half-reactions."

18. **SOA** $\quad Cl_{2(g)} + 2e^- \rightleftharpoons 2Cl^-_{(aq)}$
    $\quad\quad\quad\quad Br_{2(l)} + 2e^- \rightleftharpoons 2Br^-_{(aq)}$
    $\quad\quad\quad\quad Ag^+_{(aq)} + e^- \rightleftharpoons Ag_{(s)}$
    $\quad\quad\quad\quad I_{2(s)} + 2e^- \rightleftharpoons 2I^-_{(aq)}$
    $\quad\quad\quad\quad Cu^{2+}_{(aq)} + 2e^- \rightleftharpoons Cu_{(s)}\quad$ **SRA**

> **Lab Exercise 12C**
> **Redox Tables — Design 2 (page 359)**
>
> *Problem*
> What is the relative strength of oxidizing and reducing agents for strontium, cerium, nickel, hydrogen, platinum, and their aqueous ions?
>
> *Analysis*
> According to the evidence collected, the oxidizing agents and reducing agents are ranked from strongest to weakest in the following table of redox half-reactions.
>
> **SOA** $Pt^{4+}_{(aq)} + 4e^- \rightleftharpoons Pt_{(s)}$
> $2H^+_{(aq)} + 2e^- \rightleftharpoons H_{2(g)}$
> $Ni^{2+}_{(aq)} + 2e^- \rightleftharpoons Ni_{(s)}$
> $Ce^{3+}_{(aq)} + 3e^- \rightleftharpoons Ce_{(s)}$
> $Sr^{2+}_{(aq)} + 2e^- \rightleftharpoons Sr_{(s)}$ **SRA**

Exercise (page 359)

19. $Ag^+_{(aq)}$, $Cu^{2+}_{(aq)}$, $Pb^{2+}_{(aq)}$, $Zn^{2+}_{(aq)}$. This order agrees with the evidence collected in Investigation 12.2.

20. Metal ions, nonmetals, and a variety of acidic solutions usually behave as oxidizing agents.

21. Nonmetal ions, metals, and basic solutions of various entities usually behave as reducing agents.

22. According to a restricted version of the theory of quantum mechanics, nonmetal atoms have almost filled valence energy levels and tend to attract electrons to attain a noble gas-like electronic structure. Metal atoms, however, have few electrons in their valence energy levels and tend to lose electrons to attain a noble gas-like electronic structure. This is consistent with the empirically determined redox table, which shows that nonmetals tend to act as oxidizing agents (electron acceptors) and that metals tend to act as reducing agents (electron donors).

23. Since fluorine is the most reactive nonmetal and nonmetals are generally oxidizing agents, fluorine is expected to be the strongest oxidizing agent. Fluorine is the most reactive nonmetal because it has the greatest attraction for electrons. This reason relates the observed reactivity to the theoretical definition of an oxidizing agent.
    Some further "why" questions are as follows. Why does fluorine have the greatest attraction for electrons? Why does fluorine have the highest electronegativity? Why do atoms with almost-filled energy levels have higher electronegativities than those without? (It should rapidly become obvious that the theory presented, to this point, is limited in providing explanations of the redox table.)

24. Gold and silver (and platinum) are commonly found as metals, while sodium and potassium are never found in their elemental form. Gold and silver are listed near the top of the table of redox half-reactions and are very weak reducing agents (relatively unreactive). Sodium and potassium are listed near the bottom of the table of redox half-reactions and are very strong reducing agents (very reactive).

25. Tin(II), iron(II), and chromium(II) ions can act as both oxidizing and reducing agents. These entities can gain or lose electrons. For example, tin(II) ions can either lose electrons to produce tin(IV) ions or gain electrons to produce elemental tin. The explanation for iron(II) ions and chromium(II) ions is similar, but there is no simple explanation of the behavior of water molecules, which can also act in the same way.

26. (a) spontaneous
    (b) non-spontaneous
    (c) non-spontaneous
    The reaction is in fact spontaneous, but it is not a redox reaction.
    (d) spontaneous
    (e) spontaneous
    (f) spontaneous

27. Design 1: Mix all combinations of oxidizing agents and reducing agents from a list of elements and their ions. The oxidizing agent that reacts spontaneously with the most reducing agents is the strongest. The oxidizing agent that reacts with the next greatest number of reducing agents is the next strongest, etc. From this information, the order of reactivity and the half-reaction equations can be determined.

    Design 2: Selected mixtures of oxidizing agents and reducing agents are studied and the spontaneity rule is used to order the oxidizing and reducing agents in each reactant mixture. The individual results are combined to form a final table of redox half-reactions.

    If n half-reaction equations are to be placed in a table, a minimum of n − 1 reactions must be studied.

28. The empirical way of knowing has been most useful to this point in predicting the spontaneity of redox reactions because the construction of redox tables is based on empirical data.
    The theory presented, to this point, can barely provide an initial explanation after the fact, and has little predictive power.

Exercise (page 361)

29. (a) $Cu^{2+}_{(aq)}$ OA, $Pb_{(s)}$ RA, $SO_4^{2-}_{(aq)}$ OA, $H_2O_{(l)}$ OA/RA

    (b) $Au_{(s)}$ RA, $H^+_{(aq)}$ OA, $NO_3^-_{(aq)}$ OA, $H_2O_{(l)}$ OA/RA

    (c) $K^+_{(aq)}$ OA, $Cr_2O_7^{2-}_{(aq)}$ OA, $H^+_{(aq)}$ OA, $NO_3^-_{(aq)}$ OA, $Fe^{2+}_{(aq)}$ OA/RA, $H_2O_{(l)}$ OA/RA

    (d) $Cl_{2(aq)}$ OA, $Na^+_{(aq)}$ OA, $OH^-_{(aq)}$ RA, $H_2O_{(l)}$ OA/RA

    (e) $K^+_{(aq)}$ OA, $MnO_4^-_{(aq)}$ OA, $H^+_{(aq)}$ OA, $Sn^{2+}_{(aq)}$ OA/RA, $Cl^-_{(aq)}$ RA, $H_2O_{(l)}$ OA/RA

(f)

$\underset{\text{RA}}{\text{OH}^-_{(aq)}}, \underset{\text{RA}}{\text{SO}_3^{2-}_{(aq)}}, \quad \overset{\text{OA}}{\text{Na}^+_{(aq)}}, \overset{\text{OA}}{\underset{\text{RA}}{\text{H}_2\text{O}_{(l)}}}, \overset{\text{OA}\frown\text{OA}}{\text{I}_{2(s)}}$

Exercise (page 364)

30. (a)

$\underset{\text{SRA}}{\overset{\text{SOA}}{\text{Zn}_{(s)}}}, \text{H}^+_{(aq)}, \underset{\text{RA}}{\text{Cl}^-_{(aq)}}, \underset{\text{RA}}{\overset{\text{OA}}{\text{H}_2\text{O}_{(l)}}}$

$2\text{H}^+_{(aq)} + 2e^- \rightarrow \text{H}_{2(g)}$

$\text{Zn}_{(s)} \rightarrow \text{Zn}^{2+}_{(aq)} + 2e^-$

―――――――――――――――――――――

$2\text{H}^+_{(aq)} + \text{Zn}_{(s)} \xrightarrow{\text{spont.}} \text{H}_{2(g)} + \text{Zn}^{2+}_{(aq)}$

If a flame is inserted into a sample of the gas produced and a pop sound is heard, then hydrogen is likely present.

(b)

$\underset{\text{SRA}}{\overset{\text{SOA}}{\text{Cl}_{2(g)}}}, \underset{}{\text{I}^-_{(aq)}}, \underset{\text{RA}}{\overset{\text{OA}}{\text{H}_2\text{O}_{(l)}}}$

$\text{Cl}_{2(g)} + 2e^- \rightarrow 2\text{Cl}^-_{(aq)}$

$2\text{I}^-_{(aq)} \rightarrow \text{I}_{2(s)} + 2e^-$

―――――――――――――――――――――

$\text{Cl}_{2(g)} + 2\text{I}^-_{(aq)} \xrightarrow{\text{spont.}} 2\text{Cl}^-_{(aq)} + \text{I}_{2(s)}$

If a chlorinated hydrocarbon is added to a sample of the final mixture, and the organic layer turns violet, then iodine is likely present.
If a sample of the product solution is mixed with starch, and the starch turns dark blue, then iodine is likely present.

(c)

$\underset{\text{RA}}{\overset{\text{SOA}}{\text{Au}_{(s)}}}, \text{H}^+_{(aq)}, \text{Cl}^-_{(aq)}, \underset{\text{SRA}}{\overset{\text{OA}}{\text{H}_2\text{O}_{(l)}}}$

$2[2\text{H}^+_{(aq)} + 2e^- \rightarrow \text{H}_{2(g)}]$

$2\text{H}_2\text{O}_{(l)} \rightarrow \text{O}_{2(g)} + 4\text{H}^+_{(aq)} + 4e^-$

―――――――――――――――――――――

$2\text{H}_2\text{O}_{(l)} \xrightarrow{\text{non-spont.}} 2\text{H}_{2(g)} + \text{O}_{2(g)}$

(d)

$\overset{\text{OA}\frown\text{SOA}}{\text{H}^+_{(aq)}, \text{NO}_3^-_{(aq)}}, \underset{\text{SRA}}{\text{Cu}_{(s)}}, \underset{\text{RA}}{\overset{\text{OA}}{\text{H}_2\text{O}_{(l)}}}$

$2[\text{NO}_3^-_{(aq)} + 2\text{H}^+_{(aq)} + e^- \rightarrow \text{NO}_{2(g)} + \text{H}_2\text{O}_{(l)}]$

$\text{Cu}_{(s)} \rightarrow \text{Cu}^{2+}_{(aq)} + 2e^-$

―――――――――――――――――――――

$2\text{NO}_3^-_{(aq)} + 4\text{H}^+_{(aq)} + \text{Cu}_{(s)} \xrightarrow{\text{spont.}} 2\text{NO}_{2(g)} + 2\text{H}_2\text{O}_{(l)} + \text{Cu}^{2+}_{(aq)}$

If the color of the final solution near the copper surface is noted and the color is blue, then it is likely that copper(II) ions are produced.
If a brown gas is produced during the course of the reaction, then nitrogen dioxide is likely produced.

(e)

$\underset{\text{SRA}}{\text{Fe}_{(s)}}, \overset{\text{SOA}}{\text{O}_{2(g)}}, \underset{\text{RA}}{\overset{\text{OA}}{\text{H}_2\text{O}_{(l)}}}$

$\text{O}_{2(g)} + 2\text{H}_2\text{O}_{(l)} + 4e^- \rightarrow 4\text{OH}^-_{(aq)}$

$2[\text{Fe}_{(s)} \rightarrow \text{Fe}^{2+}_{(aq)} + 2e^-]$

―――――――――――――――――――――

$\text{O}_{2(g)} + 2\text{H}_2\text{O}_{(l)} + 2\text{Fe}_{(s)} \xrightarrow{\text{spont.}} 4\text{OH}^-_{(aq)} + \text{Fe}^{2+}_{(aq)}$

If the mixture is tested with litmus paper, and the red litmus turns blue, then hydroxide ions are likely present.
or

$$O_{2(g)} + 2H_2O_{(l)} + 2Fe_{(s)} \xrightarrow{spont.} 2Fe(OH)_{2(s)}$$

If a green solid is observed on the surface of the fender, then iron(II) hydroxide is likely produced.

(f)
$$\overset{SOA}{O_{2(g)}},\ \overset{OA}{H_2O_{(l)}},\ \overset{OA}{Na^+_{(aq)}},\ \underset{RA}{OH^-_{(aq)}},\ \underset{SRA}{SO_3^{2-}{}_{(aq)}}$$

$$O_{2(g)} + 2H_2O_{(l)} + 4e^- \rightarrow 4OH^-_{(aq)}$$
$$\underline{2[SO_3^{2-}{}_{(aq)} + 2OH^-_{(aq)} \rightarrow SO_4^{2-}{}_{(aq)} + H_2O_{(l)} + 2e^-]}$$
$$O_{2(g)} + 2SO_3^{2-}{}_{(aq)} \xrightarrow{spont.} 2SO_4^{2-}{}_{(aq)}$$

If a sample of the final solution is acidified and hydrosulfuric acid added, the formation of a yellow precipitate (of sulfur) indicates that sulfate ions are likely present.

31. Yes, because no spontaneous reaction occurs according to the redox table and spontaneity rule.

$$\overset{SOA}{H^+_{(aq)}},\ \overset{OA}{Sn^{2+}_{(aq)}},\ \underset{RA}{Cl^-_{(aq)}},\ \underset{RA}{\overset{OA}{H_2O_{(l)}}}$$
SRA   RA

$$2H^+_{(aq)} + 2e^- \rightarrow H_{2(g)}$$
$$\underline{\quad Sn^{2+}_{(aq)} \rightarrow Sn^{4+}_{(aq)} + 2e^- \quad}$$
$$2H^+_{(aq)} + Sn^{2+}_{(aq)} \xrightarrow{non\text{-}spont.} H_{2(g)} + Sn^{4+}_{(aq)}$$

32. (a) single replacement
$$2Fe_{(s)} + 3CuSO_{4(aq)} \rightarrow 3Cu_{(s)} + Fe_2(SO_4)_{3(aq)}$$
net ionic equetion
$$2Fe_{(s)} + 3Cu^{2+}_{(aq)} \rightarrow 3Cu_{(s)} + 2Fe^{3+}_{(aq)}$$

(b)
$$\overset{SOA}{Fe_{(s)}},\ \overset{}{Cu^{2+}_{(aq)}},\ \overset{OA}{SO_4^{2-}{}_{(aq)}},\ \overset{OA}{H_2O_{(l)}}$$
SRA   RA

$$Cu^{2+}_{(aq)} + 2e^- \rightarrow Cu_{(s)}$$
$$\underline{\quad Fe_{(s)} \rightarrow Fe^{2+}_{(aq)} + 2e^- \quad}$$
$$Fe_{(s)} + Cu^{2+}_{(aq)} \xrightarrow{spont.} Cu_{(s)} + Fe^{2+}_{(aq)}$$

(c) Qualitatively one could ensure complete reaction of the blue copper(II) solution and then observe the color of the solution. If the iron(III) ion is produced, the solution should be light yellow, but if the iron(II) ion is produced, the solution should be light green.

Quantitatively one could measure the mass of iron reacted and the mass of copper metal produced. Convert the masses to amounts and calculate the ratio of the amount of copper produced to the amount of iron reacted. If the ratio is 3:2, then iron(III) is produced, but if the ratio is 1:1, then iron(II) is produced.

Lab Exercise 12D
Testing Redox Concepts (page 365)

Problem
What are the products of the reaction of tin(II) chloride with an ammonium dichromate solution acidified with hydrochloric acid?

Prediction
According to redox concepts and the table of redox half-reactions, the products of the reaction are water, chromium(III) ions, and tin(IV) ions. The reasoning is shown below.

$$\begin{array}{cccccc}
\text{OA} & & \text{OA} & & \overbrace{\text{SOA}}^{} \quad \text{OA} \\
Sn^{2+}_{(aq)}, & Cl^-_{(aq)}, & H_2O_{(l)}, & NH_4^+_{(aq)}, & Cr_2O_7^{2-}_{(aq)}, & H^+_{(aq)} \\
\text{SRA} & \text{RA} \quad \text{RA} & \text{RA}
\end{array}$$

$$Cr_2O_7^{2-}{}_{(aq)} + 14H^+_{(aq)} + 6e^- \rightarrow 2Cr^{3+}_{(aq)} + 7H_2O_{(l)}$$
$$3[Sn^{2+}_{(aq)} \rightarrow Sn^{4+}_{(aq)} + 2e^-]$$

$$Cr_2O_7^{2-}{}_{(aq)} + 14H^+_{(aq)} + 3Sn^{2+}_{(aq)} \xrightarrow{spont.} 2Cr^{3+}_{(aq)} + 7H_2O_{(l)} + 3Sn^{4+}_{(aq)}$$

Experimental Design
An excess of tin(II) chloride solution is added to the acidic ammonium dichromate solution. If the color of the mixture is observed before and after the reaction, and the color changes from orange to green, then chromium(III) ions are likely produced.

INVESTIGATION 12.3

Demonstration with Sodium Metal (page 365)

Problem
What are the products of the reaction of sodium metal with water?

Prediction
According to the five-step method for predicting redox reactions, the products of the reaction are hydrogen gas and aqueous sodium hydroxide as shown below.

$$\begin{array}{cc}
 & \text{SOA} \\
Na_{(s)}, & H_2O_{(l)}, \\
\text{SRA} & \text{RA}
\end{array}$$

$$2H_2O_{(l)} + 2e^- \rightarrow H_{2(g)} + 2OH^-_{(aq)}$$
$$2[Na_{(s)} \rightarrow Na^+_{(aq)} + e^-]$$

$$2H_2O_{(l)} + 2Na_{(s)} \xrightarrow{spont.} H_{2(g)} + 2OH^-_{(aq)} + 2Na^+_{(aq)}$$

Experimental Design
A small piece of clean sodium metal is added to pure water. The following diagnostic tests are conducted first on pure water (as a control) and then on the final reaction mixture. If a flame is inserted into a sample of the gas, and a squeal or pop sound is heard, then hydrogen is likely present. If red litmus paper is immersed in the solution, and the paper turns blue, then hydroxide ions are likely present. If a flame test is conducted on the liquid, and the flame is bright yellow, then sodium ions are likely present.

Evidence

Diagnostic Test	Pure Water Control	Final Reaction Mixture
hydrogen test	no sound heard	high squeaky sound heard
litmus test	no color change	red litmus turned blue
flame test	pale yellow flame	bright yellow flame

Analysis

According to the evidence from the diagnostic tests, hydrogen gas, sodium ions, and hydroxide ions were produced in the reaction of sodium metal and water.

Evaluation

The experimental design is judged to be adequate because the problem was answered with no obvious flaws. The use of a control makes the results quite certain.

The prediction was verified because it clearly agrees with the evidence obtained. Therefore, the five-step method for predicting redox reactions is judged, with high confidence, to be acceptable.

Lab Exercise 12E
Standardizing Potassium Permanganate (page 366)

Problem

What is the concentration of the potassium permanganate solution?

Evaluation

The titration experimental design is judged to be adequate to answer the problem. The method chosen should give a low percent difference of 1–2%. The uncertainties originate from the measurements using a pipet and buret and the concentration of the tin(II) chloride solution.

The accuracy of the prediction is to within 5% as shown below.

$$\% \text{ difference} = \frac{|11.9 \text{ mmol/L} - 12.5 \text{ mmol/L}|}{12.5 \text{ mmol/L}} \times 100 = 5\%$$

This 5% difference cannot be accounted for solely by the measurement uncertainties mentioned above. Therefore, the prediction is falsified, or at best, inconclusive. It appears that the standardization of the potassium permanganate solution is important to ensure the best possible results in future titrations.

If each buret reading has an uncertainty of ±0.1 mL, then each volume of $KMnO_{4(aq)}$, which is obtained through two buret readings, will have an uncertainty of ±0.2 mL. Therefore, the uncertainty of the average volume, 16.8 mL, is ±0.2 mL. This is slightly more than 1%. The manipulation of the pipet and the concentration of the $Sn^{2+}_{(aq)}$ solution may give rise to smaller uncertainties, and hence the estimated total uncertainty of 2%. The remainder of the percent difference may be due to the spontaneous, but slow, reaction of potassium permanganate with water, a reaction that reduces the permanganate concentration.

Lab Exercise 12F
Analyzing for Tin (page 368)

Problem

What is the concentration of tin(II) ions in a solution prepared for research on toothpaste?

Analysis

$\quad$ OA $\qquad$ SOA $\quad$ OA $\qquad$ OA $\qquad$ OA
$K^+_{(aq)}$, $MnO_4^-{}_{(aq)}$, $H^+_{(aq)}$, $Sn^{2+}_{(aq)}$, $H_2O_{(l)}$
$\qquad\qquad\qquad\qquad\qquad\qquad$ SRA $\qquad$ RA

$2[MnO_4^-{}_{(aq)} + 8H^+_{(aq)} + 5e^- \rightarrow Mn^{2+}_{(aq)} + 4H_2O_{(l)}]$
$\quad 5[Sn^{2+}_{(aq)} \rightarrow Sn^{4+}_{(aq)} + 2e^-]$

$2MnO_4^-{}_{(aq)} + 16H^+_{(aq)} + 5Sn^{2+}_{(aq)} \rightarrow 2Mn^{2+}_{(aq)} + 8H_2O_{(l)} + 5Sn^{4+}_{(aq)}$

12.4 mL $\qquad\qquad\qquad$ 10.00 mL
0.0832 mol/L $\qquad\qquad\quad$ C

$n_{MnO_4^-} = 12.4 \text{ mL} \times \dfrac{0.0832 \text{ mol}}{1 \text{ L}} = 1.03 \text{ mmol}$

$n_{Sn^{2+}} = 1.03 \text{ mmol} \times \dfrac{5}{2} = 2.58 \text{ mmol}$

$C_{Sn^{2+}} = \dfrac{2.58 \text{ mmol}}{10.00 \text{ mL}} = 0.258 \text{ mol/L}$

According to the evidence and the stoichiometric analysis, the concentration of tin(II) ions in the solution is 0.258 mol/L.

Lab Exercise 12G
Analysis of Chromium in Steel (page 368)

Problem

What is the concentration of chromium(II) ions in a solution obtained in the analysis of a stainless steel alloy?

Analysis

OA $\qquad$ OA $\qquad\qquad$ SOA $\quad$ OA $\qquad$ OA
$Cr^{2+}_{(aq)}$, $K^+_{(aq)}$, $Cr_2O_7^{2-}{}_{(aq)}$, $H^+_{(aq)}$, $H_2O_{(l)}$
SRA $\qquad\qquad\qquad\qquad\qquad\qquad\qquad$ RA

$Cr_2O_7^{2-}{}_{(aq)} + 14H^+_{(aq)} + 6e^- \rightarrow 2Cr^{3+}_{(aq)} + 7H_2O_{(l)}$
$\quad 6[Cr^{2+}_{(aq)} \rightarrow Cr^{3+}_{(aq)} + e^-]$

$Cr_2O_7^{2-}{}_{(aq)} + 14H^+_{(aq)} + 6Cr^{2+}_{(aq)} \rightarrow 8Cr^{3+}_{(aq)} + 7H_2O_{(l)}$
17.4 mL $\qquad\qquad\qquad\qquad$ 10.00 mL
0.125 mol/L $\qquad\qquad\qquad\quad$ C

$$n_{Cr_2O_7^{2-}} = 17.4 \text{ mL} \times \frac{0.125 \text{ mol}}{1 \text{ L}} = 2.18 \text{ mmol}$$

$$n_{Cr^{2+}} = 2.18 \text{ mmol} \times \frac{6}{1} = 13.1 \text{ mmol}$$

$$C_{Cr^{2+}} = \frac{13.1 \text{ mmol}}{10.00 \text{ mL}} = 1.31 \text{ mol/L}$$

According to the evidence and the stoichiometric analysis, the concentration of chromium(II) ions is 1.31 mol/L.

Lab Exercise 12H
Analyzing for Iron (page 369)

Problem

What is the concentration of iron(II) ions in a solution obtained in an iron ore analysis?

Analysis

SOA OA OA
$Ce^{4+}_{(aq)}$, $Fe^{2+}_{(aq)}$, $H_2O_{(l)}$
    SRA RA

$$Ce^{4+}_{(aq)} + e^- \rightarrow Ce^{3+}_{(aq)}$$
$$Fe^{2+}_{(aq)} \rightarrow Fe^{3+}_{(aq)} + e^-$$

$Ce^{4+}_{(aq)}$ + $Fe^{2+}_{(aq)}$ $\rightarrow$ $Ce^{3+}_{(aq)}$ + $Fe^{3+}_{(aq)}$
15.0 mL   25.0 mL
0.125 mol/L C

$$n_{Ce^{4+}} = 15.0 \text{ mL} \times \frac{0.125 \text{ mol}}{1 \text{ L}} = 1.87 \text{ mmol}$$

$$n_{Fe^{2+}} = 1.87 \text{ mmol} \times \frac{1}{1} = 1.87 \text{ mmol}$$

$$C_{Fe^{2+}} = \frac{1.87 \text{ mmol}}{25.0 \text{ mL}} = 74.9 \text{ mmol/L}$$

According to the evidence and the stoichiometric analysis, the concentration of iron(II) ions is 74.9 mmol/L.

Evaluation

The prediction is accurate to within 6.4%.

$$\% \text{ difference} = \frac{|74.9 \text{ mmol/L} - 80.0 \text{ mmol/L}|}{80.0 \text{ mmol/L}} \times 100 = 6.4\%$$

The prediction is judged to be inconclusive because it is uncertain whether the percent difference can be accounted for by the uncertainties in the experiment. The metallurgical process must also have allowed for an acceptable tolerance of difference in results which is not known. Therefore, the process is still deemed to be acceptable.

A case could also be made for a falsified prediction and an unacceptable process.

Exercise (page 369)

33. Four assumptions for using titration evidence to do stoichiometric calculations are that the reaction is spontaneous, fast, quantitative, and stoichiometric.

34. Alternative designs include crystallization, filtration, and gas collection (page 193).
    Other methods include colorimetry and a variety of electrical methods.

35. SOA OA
$Ni_{(s)}, Ag^+_{(aq)}, H_2O_{(l)}$
 SRA RA

$$2[Ag^+_{(aq)} + e^- \rightarrow Ag_{(s)}]$$
$$Ni_{(s)} \rightarrow Ni^{2+}_{(aq)} + 2e^-$$
$$\overline{2Ag^+_{(aq)} + Ni_{(s)} \rightarrow 2Ag_{(s)} + Ni^{2+}_{(aq)}}$$

0.10 mol/L 25.0 g
v 58.69 g/mol

$$n_{Ni} = 25.0 \text{ g} \times \frac{1 \text{ mol}}{58.69 \text{ g}} = 0.426 \text{ mol}$$

$$n_{Ag^+} = 0.426 \text{ mol} \times \frac{2}{1} = 0.852 \text{ mol}$$

$$v_{Ag^+} = 0.852 \text{ mol} \times \frac{1 \text{ L}}{0.10 \text{ mol}} = 8.5 \text{ L}$$

36. SOA OA
$Cl_{2(g)}, H_2O_{(l)}, Br^-_{(aq)}$
 RA SRA

$$Cl_{2(g)} + 2e^- \rightarrow 2Cl^-_{(aq)}$$
$$2Br^-_{(aq)} \rightarrow Br_{2(l)} + 2e^-$$
$$\overline{Cl_{2(g)} + 2Br^-_{(aq)} \rightarrow Br_{2(l)} + 2Cl^-_{(aq)}}$$

m 3.00 kL
70.90 g/mol 0.40 mmol/L

$$n_{Br^-} = 3.00 \text{ kL} \times \frac{0.40 \text{ mmol}}{1 \text{ L}} = 1.2 \text{ mol}$$

$$n_{Cl_2} = 1.2 \text{ mol} \times \frac{1}{2} = 0.60 \text{ mol}$$

$$m_{Cl_2} = 0.60 \text{ mol} \times \frac{70.90 \text{ g}}{1 \text{ mol}} = 43 \text{ g}$$

37. $n_{Fe} = 1.08 \text{ g} \times \dfrac{1 \text{ mol}}{55.85 \text{ g}} = 19.3 \text{ mmol}$

$$C_{Fe^{2+}} = \frac{19.3 \text{ mmol}}{250.0 \text{ mL}} = 0.0774 \text{ mol/L}$$

```
         OA           SOA   OA       OA         OA
         K⁺(aq),  MnO₄⁻(aq),   H⁺(aq),   Fe²⁺(aq),   H₂O(l)
                                         SRA        RA
```

$$MnO_4^-{}_{(aq)} + 8H^+{}_{(aq)} + 5e^- \rightarrow Mn^{2+}{}_{(aq)} + 4H_2O_{(l)}$$
$$5[Fe^{2+}{}_{(aq)} \rightarrow Fe^{3+}{}_{(aq)} + e^-]$$

$$MnO_4^-{}_{(aq)} + 8H^+{}_{(aq)} + 5Fe^{2+}{}_{(aq)} \rightarrow Mn^{2+}{}_{(aq)} + 4H_2O_{(l)} + 5Fe^{3+}{}_{(aq)}$$

13.6 mL   10.0 mL
C   0.0774 mol/L

$$n_{Fe^{2+}} = 10.0 \text{ mL} \times \frac{0.0774 \text{ mol}}{1 \text{ L}} = 0.774 \text{ mmol}$$

$$n_{MnO_4^-} = 0.774 \text{ mmol} \times \frac{1}{5} = 0.155 \text{ mmol}$$

$$C_{MnO_4^-} = \frac{0.155 \text{ mmol}}{13.6 \text{ mL}} = 11.4 \text{ mmol/L}$$

38. In the following calculations, P represents both the percentage of iron in the iron ore sample and the volume of potassium dichromate used in the titration. Alternatively, a specific percentage and volume (e.g., 5% and 5 mL) could be chosen and the concentration of the potassium dichromate calculated stoichiometrically from these values. The same concentration should be predicted regardless of the values chosen.

```
  OA         OA             SOA    OA        OA
  Fe²⁺(aq),  K⁺(aq),   Cr₂O₇²⁻(aq),   H⁺(aq),   H₂O(l)
  SRA                                           RA
```

$$Cr_2O_7^{2-}{}_{(aq)} + 14H^+{}_{(aq)} + 6e^- \rightarrow 2Cr^{3+}{}_{(aq)} + 7H_2O_{(l)}$$
$$6[Fe^{2+}{}_{(aq)} \rightarrow Fe^{3+}{}_{(aq)} + e^-]$$

$$Cr_2O_7^{2-}{}_{(aq)} + 14H^+{}_{(aq)} + 6Fe^{2+}{}_{(aq)} \rightarrow 2Cr^{3+}{}_{(aq)} + 7H_2O_{(l)} + 6Fe^{3+}{}_{(aq)}$$

P mL   P% of 1.00 g
C   55.85 g/mol

$$n_{Fe^{2+}} = \frac{P}{100} \times 1.00 \text{ g} \times \frac{1 \text{ mol}}{55.85 \text{ g}} = 0.179P \text{ mmol}$$

$$n_{Cr_2O_7^{2-}} = 0.179P \text{ mmol} \times \frac{1}{6} = 0.0298P \text{ mmol}$$

$$C_{Cr_2O_7^{2-}} = \frac{0.0298P \text{ mmol}}{P \text{ mL}} = 29.8 \text{ mmol/L}$$

Lab Exercise 12I
Analysis for Tin(II) Chloride (page 370)

Problem

What is the concentration of a tin(II) chloride solution prepared from a sample of tin ore?

Analysis

The first endpoint was overshot and the result was discarded. The average volume of $K_2Cr_2O_{7(aq)}$ used in trials 2 to 4 was 10.7 mL.

$$\underset{SRA}{\overset{OA}{Fe^{2+}_{(aq)},}} \quad \overset{OA}{K^+_{(aq)},} \quad \overset{SOA}{Cr_2O_7^{2-}_{(aq)},} \quad \overset{OA}{H^+_{(aq)},} \quad \underset{RA}{\overset{OA}{H_2O_{(l)}}} \quad \overset{OA}{SO_4^{2-}_{(aq)},} \quad NH_4^+_{(aq)}$$

$$Cr_2O_7^{2-}{}_{(aq)} + 14H^+_{(aq)} + 6e^- \rightarrow 2Cr^{3+}_{(aq)} + 7H_2O_{(l)}$$
$$6[Fe^{2+}_{(aq)} \rightarrow Fe^{3+}_{(aq)} + e^-]$$

$$Cr_2O_7^{2-}{}_{(aq)} + 14H^+_{(aq)} + 6Fe^{2+}_{(aq)} \rightarrow 2Cr^{3+}_{(aq)} + 7H_2O_{(l)} + 6Fe^{3+}_{(aq)}$$
10.7 mL   10.00 mL
C   0.0500 mol/L

$$n_{Fe^{2+}} = 10.00 \text{ mL} \times \frac{0.0500 \text{ mol}}{1 \text{ L}} = 0.500 \text{ mmol}$$

$$n_{Cr_2O_7^{2-}} = 0.500 \text{ mmol} \times \frac{1}{6} = 0.0833 \text{ mmol}$$

$$C_{Cr_2O_7^{2-}} = \frac{0.0833 \text{ mmol}}{10.7 \text{ mL}} = 0.00779 \text{ mol/L}$$

The average volume of $K_2Cr_2O_{7(aq)}$ used in the second titration was 11.1 mL, based on trials 2 to 4.

$$\underset{SRA}{\overset{OA}{Sn^{2+}_{(aq)},}} \quad \overset{OA}{K^+_{(aq)},} \quad \overset{SOA}{Cr_2O_7^{2-}_{(aq)},} \quad \overset{OA}{H^+_{(aq)},} \quad \underset{RA}{\overset{OA}{H_2O_{(l)},}} \quad \underset{RA}{Cl^-_{(aq)}}$$

$$Cr_2O_7^{2-}{}_{(aq)} + 14H^+_{(aq)} + 6e^- \rightarrow 2Cr^{3+}_{(aq)} + 7H_2O_{(l)}$$
$$3[Sn^{2+}_{(aq)} \rightarrow Sn^{4+}_{(aq)} + 2e^-]$$

$$Cr_2O_7^{2-}{}_{(aq)} + 14H^+_{(aq)} + 3Sn^{2+}_{(aq)} \rightarrow 2Cr^{3+}_{(aq)} + 7H_2O_{(l)} + 3Sn^{4+}_{(aq)}$$
11.1 mL   10.00 mL
0.00779 mol/L   C

$$n_{Cr_2O_7^{2-}} = 11.1 \text{ mL} \times \frac{0.00779 \text{ mol}}{1 \text{ L}} = 0.0862 \text{ mmol}$$

$$n_{Sn^{2+}} = 0.0862 \text{ mmol/L} \times \frac{3}{1} = 0.259 \text{ mmol}$$

$$C_{Sn^{2+}} = \frac{0.259 \text{ mmol}}{10.00 \text{ mL}} = 0.0259 \text{ mol/L}$$

According to the evidence and the stoichiometric analysis, the concentration of tin(II) chloride is 25.9 mmol/L.

INVESTIGATION 12.4

Analysis of a Hydrogen Peroxide Solution (page 371)

Problem

What is the percent concentration of hydrogen peroxide in a consumer product?

Prediction

According to the manufacturer's label, the concentration of hydrogen peroxide is 3%.

Evidence

mass of $FeSO_4 \cdot (NH_4)_2SO_4 \cdot 6H_2O_{(s)}$ to prepare 100.0 mL of solution = 1.96 g

TITRATION OF 10.00 mL OF ACIDIC IRON(II) STANDARD WITH $KMnO_{4(aq)}$				
Trial	1	2	3	4
Final buret reading (mL)	13.6	26.6	39.6	13.9
Initial buret reading (mL)	0.2	13.6	26.6	0.8
Volume of $KMnO_{4(aq)}$ (mL)	13.4	13.0	13.0	13.1
Endpoint color	red	pink	pink	pink

- dilution of consumer $H_2O_{2(aq)}$, 25.0 mL to 1.00 L
- 5 mL of $H_2SO_{4(aq)}$ added to $H_2O_{2(aq)}$ in each trial

TITRATION OF 10.00 mL OF ACIDIFIED, DILUTED $H_2O_{2(aq)}$ WITH $KMnO_{4(aq)}$			
Trial	1	2	3
Final buret reading (mL)	11.5	22.7	33.9
Initial buret reading (mL)	0.2	11.5	22.7
Volume of $KMnO_{4(aq)}$ (mL)	11.3	11.2	11.2
Endpoint color	pink	pink	pink

Analysis

Standardization of the potassium permanganate solution:

$$\text{OA} \quad \text{SOA} \quad \text{OA} \quad \text{OA} \quad \text{OA} \quad \text{OA}$$
$$K^+_{(aq)}, \; MnO_4^-{}_{(aq)}, \; H^+_{(aq)}, \; Fe^{2+}_{(aq)}, \; H_2O_{(l)}, \; SO_4^{2-}{}_{(aq)}, \; NH_4^+{}_{(aq)}$$
$$\quad\quad\quad\quad \text{SRA} \quad\quad \text{RA}$$

$$MnO_4^-{}_{(aq)} + 8H^+_{(aq)} + 5e^- \rightarrow Mn^{2+}_{(aq)} + 4H_2O_{(l)}$$
$$5[Fe^{2+}_{(aq)} \rightarrow Fe^{3+}_{(aq)} + e^-]$$

$$MnO_4^-{}_{(aq)} + 8H^+_{(aq)} + 5Fe^{2+}_{(aq)} \rightarrow Mn^{2+}_{(aq)} + 4H_2O_{(l)} + 5Fe^{3+}_{(aq)}$$
13.0 mL 10.00 mL
C 0.0500 mol/L

$$n_{Fe^{2+}} = 10.00 \text{ mL} \times \frac{0.0500 \text{ mol}}{1 \text{ L}} = 0.500 \text{ mmol}$$

$$n_{MnO_4^-} = 0.500 \text{ mmol} \times \frac{1}{5} = 0.100 \text{ mmol}$$

$$C_{MnO_4^-} = \frac{0.100 \text{ mmol}}{13.0 \text{ mL}} = 0.007\ 67 \text{ mol/L}$$

Analysis of the hydrogen peroxide solution:

$$\text{OA} \quad \text{SOA} \quad \text{OA} \quad\quad \text{OA} \quad \text{OA}$$
$$K^+_{(aq)}, \; MnO_4^-{}_{(aq)}, \; H^+_{(aq)}, \; H_2O_{2(aq)}, \; H_2O_{(l)}, \; SO_4^{2-}{}_{(aq)}$$
$$\quad\quad\quad\quad \text{SRA} \quad\quad \text{RA}$$

$$2[MnO_4^-{}_{(aq)} + 8H^+_{(aq)} + 5e^- \rightarrow Mn^{2+}_{(aq)} + 4H_2O_{(l)}]$$
$$5[H_2O_{2(aq)} \rightarrow O_{2(g)} + 2H^+_{(aq)} + 2e^-]$$

$$2MnO_4^-{}_{(aq)} + 6H^+_{(aq)} + 5H_2O_{2(aq)} \rightarrow 2Mn^{2+}_{(aq)} + 8H_2O_{(l)} + 5O_{2(g)}$$
11.2 mL 10.00 mL
0.007 67 mol/L C

$$n_{MnO_4^-} = 11.2 \text{ mL} \times \frac{0.007\ 67 \text{ mol}}{1 \text{ L}} = 0.0862 \text{ mmol}$$

$$n_{H_2O_2} = 0.0862 \text{ mmol} \times \frac{5}{2} = 0.215 \text{ mmol}$$

$$C_{H_2O_2} = \frac{0.215 \text{ mmol}}{10.00 \text{ mL}} = 0.0215 \text{ mol/L (diluted sample)}$$

Since the original hydrogen peroxide was diluted by a factor of 40, the molar concentration of the orginal solution was 0.0215 mol/L × 40 = 0.862 mol/L. This corresponds to a percent concentration of 2.94% according to the graph.

According to the evidence and the stoichiometric analysis, the percent concentration of hydrogen peroxide is 2.94%.

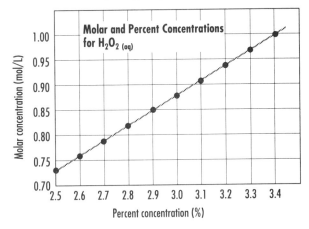

Evaluation

The titration design is judged to be adequate and is the best design for this type of chemical analysis. Based on previous experience, I am quite confident in this design. The procedure allowed sufficient evidence to be collected and was, therefore, adequate. Technological skills were routine and adequate as shown by the consistent results among the trials. Preparation of the iron(II) solution and judgment of the endpoint color, if not done correctly, are parts that could have a noticeable effect on the results.

Based upon my evaluation of this experiment, I am very certain of the results and I would expect a 1–2% difference. This difference would be due to small measurement uncertainties.

The prediction was verified with a percent difference of 2%.

$$\% \text{ difference} = \frac{|2.94\% - 3\%|}{3\%} \times 100 = 2\%$$

The prediction clearly agrees with the experimental result taking into account expected uncertainties. The claim by the manufacturer is judged to be acceptable because the prediction was verified. I am very confident in this judgment.

Exercise (page 374)

39. (a) $x + 2(-2) = 0$, $\quad x = +4$
    (b) $+1 + x + 4(-2) = 0$, $\quad x = +7$
    (c) $x + 4(-2) = -2$, $\quad x = +6$
    (d) $2x + 7(-2) = -2$, $\quad x = +6$
    (e) $+2 + 2x = 0$, $\quad x = -1$

40. (a) $2x + (-2) = 0$, $\quad x = +1$
    (b) $x + (-2) = 0$, $\quad x = +2$
    (c) $x + 2(-2) = 0$, $\quad x = +4$
    (d) $x + 3(+1) = 0$, $\quad x = -3$
    (e) $2x + 4(+1) = 0$, $\quad x = -2$
    (f) $+1 + x + 3(-2) = 0$, $\quad x = +5$
    (g) $2x = 0$, $\quad x = 0$
    (h) $x + 4(+1) + (-1) = 0$, $\quad x = -3$

41. (a) C, $\quad x = 0$
    (b) $C_6H_{12}O_6$, $\quad 6x + 12(+1) + 6(-2) = 0$, $\quad x = 0$
    (c) Na_2CO_3, $\quad 2(+1) + x + 3(-2) = 0$, $\quad x = +4$
    (d) CO, $\quad x + (-2) = 0$, $\quad x = +2$

42. methane, $\quad CH_4$, $\quad x + 4(+1) = 0$, $\quad x = -4$
    methanol, $\quad CH_3OH$, $\quad x + 4(+1) + (-2) = 0$, $\quad x = -2$
    methanal, $\quad CH_2O$, $\quad x + 2(+1) + (-2) = 0$, $\quad x = 0$
    methanoic acid, $\quad HCOOH$, $\quad 2(+1) + x + 2(-2) = 0$, $\quad x = +2$
    carbon dioxide, $\quad CO_2$, $\quad x + 2(-2) = 0$, $\quad x = +4$

Exercise (page 375)

43. (a) $\overset{0}{Cu}_{(s)} + 2\overset{+1\ +5\ -2}{AgNO_3}{}_{(aq)} \rightarrow 2\overset{0}{Ag}_{(s)} + \overset{+2\ +5-2}{Cu(NO_3)_2}{}_{(aq)}$ (redox)
    RA   OA

(b) $\overset{+2\ +5-2}{Pb(NO_3)_2}{}_{(aq)} + 2\overset{+1-1}{KI}_{(aq)} \rightarrow \overset{+2-1}{PbI_2}{}_{(s)} + 2\overset{+1+5-2}{KNO_3}{}_{(aq)}$ (not redox)

(c) $\overset{0}{Cl_2}{}_{(aq)} + 2\overset{+1-1}{KI}_{(aq)} \rightarrow \overset{0}{I_2}{}_{(s)} + 2\overset{+1-1}{KCl}_{(aq)}$ (redox)
    OA   RA

(d) $2\overset{+1\ -1}{NaCl}_{(l)} \rightarrow 2\overset{0}{Na}_{(l)} + \overset{0}{Cl_2}{}_{(g)}$ (redox)
    OA RA

(e) $\overset{+1-1}{HCl}_{(aq)} + \overset{+1\ -2+1}{NaOH}_{(aq)} \rightarrow \overset{+1\ -2}{H_2O}_{(l)} + \overset{+1\ -1}{NaCl}_{(aq)}$ (not redox)

(f) $2\overset{0}{Al}_{(s)} + 3\overset{0}{Cl_2}{}_{(g)} \rightarrow 2\overset{+3\ -1}{AlCl_3}{}_{(s)}$ (redox)
    RA   OA

(g) $2\overset{-2.5\ +1}{C_4H_{10}}{}_{(g)} + 13\overset{0}{O_2}{}_{(g)} \rightarrow 4\overset{+4-2}{CO_2}{}_{(g)} + 5\overset{+1\ -2}{H_2O}_{(l)}$ (redox)
    RA   OA

(h) $5\overset{-2+1-2\ +1}{CH_3OH}_{(l)} + 2\overset{+7\ -2}{MnO_4^-}{}_{(aq)} + 6\overset{+1}{H^+}_{(aq)} \rightarrow 5\overset{0\ +1\ -2}{CH_2O}_{(l)} + 2\overset{+2}{Mn^{2+}}{}_{(aq)} + 8\overset{+1-2}{H_2O}_{(l)}$ (redox)
    RA   OA

44. (a) single replacement
    (b) double replacement
    (c) single replacement
    (d) simple decomposition
    (e) double replacement (neutralization)
    (f) formation
    (g) combustion
    (h) other

Double replacement reactions do not appear to be redox reactions.

45. Methane and hydrogen are both reducing agents. The atmosphere on Saturn would tend to reduce carbon and its compounds, causing a decrease in oxidation number of carbon. For example, carbon can be reduced to methane.

$$\overset{0}{C_{(s)}} \rightarrow \overset{-4}{CH_{4(g)}}$$

On Earth, the oxidizing atmosphere oxidizes carbon and its compounds, causing an increase in oxidation number of carbon. For example, carbon is oxidized to carbon dioxide.

$$\overset{0}{C_{(s)}} \rightarrow \overset{+4}{CO_{2(g)}}$$

Exercise (page 379)

46. (a) $\overset{0}{H_{2(g)}} + \overset{+3\ -2}{Fe_2O_{3(aq)}} \rightarrow \overset{+2-2}{2FeO_{(s)}} + \overset{+1-2}{H_2O_{(l)}}$

        0   +3   +2   +1

    1e⁻/H   1e⁻/Fe
    2e⁻/H₂ 2e⁻/Fe₂O₃

(b) $\overset{+1-1}{2HBr_{(aq)}} + \overset{+1\ +6-2}{H_2SO_{4(aq)}} \rightarrow \overset{+4-2}{SO_{2(g)}} + \overset{0}{Br_{2(aq)}} + \overset{+1\ -2}{2H_2O_{(l)}}$

       -1   +6   +4   0

    1e⁻/Br   2e⁻/S
    1e⁻/HBr   2e⁻/H₂SO₄

(c) $\overset{+7\ -2}{2MnO_4^-{}_{(aq)}} + \overset{+1}{6H^+{}_{(aq)}} + \overset{-2+1\ -2+1}{5CH_3OH_{(l)}} \rightarrow \overset{+2}{2Mn^{2+}{}_{(aq)}} + \overset{+1-2}{8H_2O_{(l)}} + \overset{0\ +1-2}{5CH_2O_{(aq)}}$

    +7   -2   +2   0

    5e⁻/Mn   2e⁻/C
    5e⁻/MnO₄⁻   2e⁻/CH₃OH

(d) $\overset{+5-2}{2IO_3^-{}_{(aq)}} + \overset{+1\ +4-2}{5HSO_3^-{}_{(aq)}} \rightarrow \overset{+6-2}{5SO_4^{2-}{}_{(aq)}} + \overset{+1}{3H^+{}_{(aq)}} + \overset{0}{I_{2(aq)}} + \overset{+1\ -2}{H_2O_{(l)}}$

     +5   +4   +6   0

    5e⁻/I   2e⁻/S
    5e⁻/IO₃⁻ 2e⁻/HSO₃⁻

(e) $\overset{0}{I_{2(aq)}} + \overset{+1+4-2}{HSO_3^-{}_{(aq)}} + \overset{+1-2}{H_2O_{(l)}} \rightarrow \overset{-1}{2I^-{}_{(aq)}} + \overset{+6-2}{SO_4^{2-}{}_{(aq)}} + \overset{+1}{3H^+{}_{(aq)}}$

    0   +4   -1   +6

    1e⁻/I 2e⁻/S
    2e⁻/I₂ 2e⁻/HSO₃⁻

(f) $\overset{-3+1}{4NH_{3(g)}} + \overset{0}{7O_{2(g)}} \rightarrow \overset{+4-2}{4NO_{2(g)}} + \overset{+1\ -2}{6H_2O_{(g)}}$

     -3   0   +4   -2

    7e⁻/N   2e⁻/O
    7e⁻/NH₃ 4e⁻/O₂

(g) $\overset{-2+1\ -2+1}{C_2H_5OH_{(aq)}} + \overset{+1}{4H^+{}_{(aq)}} + \overset{+5-2}{4NO_3^-{}_{(aq)}} \rightarrow \overset{0+1\ 0-2\ +1}{CH_3COOH_{(aq)}} + \overset{+4-2}{4NO_{2(g)}} + \overset{+1\ -2}{3H_2O_{(l)}}$

    -2   +5   0   +4

    2e⁻/C   1e⁻/N
    4e⁻/C₂H₅OH   1e⁻/NO₃⁻

(h)
$$\overset{+1\;-2}{ClO^-_{(aq)}} + 2\overset{+1\;-2}{ClO^-_{(aq)}} \rightarrow \overset{+5\;-2}{ClO_3^-_{(aq)}} + 2\overset{-1}{Cl^-_{(aq)}}$$

+1   +1   +5   −1
4e⁻/Cl   2e⁻/Cl
4e⁻/ClO⁻ 2e⁻/ClO⁻

$$3ClO^-_{(aq)} \rightarrow ClO_3^-_{(aq)} + 2Cl^-_{(aq)}$$

47. The three methods of balancing oxidation-reduction reaction equations are by inspection (trial and error), by half-reactions, and by oxidation numbers.
In addition to discussing preferences, also discuss appropriate choice of method.

Lab Exercise 12J
Analyzing Blood for Alcohol Content (page 379)

Problem

What is the blood alcohol content in a blood sample?

Analysis

OA   OA   ⎯SOA⎯ OA   OA   ⎯OA
$Fe^{2+}_{(aq)}$, $K^+_{(aq)}$, $Cr_2O_7^{2-}_{(aq)}$, $H^+_{(aq)}$, $H_2O_{(l)}$, $SO_4^{2-}_{(aq)}$, $NH_4^+_{(aq)}$
SRA   RA

$$Cr_2O_7^{2-}_{(aq)} + 14H^+_{(aq)} + 6e^- \rightarrow 2Cr^{3+}_{(aq)} + 7H_2O_{(l)}$$
$$6[Fe^{2+}_{(aq)} \rightarrow Fe^{3+}_{(aq)} + e^-]$$

$$Cr_2O_7^{2-}_{(aq)} + 14H^+_{(aq)} + 6Fe^{2+}_{(aq)} \rightarrow 2Cr^{3+}_{(aq)} + 7H_2O_{(l)} + 6Fe^{3+}_{(aq)}$$

n 6.14 mL
           20.40 mmol/L

$$n_{Cr_2O_7^{2-}} = 2.70\text{ mL} \times \frac{0.01062\text{ mol}}{1\text{ L}} = 0.0287\text{ mmol (initially added)}$$

$$n_{Fe^{2+}} = 6.14\text{ mL} \times \frac{0.02040\text{ mol}}{1\text{ L}} = 0.125\text{ mmol}$$

$$n_{Cr_2O_7^{2-}} = 0.125\text{ mmol} \times \frac{1}{6} = 0.0209\text{ mmol (excess left after heating)}$$

$$n_{Cr_2O_7^{2-}} = 0.0287\text{ mmol} - 0.0209\text{ mmol} = 0.0078\text{ mmol (reduced by ethanol)}$$

+6 −2   +1   −2 +1 −2+1   +3   +1−2   0 +1 0 −2 +1
$$2Cr_2O_7^{2-}_{(aq)} + 16H^+_{(aq)} + 3C_2H_5OH_{(l)} \rightarrow 4Cr^{3+}_{(aq)} + 11H_2O_{(l)} + 3CH_3COOH_{(aq)}$$
+6   −2   +3   0

3e⁻/Cr   2e⁻/C
6e⁻/Cr₂O₇²⁻   4e⁻/C₂H₅OH
(× 2)   (× 3)

$$n_{C_2H_5OH} = 0.0078\text{ mmol} \times \frac{3}{2} = 0.0117\text{ mmol}$$

$$C_{C_2H_5OH} = \frac{0.0117\text{ mmol}}{0.500\text{ mL}} = 23.4\text{ mmol/L}$$

According to the evidence, the stoichiometric analysis, and the following graph, the blood alcohol content is determined to be 0.108 mg/100 mL.

Evaluation

The roadside breathalyzer screening test indicated a blood alcohol content greater than 0.10 mg/100 mL. This prediction is judged to be verified since the blood alcohol content determined by titration was 8% greater than 0.10 mg/100 mL. Total uncertainties from this titration experiment would be expected to be low and much less than 8%. The breathalyzer screening test appears to be acceptable.

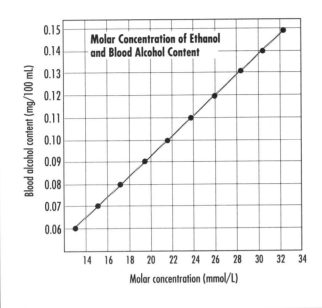

Exercise (page 381)

48. (a)

Menten	scientific	Herzberg	scientific
Banting	technological	Bartlett	scientific
Taylor	scientific	Guillet	technological
AEC	technological	McPherson	technological
Taube	scientific	Franklin	scientific
Bombardier	technological	Bell	technological
Hillier	technological	Newton	technological
Spar	technological	Battle	scientific
Lemieux	scientific	Fenerty	technological
Milner	scientific	Polanyi	scientific
Gibbs	technological	Gesner	technological
Wilson	scientific	Huntsman	technological

(b) *This could be a presentation, perhaps in costume, instead of a written report.*

Overview (page 382)

Review

1. Historically, oxidation referred to reactions involving oxygen, while reduction referred to the reduction in mass of a metal ore when a metal is produced.
2. A redox reaction is the transfer of electrons from a reducing agent to an oxidizing agent.
3. (a) An oxidizing agent accepts electrons from another substance (a reducing agent), causing that substance to be oxidized.

(b) A reducing agent donates electrons to another substance (an oxidizing agent), causing that substance to be reduced.

4. Possible evidence of a spontaneous redox reaction includes the formation of a precipitate or gas, a color or odor change, or an energy change.

5. In a table of redox half-reactions, if the oxidizing agent is listed above the reducing agent, the reaction is predicted to be spontaneous. If the oxidizing agent is listed below the reducing agent, the reaction is predicted to be non-spontaneous.

6. (a) spontaneous
   (b) non-spontaneous
   (c) non-spontaneous
   (d) spontaneous

7. The term standardized means the molar concentration has been determined empirically using a primary standard.

8. (a) 0
   (b) −1
   (c) +1

9. (a) Oxidation is defined as a loss of electrons involving an increase in oxidation number.
   (b) Reduction is defined as a gain of electrons involving a decrease in oxidation number.
   (c) In a redox reaction, an oxidizing agent gains electrons from a reducing agent. The total decrease in the oxidation number of the atoms/ions in the oxidizing agent is balanced by the total increase in the oxidation number of the atoms/ions in the reducing agent.

10. A net ionic equation is balanced in terms of the numbers of different kinds of atoms or ions, and total charge.

Applications

11. (a) $2\ Fe^{3+}_{(aq)} + 2e^- \rightarrow 2Fe^{2+}_{(aq)}$
    $Ni_{(s)} \rightarrow Ni^{2+}_{(aq)} + 2e^-$
    OA is $Fe^{3+}_{(aq)}$; RA is $Ni_{(s)}$

    (b) $Br_{2(aq)} + 2e^- \rightarrow 2Br^-_{(aq)}$
    $2I^-_{(aq)} \rightarrow I_{2(s)} + 2e^-$
    OA is $Br_{2(aq)}$; RA is $I^-_{(aq)}$

    (c) $Pd^{2+}_{(aq)} + 2e^- \rightarrow Pd_{(s)}$
    $Sn^{2+}_{(aq)} \rightarrow Sn^{4+}_{(aq)} + 2e^-$
    OA is $Pd^{2+}_{(aq)}$; RA is $Sn^{2+}_{(aq)}$

12. (a)   OA   OA
    $Cl_{2(aq)},\ H_2O_{(l)}$
      RA

    (b) OA   OA
    $Sn^{2+}_{(aq)},\ NO_3^-{}_{(aq)},\ H_2O_{(l)}$
    RA   RA

    (c) OA   OA   OA   OA
    $K^+_{(aq)},\ H^+_{(aq)},\ IO_3^-{}_{(aq)},\ H_2O_{(l)}$
                            RA

13. SOA $Tl^+_{(aq)} + e^- \rightleftharpoons Tl_{(s)}$
$In^{3+}_{(aq)} + 3e^- \rightleftharpoons In_{(s)}$
$Ga^{3+}_{(aq)} + 3e^- \rightleftharpoons Ga_{(s)}$
$Al^{3+}_{(aq)} + 3e^- \rightleftharpoons Al_{(s)}$ SRA

14. SOA $Cd^{2+}_{(aq)} + 2e^- \rightleftharpoons Cd_{(s)}$
$Ga^{3+}_{(aq)} + 3e^- \rightleftharpoons Ga_{(s)}$
$Mn^{2+}_{(aq)} + 2e^- \rightleftharpoons Mn_{(s)}$
$Ce^{3+}_{(aq)} + 3e^- \rightleftharpoons Ce_{(s)}$ SRA

15. According to the kinetic molecular theory, aqueous silver ions are in constant motion. According to collision theory, some of these silver ions collide with atoms of copper. According to redox theory, a competition for electrons results in silver ions removing electrons from the copper. This results in the silver ions being reduced to solid silver and the solid copper being oxidized to aqueous copper(II) ions.

16. (a) SOA   OA   OA   OA
    $Cl_{2(g)}$, $Fe^{2+}_{(aq)}$, $SO_4^{2-}_{(aq)}$ $H_2O_{(l)}$
    SRA   RA

$Cl_{2(g)} + 2e^- \rightarrow 2Cl^-_{(aq)}$
$2[Fe^{2+}_{(aq)} \rightarrow Fe^{3+}_{(aq)} + e^-]$
─────────────────────────────
$Cl_{2(g)} + 2Fe^{2+}_{(aq)} \rightarrow 2Cl^-_{(aq)} + 2Fe^{3+}_{(aq)}$
spontaneous

(b) OA   SOA   OA   OA
 $Ni^{2+}_{(aq)}$, $NO_3^-_{(aq)}$, $Sn^{2+}_{(aq)}$, $SO_4^{2-}_{(aq)}$ $H_2O_{(l)}$
                    SRA   RA

$Sn^{2+}_{(aq)} + 2e^- \rightarrow Sn_{(s)}$
$Sn^{2+}_{(aq)} \rightarrow Sn^{4+}_{(aq)} + 2e^-$
─────────────────────────────
$2Sn^{2+}_{(aq)} \rightarrow Sn_{(s)} + Sn^{4+}_{(aq)}$
non-spontaneous

(c)   OA   SOA
 $Zn_{(s)}$, $H_2O_{(l)}$ $O_{2(g)}$
 SRA   RA

$O_{2(g)} + 2H_2O_{(l)} + 4e^- \rightarrow 4OH^-_{(aq)}$
$2[Zn_{(s)} \rightarrow Zn^{2+}_{(aq)} + 2e^-]$
─────────────────────────────
$O_{2(g)} + 2H_2O_{(l)} + 2Zn_{(s)} \rightarrow 4OH^-_{(aq)} + 2Zn^{2+}_{(aq)}$
$O_{2(g)} + 2H_2O_{(l)} + 2Zn_{(s)} \rightarrow 2Zn(OH)_{2(s)}$
spontaneous

(d) OA   SOA   OA   OA   OA
 $H^+_{(aq)}$, $SO_4^{2-}_{(aq)}$, $H_2O_{(l)}$, $Fe_{(s)}$, $Na^+_{(aq)}$
                RA   SRA

$SO_4^{2-}_{(aq)} + 4H^+_{(aq)} + 2e^- \rightarrow H_2SO_{3(aq)} + H_2O_{(l)}$
$Fe_{(s)} \rightarrow Fe^{2+}_{(aq)} + 2e^-$
─────────────────────────────
$Fe_{(s)} + SO_4^{2-}_{(aq)} + 4H^+_{(aq)} \rightarrow Fe^{2+}_{(aq)} + H_2SO_{3(aq)} + H_2O_{(l)}$
spontaneous

(e)   OA   SOA   OA

$Na^+_{(aq)}$, $H_2O_{(l)}$, $K^+_{(aq)}$, $SO_3^{2-}_{(aq)}$, $OH^-_{(aq)}$
                         SRA   RA

$$2H_2O_{(l)} + 2e^- \rightarrow H_{2(g)} + 2OH^-_{(aq)}$$
$$SO_3^{2-}_{(aq)} + 2OH^-_{(aq)} \rightarrow SO_4^{2-}_{(aq)} + H_2O_{(l)} + 2e^-$$

$$SO_3^{2-}_{(aq)} + H_2O_{(l)} \rightarrow H_{2(g)} + SO_4^{2-}_{(aq)}$$

spontaneous

17.   SOA

$Mg_{(s)}$, $H_2O_{(l)}$

SRA   RA

$$2H_2O_{(l)} + 2e^- \rightarrow H_{2(g)} + 2OH^-_{(aq)}$$
$$Mg_{(s)} \rightarrow Mg^{2+}_{(aq)} + 2e^-$$

$$2H_2O_{(l)} + Mg_{(s)} \rightarrow H_{2(g)} + Mg(OH)_{2(s)}$$
                2.0 g m
              24.31 g/mol   58.33 g/mol

$$n_{Mg} = 2.0 \text{ g} \times \frac{1 \text{ mol}}{24.31 \text{ g}} = 0.082 \text{ mol}$$

$$n_{Mg(OH)_2} = 0.082 \text{ mol} \times \frac{1}{1} = 0.082 \text{ mol}$$

$$m_{Mg(OH)_2} = 0.082 \text{ mol} \times \frac{58.33 \text{ g}}{1 \text{ mol}} = 4.8 \text{ g}$$

18. OA   OA   SOA   OA   OA

$Sn^{2+}_{(aq)}$, $Cl^-_{(aq)}$, $K^+_{(aq)}$, $Cr_2O_7^{2-}_{(aq)}$, $H^+_{(aq)}$, $H_2O_{(l)}$

SRA   RA   RA

$$Cr_2O_7^{2-}_{(aq)} + 14H^+_{(aq)} + 6e^- \rightarrow 2Cr^{3+}_{(aq)} + 7H_2O_{(l)}$$
$$3[Sn^{2+}_{(aq)} \rightarrow Sn^{4+}_{(aq)} + 2e^-]$$

$$Cr_2O_7^{2-}_{(aq)} + 14H^+_{(aq)} + 3Sn^{2+}_{(aq)} \rightarrow 2Cr^{3+}_{(aq)} + 3Sn^{4+}_{(aq)} + 7H_2O_{(l)}$$
12.7 mL   25.0 mL
C 0.100 mol/L

$$n_{Sn^{2+}} = 25.0 \text{ mL} \times \frac{0.100 \text{ mol}}{1 \text{ L}} = 2.50 \text{ mmol}$$

$$n_{Cr_2O_7^{2-}} = 2.50 \text{ mmol} \times \frac{1}{3} = 0.833 \text{ mmol}$$

$$C_{Cr_2O_7^{2-}} = \frac{0.833 \text{ mmol}}{12.7 \text{ mL}} = 65.6 \text{ mmol/L}$$

19. OA   OA   SOA   OA   OA

$Fe^{2+}_{(aq)}$, $K^+_{(aq)}$, $MnO_4^-_{(aq)}$, $H^+_{(aq)}$, $H_2O_{(l)}$

SRA   RA

$$MnO_4^-_{(aq)} + 8H^+_{(aq)} + 5e^- \rightarrow Mn^{2+}_{(aq)} + 4H_2O_{(l)}$$
$$5[Fe^{2+}_{(aq)} \rightarrow Fe^{3+}_{(aq)} + e^-]$$

$$MnO_4^-_{(aq)} + 8H^+_{(aq)} + 5Fe^{2+}_{(aq)} \rightarrow Mn^{2+}_{(aq)} + 4H_2O_{(l)} + 5Fe^{3+}_{(aq)}$$
15.0 mL   10.0 mL
7.50 mmol/L C

$$n_{MnO_4^-} = 15.0 \text{ mL} \times \frac{7.50 \text{ mmol}}{1 \text{ L}} = 0.113 \text{ mmol}$$

$$n_{Fe^{2+}} = 0.113 \text{ mmol} \times \frac{5}{1} = 0.563 \text{ mmol}$$

$$C_{Fe^{2+}} = \frac{0.563 \text{ mmol}}{10.0 \text{ mL}} = 56.3 \text{ mmol/L}$$

20. (a)   SOA
    $K_{(s)}$, $H_2O_{(l)}$
    SRA RA

$$2H_2O_{(l)} + 2e^- \rightarrow H_{2(g)} + 2OH^-_{(aq)}$$
$$2[K_{(s)} \rightarrow K^+_{(aq)} + e^-]$$
$$\overline{2K_{(s)} + 2H_2O_{(l)} \rightarrow H_{2(g)} + 2OH^-_{(aq)} + 2K^+_{(aq)}}$$

(b) If a gas is collected and exposed to a flame, and a pop sound is heard, then hydrogen gas was likely produced. If a piece of red litmus paper is placed into the reaction mixture and the litmus paper turns blue, then hydroxide ions were likely produced. If a sample of the final solution is placed into a burner flame and a pale violet color is produced, then potassium ions were likely produced.

21.
- Clean three strips of magnesium metal with steel wool.
- Place a strip of magnesium metal into each solution and record evidence of reaction.
- Add drops of sodium carbonate solution to each of the solutions that did not react spontaneously with magnesium metal.

Evidence and Analysis: The solution that produced a precipitate with magnesium metal can be identified as lead(II) nitrate. The solution that produced a precipitate with sodium carbonate solution can be identified as calcium nitrate.

22. Sulfur is oxidized in each case.

$$\overset{-2}{H_2}S_{(g)} \rightarrow \overset{+4}{S}O_{2(g)}$$

$$\overset{+4}{S}O_{2(g)} \rightarrow \overset{+6}{S}O_{3(g)}$$

23. (a) $\overset{+2}{Ag^{2+}}_{(aq)} + \overset{+1\ -2}{H_2O}_{(l)} \rightarrow \overset{+1}{Ag^+}_{(aq)} + \overset{0}{O}_{2(g)} + \overset{+1}{H^+}_{(aq)}$

(b)   OA   RA
    $1e^-/Ag$ $2e^-/O$
    $1e^-/Ag^{2+}$ $2e^-/H_2O$

(c) $2Ag^{2+}_{(aq)} + H_2O_{(l)} \rightarrow 2Ag^+_{(aq)} + \frac{1}{2}O_{2(g)} + 2H^+_{(aq)}$

or

$4Ag^{2+}_{(aq)} + 2H_2O_{(l)} \rightarrow 4Ag^+_{(aq)} + O_{2(g)} + 4H^+_{(aq)}$

24. (a) $\overset{0}{C_6}\overset{+1\ -2}{H_{12}O_6}_{(s)} + \overset{0}{6O}_{2(g)} \rightarrow \overset{+4\ -2}{6CO}_{2(g)} + \overset{+1\ -2}{6H_2O}_{(l)}$
     0   0   +4 -2   -2

    $4e^-/C$ $2e^-/O$
    $24e^-/C_6H_{12}O_6$ $4e^-/O_2$

(b) $\overset{+3\ -1}{2AuBr_{3(aq)}} + \overset{+4-2}{3SO_{2(g)}} + \overset{+1\ -2}{6H_2O_{(l)}} \rightarrow \overset{+1\ +6-2}{3H_2SO_{4(aq)}} + \overset{+1-1}{6HBr_{(aq)}} + \overset{0}{2Au_{(s)}}$
$\quad\quad +3 \quad\quad\quad\quad +4 \quad\quad\quad\quad\quad\quad\quad\quad\quad\quad +6 \quad\quad\quad\quad\quad\quad\quad\quad 0$

$3e^-/Au \quad 2e^-/S$
$3e^-/AuBr_3 \quad 2e^-/SO_2$

(c) $\overset{+5-2}{2BrO_3^-_{(aq)}} + \overset{-2\ +1\ -2}{C_2H_6O_{(aq)}} \rightarrow \overset{+4-2}{2CO_{2(g)}} + \overset{-1}{2Br^-_{(aq)}} + \overset{+1\ -2}{3H_2O_{(l)}}$
$\quad\quad +5 \quad\quad\quad\quad\quad -2 \quad\quad\quad\quad\quad\quad +4 \quad\quad\quad\quad -1$

$6e^-/Br \quad\quad 6e^-/C$
$6e^-/BrO_3^- \quad 12e^-/C_2H_6O$

(d) $\overset{0}{3Ag_{(s)}} + \overset{+5-2}{NO_3^-_{(aq)}} + \overset{+1}{4H^+_{(aq)}} \rightarrow \overset{+1}{3Ag^+_{(aq)}} + \overset{+2-2}{NO_{(g)}} + \overset{+1-2}{2H_2O_{(l)}}$
$\quad\quad 0 \quad\quad\quad\quad +5 \quad\quad\quad\quad\quad\quad\quad\quad\quad\quad +1 \quad\quad\quad +2$

$1e^-/Ag \quad 3e^-/N$
$1e^-/Ag \quad 3e^-/NO_3^-$

(e) $\overset{+1\ +5-2}{2HNO_{3(aq)}} + \overset{+4-2}{3SO_{2(g)}} + \overset{+1\ -2}{2H_2O_{(l)}} \rightarrow \overset{+1\ +6-2}{3H_2SO_{4(aq)}} + \overset{+2\ -2}{2NO_{(g)}}$
$\quad\quad\quad +5 \quad\quad\quad\quad +4 \quad\quad\quad\quad\quad\quad\quad\quad\quad\quad +6 \quad\quad\quad\quad +2$

$3e^-/N \quad 2e^-/S$
$3e^-/NO_3^- \quad 2e^-/SO_2$

(f) $\overset{0}{4Al_{(s)}} + \overset{0}{3O_{2(g)}} \rightarrow \overset{+3\ -2}{2Al_2O_{3(s)}}$
$\quad\quad 0 \quad\quad\quad\quad 0 \quad\quad\quad\quad +3\ -2$

$3e^-/Al \quad 2e^-/O$
$3e^-/Al \quad 4e^-/O_2$

25. (a) $Zn_{(s)} + Ag_2S_{(s)} \rightarrow ZnS_{(s)} + 2Ag_{(s)}$
(balanced by inspection)

(b) $2e^-/Zn \quad 1e^-/Ag$
$2e^-/Zn \quad 2e^-/Ag_2S$

(c) reduction $\quad\quad Ag^+_{(s)} + e^- \rightarrow Ag_{(s)}$
oxidation $\quad\quad\quad Zn_{(s)} \rightarrow Zn^{2+}_{(s)} + 2e^-$

Extension

26. (a) +5 in VF_5 and V_2O_5
+4 in VCl_4
+3 in VBr_3
+2 in VI_2 and VCl_2

(b) The highest oxidation state (+5) is reached with fluorine or oxygen acting as the oxidizing agent. Therefore, these oxidizing agents are the strongest. The next highest oxidation state is reached with chlorine, which means chlorine is the next strongest oxidizing agent. Using the same argument, bromine is the next strongest while iodine is the least strong oxidizing agent.

(c) According to the *CRC Handbook of Chemistry and Physics*, compounds of vanadium exhibit different colors depending upon the oxidation state of vanadium.

  +5   yellow-red in $V_2O_{5(s)}$
  +4   blue in $VO_{2(s)}$, and red-brown in $VCl_{4(l)}$
  +3   pink in $VCl_{3(s)}$, green-black in $VBr_{3(s)}$, and green in $VF_3 \cdot 6H_2O_{(s)}$
  +2   green in $VCl_{2(s)}$, and violet-rose in $VI_{2(s)}$

(d) *Answers may vary.* Vanadium is widespread in nature, occurring in over 65 different minerals. Vanadium and its compounds are used in the manufacture of rust-resistant, spring, and high-speed tool steels; nuclear applications; ceramics; catalysts; superconducting materials.

27. *This is a widely reported topic. Considerable information is available from reports in popular magazines, from government environmental agencies, and from industries. For a report, it is best to limit this topic to a comparison of the common chlorine process with one alternative. See the article on page 376.*

28. For qualitative results, samples of iron are placed in beakers containing varying concentrations of calcium chloride solutions or containing equal concentrations of electrolytic and non-electrolytic solutions. The beakers are covered with watch glasses and the iron samples are visually inspected every 24 h.

29. A successful sacrificial anode would be a stronger reducing agent than iron. Therefore, common metals such as magnesium or zinc would be useful.

30. • Silver ions in a measured volume of the solution are precipitated using an excess of sodium chloride solution and the mass of the precipitate is determined.
    • Excess copper metal is added to a measured volume of the solution and the silver metal precipitate is collected and weighed.
    • An electrode, similar to a pH electrode but sensitive to silver ion concentration, is used to directly measure the concentration of the silver ions in the solution.

31. Some methods of determining or approximating the position of the beryllium half-reaction in a table of half-reactions include:
    • Reference the information from a source such as the *CRC Handbook of Chemistry and Physics*.
    • Look for evidence of reaction between beryllium metal and a large number of oxidizing agents. Place the beryllium half-reaction below all of the oxidizing agents with which it reacts spontaneously.
    • Look for evidence of reaction between aqueous beryllium ions and a large number of reducing agents. Place the beryllium half-reaction above all of the reducing agents with which it reacts spontaneously.
    • Construct a standard voltaic cell with $Be_{(s)}/Be^{2+}_{(aq)}$ as one half-cell and $H^+_{(aq)}/H_{2(g)}$ as the other half-cell. The measured voltage provides the $E°$ value for $Be_{(s)}/Be^{2+}_{(aq)}$, which allows this half-reaction to be positioned in a table of half-reactions.
    • Determine the minimum voltage required to cause a standard solution of $Be^{2+}_{(aq)}$ to react in an electrolytic cell. Then determine the $E°$ for the beryllium half-reaction and place it in the table of half-reactions.

Lab Exercise 12K
Analyzing Antifreeze (page 385)

Problem
What is the freezing point of a sample of windshield washer fluid?

Analysis
Standardization of $KMnO_{4(aq)}$:

$$\overset{\text{OA}}{K^+_{(aq)}}, \overset{\overset{\frown}{\text{SOA}}}{MnO_4^-{}_{(aq)}}, \overset{\text{OA}}{H^+_{(aq)}}, \overset{\text{OA}}{Fe^{2+}_{(aq)}}, NH_4^+{}_{(aq)}, SO_4^{2-}{}_{(aq)}, \overset{\overset{\frown}{\text{OA}}}{H_2O_{(l)}}$$
$$\text{SRA} \qquad\qquad\qquad\qquad\qquad\qquad \text{RA}$$

$$MnO_4^-{}_{(aq)} + 8H^+_{(aq)} + 5e^- \rightarrow Mn^{2+}_{(aq)} + 4H_2O_{(l)}$$
$$5[Fe^{2+}_{(aq)} \rightarrow Fe^{3+}_{(aq)} + e^-]$$

$MnO_4^-{}_{(aq)} + 8H^+_{(aq)} + 5Fe^{2+}_{(aq)} \rightarrow Mn^{2+}_{(aq)} + 4H_2O_{(l)} + 5Fe^{3+}_{(aq)}$
12.4 mL   10.00 mL
C 0.331 mol/L

$$n_{Fe^{2+}} = 10.00 \text{ mL} \times \frac{0.331 \text{ mol}}{1 \text{ L}} = 3.31 \text{ mmol}$$

$$n_{MnO_4^-} = 3.31 \text{ mmol} \times \frac{1}{5} = 0.662 \text{ mmol}$$

$$C_{MnO_4^-} = \frac{0.662 \text{ mmol}}{12.4 \text{ mL}} = 0.0534 \text{ mol/L}$$

Chemical analysis of basic methanol:

$$\overset{-2+1\ -2+1}{CH_3OH_{(aq)}} + \overset{+7\ -2}{6MnO_4^-{}_{(aq)}} + \overset{-2+1}{8OH^-_{(aq)}} \rightarrow \overset{+4-2}{CO_3^{2-}{}_{(aq)}} + \overset{+6\ -2}{6MnO_4^{2-}{}_{(aq)}} + \overset{+1\ -2}{6H_2O_{(l)}}$$
$\ \ -2 \qquad\qquad\qquad +7 \qquad\qquad\qquad\qquad\qquad\qquad +4 \qquad\qquad +6$

6e⁻/C   1e⁻/Mn
6e⁻/CH₃OH   1e⁻/MnO₄⁻
(× 1)   (× 6)
10.00 mL   11.7 mL
C 0.0534 mol/L

$$n_{MnO_4^-} = 11.7 \text{ mL} \times \frac{0.0534 \text{ mol}}{1 \text{ L}} = 0.625 \text{ mmol}$$

$$n_{CH_3OH} = 0.625 \text{ mmol} \times \frac{1}{6} = 0.104 \text{ mmol}$$

$$C_{CH_3OH} = \frac{0.104 \text{ mmol}}{10.00 \text{ mL}} = 0.0104 \text{ mol/L}$$

Since the windshield washer fluid has a concentration 1000 times the laboratory sample, its concentration is 10.4 mol/L according to the evidence presented.

According to the graph shown below, the freezing point of a 10.4 mol/L solution of methanol is approximately –33°C.

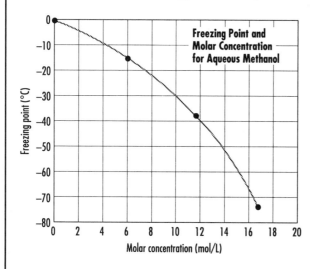

Freezing Point and Molar Concentration for Aqueous Methanol

Lab Exercise 12L
Redox Indicators (page 385)

Problem
Where do the redox indicators in the given list fit in a table of oxidizing and reducing agents?

Analysis

Reaction 1
Since red-yellow eriogreen reacts spontaneously with $IO_3^-{}_{(aq)} + H^+{}_{(aq)}$, it is listed below $IO_3^-{}_{(aq)} + H^+{}_{(aq)}$.
$2IO_3^-{}_{(aq)} + 12H^+{}_{(aq)} + 10e^- \rightleftharpoons I_{2(s)} + 6H_2O_{(l)}$
eriogreen(rose) + $ne^- \rightleftharpoons$ eriogreen (red-yellow)

Reaction 2
Since red nitroferrion does not react spontaneously with $IO_3^-{}_{(aq)} + H^+{}_{(aq)}$, it is listed above $IO_3^-{}_{(aq)} + H^+{}_{(aq)}$.
nitroferrion (faint blue) + $ne^- \rightleftharpoons$ nitroferrion (red)
$2IO_3^-{}_{(aq)} + 12H^+{}_{(aq)} + 10e^- \rightleftharpoons I_{2(s)} + 6H_2O_{(l)}$
eriogreen (rose) + $ne^- \rightleftharpoons$ eriogreen (red-yellow)

Reaction 3
Since red-yellow eriogreen does not react spontaneously with $Ag^+{}_{(aq)}$, it is listed above $Ag^+{}_{(aq)}$.
nitroferrion (faint blue) + $ne^- \rightleftharpoons$ nitroferrion (red)
$2IO_3^-{}_{(aq)} + 12H^+{}_{(aq)} + 10e^- \rightleftharpoons I_{2(s)} + 6H_2O_{(l)}$
eriogreen (rose) + $ne^- \rightleftharpoons$ eriogreen (red-yellow)
$Ag^+{}_{(aq)} + e^- \rightleftharpoons Ag_{(s)}$

Reaction 4
Since colorless diphenylamine reacts spontaneously with $Ag^+{}_{(aq)}$, it is listed below $Ag^+{}_{(aq)}$.
nitroferrion (faint blue) + $ne^- \rightleftharpoons$ nitroferrion (red)
$2IO_3^-{}_{(aq)} + 12H^+{}_{(aq)} + 10e^- \rightleftharpoons I_{2(s)} + 6H_2O_{(l)}$
eriogreen (rose) + $ne^- \rightleftharpoons$ eriogreen (red-yellow)
$Ag^+{}_{(aq)} + e^- \rightleftharpoons Ag_{(s)}$
diphenylamine (violet) + $ne^- \rightleftharpoons$ diphenylamine (colorless)

Reaction 5
Since red nitroferrion reacts spontaneously with $Au^{3+}{}_{(aq)}$, it is listed below $Au^{3+}{}_{(aq)}$.
$Au^{3+}{}_{(aq)} + 3e^- \rightleftharpoons Au_{(s)}$
nitroferrion (faint blue) + $ne^- \rightleftharpoons$ nitroferrion (red)
$2IO_3^-{}_{(aq)} + 12H^+{}_{(aq)} + 10e^- \rightleftharpoons I_{2(s)} + 6H_2O_{(l)}$
eriogreen (rose) + $ne^- \rightleftharpoons$ eriogreen (red-yellow)
$Ag^+{}_{(aq)} + e^- \rightleftharpoons Ag_{(s)}$
diphenylamine (violet) + $ne^- \rightleftharpoons$ diphenylamine (colorless)

Reaction 6
Since blue methylene blue reacts spontaneously with violet diphenylamine, it is listed below diphenylamine.
$Au^{3+}{}_{(aq)} + 3e^- \rightleftharpoons Au_{(s)}$
nitroferrion (faint blue) + $ne^- \rightleftharpoons$ nitroferrion (red)
$2IO_3^-{}_{(aq)} + 12H^+{}_{(aq)} + 10e^- \rightleftharpoons I_{2(s)} + 6H_2O_{(l)}$

eriogreen (rose) + ne⁻ ⇌ eriogreen (red-yellow)
Ag⁺$_{(aq)}$ + e⁻ ⇌ Ag$_{(s)}$
diphenylamine (violet) + ne⁻ ⇌ diphenylamine (colorless)
methylene blue (colorless) + ne⁻ ⇌ methylene blue (blue)

Reaction 7
Since blue methylene blue does not react spontaneously with $Cu^{2+}_{(aq)}$, it is listed above $Cu^{2+}_{(aq)}$.

$Au^{3+}_{(aq)}$ + 3e⁻ ⇌ Au$_{(s)}$
nitroferrion (faint blue) + ne⁻ ⇌ nitroferrion (red)
$2IO_3^-{}_{(aq)}$ + $12H^+{}_{(aq)}$ + 10e⁻ ⇌ $I_{2(s)}$ + $6H_2O_{(l)}$
eriogreen (rose) + ne⁻ ⇌ eriogreen (red-yellow)
Ag⁺$_{(aq)}$ + e⁻ ⇌ Ag$_{(s)}$
diphenylamine (violet) + ne⁻ ⇌ diphenylamine (colorless)
methylene blue (colorless) + ne⁻ ⇌ methylene blue (blue)
$Cu^{2+}_{(aq)}$ + 2e⁻ ⇌ Cu$_{(s)}$

Based on the evidence gathered and on the redox reaction spontaneity generalization, the position of the redox indicators in a table of oxidizing and reducing agents is, in order of decreasing strength of oxidizing agent, $Au^{3+}_{(aq)}$, nitroferrion, $IO_3^-{}_{(aq)}$ + $H^+{}_{(aq)}$, eriogreen, $Ag^+{}_{(aq)}$, diphenylamine, methylene blue, $Cu^{2+}_{(aq)}$.

13 Voltaic and Electrolytic Cells

INVESTIGATION 13.1

Demonstration of a Simple Electric Cell (page 387)

Problem

What electrical properties are observed when two metals come in contact with a conducting solution?

Evidence

- Metal strips used were approximately 2 cm by 3 cm in size.
- Strips were cleaned with steel wool before each measurement since some of the metals, particularly lead, corroded noticeably.
- The readings were not very stable in most cases, especially in the measurement of current.

Metals	Paper Towel Soaked in Salt Water		Apple		Lemon	
	Current (mA)	Voltage (V)	Current (µA)	Voltage (V)	Current (µA)	Voltage (V)
Zn, Cu	2.4	0.80	170	1.01	179	0.96
Zn, Pb	4.9	0.48	107	0.57	125	0.50
Zn, Ag	2.2	0.76	124	0.93	146	0.75
Cu, Pb	0.9	0.31	65	0.41	60	0.45
Cu, Ag	0.7	0.28	16	0.17	41	0.22
Pb, Ag	0.4	0.27	49	0.39	48	0.20

Analysis

All combinations of metals in contact with a salt solution, apple juice, or lemon juice produced different electric currents and voltages.

Evaluation

The experimental design is judged to be adequate to answer the problem since only a general answer to the question was required. Better control of variables would be needed to obtain reliable measurements. Both the procedure and the technological skills were adequate for the design used. The positioning of the metal strips and the time used to obtain a measurement were two factors that had a significant influence on the results. There appear to be many possible uncertainties in the measurements made.

The prediction based on Galvani's hypothesis was falsified since all solutions tested were found to produce electricity and none of the solutions used contained animal tissue. The hypothesis is, therefore, judged to be unacceptable because the prediction was falsified.

Exercise (page 390)

1. A simple electric cell contains two solid conductors (electrodes) and an aqueous conductor (electrolyte).

2. An electrode is a solid conductor. An electrolyte is an aqueous conductor. A cathode is the positive electrode in an electric cell. An anode is the negative electrode in an electric cell.

3. $\dfrac{6\text{ V}}{1.5\text{ V/cell}} = 4\text{ cells}$ or $6\text{ V} \times \dfrac{1\text{ cell}}{1.5\text{ V}} = 4\text{ cells}$

Connecting cells in series does not increase the current of the resulting battery. The voltage (energy) of the circuit is increased by series arrangements. However, for cells connected in parallel, the current increases and the voltage remains constant.

4. From a scientific perspective, a diagram shows where the electrons are believed to enter the device and where the electrons are believed to exit the device.
   From a technological perspective, a diagram shows how the batteries should be placed in the device to make it work and to avoid damaging certain components.

5. Electric current is a measure of the rate of flow of charge past a point in an electric circuit. Voltage is a measure of the energy difference per unit electric charge.

INVESTIGATION 13.2

Designing an Electric Cell (page 390)

Problem

What combination of electrodes and electrolyte gives the largest voltage for an aluminum-can cell?

Prediction

According to the table of redox half-reactions, copper and aluminum are quite far apart so this combination might give a significant voltage. Salt water will be tried first as the electrolyte because it is common and inexpensive. This initial prediction will be revised by trial and error to answer the problem.

Procedure

1. Using steel wool, scrape an area near the top of the aluminum can to expose a clean, shiny surface.
2. Half fill the can with 0.5 mol/L $NaCl_{(aq)}$.
3. Clean the end of a piece of copper metal with steel wool.
4. Attach the negative (black) terminal of the voltmeter to the aluminum and the positive (red) terminal to the copper.
5. While holding the copper in the electrolyte, avoiding the sides of the can, measure the voltage.
6. Repeat the measurement after 15–30 min.
7. Repeat steps 2 to 6 using a carbon rod in place of the copper metal.
8. Repeat steps 2 to 7 using a 0.5 mol/L $NaOH_{(aq)}$ and then a 0.5 mol/L $HCl_{(aq)}$ in place of the $NaCl_{(aq)}$.

Evidence/Analysis

Electrode	INITIAL VOLTAGES (V)			FINAL VOLTAGES (V)		
	$NaCl_{(aq)}$	$NaOH_{(aq)}$	$HCl_{(aq)}$	$NaCl_{(aq)}$	$NaOH_{(aq)}$	$HCl_{(aq)}$
copper	0.5	0.4	0.6	1.1	2.4	1.2
carbon	0.4	1.5	0.9	1.5	3.8	2.4

According to the evidence collected, the largest voltage of an aluminum-can cell is 3.8 V. This voltage is obtained with a carbon electrode in a 0.5 mol/L sodium hydroxide electrolyte left sitting in the can for a period of time.

Evaluation

The trial and error design is judged to be adequate to answer the problem. There are no obvious flaws and no better alternatives appear to exist. The procedure was adequate, but for more reliable results, the time the electrolyte is left in the can needs to be considered. The results changed when the electrolyte remained in the can for a longer time. Improvements could include cleaning and scraping the inside of the aluminum can before use. Technological skills are not likely responsible for any uncertainties since no special skills are required. Overall, I am quite certain that I found the best combination but I am not completely certain about the specific voltage value.

The $C_{(s)}/NaOH_{(aq)}/Al_{(s)}$ electric cell is judged according to these criteria: reliability, economy, and simplicity. The cell appears reliable since it produced a voltage quickly. However, if a constant voltage is required, this cell may not function reliably. More trials are required to investigate long-term reliability and the time it takes the thin wall of the aluminum can to disintegrate. The economics of the cell would depend largely on the cost of aluminum and the cell's ability to be recharged. Recycled aluminum would be a significant economic advantage. Certainly, there does not appear to be a shortage of empty aluminum cans. The cell is simple as it does not have any complex parts or technical design. Further trials are needed using a moist, basic electrolyte paste to obtain a cell that will not easily leak its contents.

Exercise (page 393)

6. Scientific knowledge may be used to describe, explain, or predict the parts or operation of a device. Technological problem solving is used to develop a device that works based on established criteria.
7. The steps involved in technological problem solving are:
   - Develop a general design; for example, select variables to manipulate and control.
   - Follow several prediction-procedure-evidence-analysis cycles, manipulating and systematically studying one variable at a time.
   - Complete an evaluation based on criteria such as efficiency, reliability, cost, and simplicity.
8. A technical and safety problem is the possible leakage of the liquid electrolyte. The electrolyte should be changed to a moist paste to reduce this problem.
9. Advantages include simplicity, reliability, and relatively low cost. The main disadvantage is that the cell is not rechargeable and must be discarded when depleted. Other possible disadvantages for certain uses include short shelf life, relatively low current produced, and constancy of voltage.
10. The Molicel contains an anode and a cathode in a layered design analogous to a jelly-roll. This creative design gives the Molicel a large surface area, allowing for more electrons to be exchanged per second. It allows the cell to output relatively large currents compared with other batteries of about the same size.
11. The development of the fuel cell solves the technological problem of the limited life of a cell caused by the depletion of the reactants as the cell operates.

12. The AA, C, and D cells differ in size, current produced, and longevity of the cell, but the cell potential remains the same in each size. The smaller the cell, the smaller the electrodes and the smaller the current produced.
13. Both have containers which are consumed as they are used.

Lab Exercise 13A
Evaluating Batteries (page 393)

Problem

Taking all of the preceding criteria into account, what is the best cell or battery for a portable radio, cassette player, or CD player?

Evidence/Analysis

Table 13.2, page 392, can be used as a starting point and further information can be gathered from popular science and technology magazines. Some library books on general electricity and "Consumer Reports" are also good sources of information. The problem could be limited to a decision between a common dry cell (zinc chloride) and a common secondary cell (Ni-Cad). An interesting comparison that has been reported in consumer magazines and TV programs is the cost comparison between disposable dry cells and rechargeable cells. Note that there is a variety of "trade-offs" which will be part of the evaluation process.

INVESTIGATION 13.3

Demonstration of a Voltaic Cell (page 396)

Problem

What is the design and operation of a voltaic cell?

Evidence

VOLTAGE FOR VARIOUS Ag-Cu CELL DESIGNS			
Cell Design	Positive Electrode	Negative Electrode	Voltage (V)
(a) single electrolyte	$Ag_{(s)}$	$Cu_{(s)}$	0.15
(b) salt bridge	$Ag_{(s)}$	$Cu_{(s)}$	0.42
(c) porous cup	$Ag_{(s)}$	$Cu_{(s)}$	0.42

- Removing either electrode from the solution, removing the salt bridge, or removing the porous cup immediately produced a zero volt reading. Replacing the removed part restored the original voltmeter reading.
- No evidence of reaction was visible after several minutes. After several days, silvery crystals formed on the silver electrode and some blue colored solution had moved toward the silver side of the cell. The copper electrode appeared smaller in size.

Analysis

A voltaic cell requires two different electrodes, both in contact with an electrolyte. All parts, external and internal, must be connected for the cell to operate (produce a voltage). The salt bridge and porous cup designs functioned in very much the same way.

Exercise (page 399)

14. A voltaic cell is an arrangement of two half-cells that spontaneously produces electricity.
    A half-cell is an electrode-electrolyte combination forming one-half of a complete cell.
    A porous boundary is a barrier that separates electrolytes while still permitting the movement of ions.
    An inert electrode is a solid conductor that serves as an anode or a cathode in a voltaic cell, but is chemically unreactive.

15. A cathode is the electrode where reduction occurs. An anode is the electrode where oxidation occurs.

16. (a) cathode
    (b) anode
    (c) anode
    (d) cathode

17. (a) Ions move to maintain electrical neutrality at each electrode. In a simple voltaic cell composed of metals and metal ions, reduction removes positive ions from the solution and oxidation adds positive ions to the solution around the electrodes. As electrons move externally from the anode to the cathode, anions move toward the anode and take away the excess positive charge.
    (b) Cations move toward the cathode and anions move toward the anode.

18. An inert electrode is a conducting solid which will not react with an electrolyte in a cell. It is used as the anode or cathode in a half-cell in which no solid is involved in the half-reaction equation.

19. The solution in a salt bridge must be an inert electrolyte; that is, it must not react with the ions which pass through. Sodium sulfate solution is an example.

20. (a) $Ag_{(s)} | Ag^+_{(aq)} \| Zn^{2+}_{(aq)} | Zn_{(s)}$
    $\phantom{Ag_{(s)} |}$SOA$\phantom{Ag^+_{(aq)} \|}$SRA

    cathode $\quad 2[Ag^+_{(aq)} + e^- \rightarrow Ag_{(s)}]$
    anode $\quad\quad\quad Zn_{(s)} \rightarrow Zn^{2+}_{(aq)} + 2e^-$

    net $\quad\quad 2Ag^+_{(aq)} + Zn_{(s)} \rightarrow 2Ag_{(s)} + Zn^{2+}_{(aq)}$

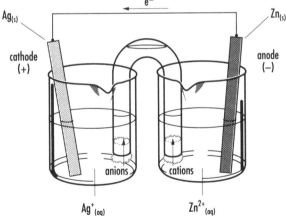

(b) $Cu_{(s)} | Cu^{2+}_{(aq)} \| Zn^{2+}_{(aq)} | Zn_{(s)}$
    $\phantom{Cu_{(s)} |}$SOA$\phantom{Cu^{2+}_{(aq)} \|}$SRA

    cathode $\quad Cu^{2+}_{(aq)} + 2e^- \rightarrow Cu_{(s)}$
    anode $\quad\quad\quad Zn_{(s)} \rightarrow Zn^{2+}_{(aq)} + 2e^-$

    net $\quad\quad Cu^{2+}_{(aq)} + Zn_{(s)} \rightarrow Cu_{(s)} + Zn^{2+}_{(aq)}$

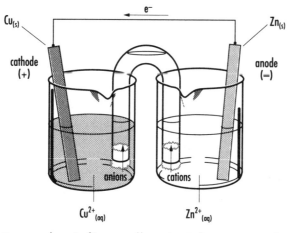

(c) $Sn_{(s)} \mid Sn^{2+}_{(aq)}, \parallel Cr_2O_7^{2-}_{(aq)}, \quad H^+_{(aq)}, \mid C_{(s)}$
SRA   SOA

cathode $Cr_2O_7^{2-}_{(aq)} + 14H^+_{(aq)} + 6e^- \rightarrow 2Cr^{3+}_{(aq)} + 7H_2O_{(l)}$
anode $3[Sn_{(s)} \rightarrow Sn^{2+}_{(aq)} + 2e^-]$

net $Cr_2O_7^{2-}_{(aq)} + 14H^+_{(aq)} + 3Sn_{(s)} \rightarrow 2Cr^{3+}_{(aq)} + 7H_2O_{(l)} + Sn^{2+}_{(aq)}$

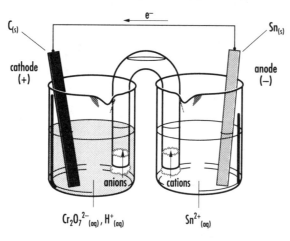

(d) $Al_{(s)} \mid Al^{3+}_{(aq)} \parallel Na^+_{(aq)}, Cl^-_{(aq)}, O_{2(g)} \mid Pt_{(s)}$
SRA   SOA

cathode $3[O_{2(g)} + 2H_2O_{(l)} + 4e^- \rightarrow 4OH^-_{(aq)}]$
anode $4[Al_{(s)} \rightarrow Al^{3+}_{(aq)} + 3e^-]$

net $3O_{2(g)} + 6H_2O_{(l)} + 4Al_{(s)} \rightarrow 4Al(OH)_{3(s)}$

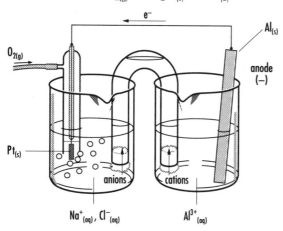

CHANGE AND SYSTEMS

21. cathode $Ni^{2+}_{(aq)} + 2e^- \rightarrow Ni_{(s)}$
    anode $Cd_{(s)} \rightarrow Cd^{2+}_{(aq)} + 2e^-$

    net $Ni^{2+}_{(aq)} + Cd_{(s)} \rightarrow Ni_{(s)} + Cd^{2+}_{(aq)}$

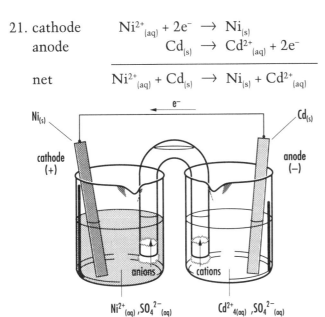

22. (Extension) The nickel cathode could be replaced by any inert conducting material such as carbon or platinum. The anode solution could be replaced with an inert aqueous electrolyte such as aqueous potassium sulfate.

Exercise (page 405)

23. (a) cathode $Sn^{2+}_{(aq)} + 2e^- \rightarrow Sn_{(s)}$ $E_r° = -0.14$ V
        anode $Cr_{(s)} \rightarrow Cr^{2+}_{(aq)} + 2e^-$ $E_r° = -0.91$ V

        net $Sn^{2+}_{(aq)} + Cr_{(s)} \rightarrow Sn_{(s)} + Cr^{2+}_{(aq)}$ $\Delta E° = +0.77$ V

    (b) cathode $SO_4^{2-}{}_{(aq)} + 4H^+_{(aq)} + 2e^- \rightarrow H_2SO_{3(aq)} + H_2O_{(l)}$ $E_r° = +0.17$ V
        anode $Co_{(s)} \rightarrow Co^{2+}_{(aq)} + 2e^-$ $E_r° = -0.28$ V

        net $SO_4^{2-}{}_{(aq)} + 4H^+_{(aq)} + Co_{(s)} \rightarrow$
                                      $H_2SO_{3(aq)} + H_2O_{(l)} + Co^{2+}_{(aq)}$ $\Delta E° = +0.45$ V

    (c) cathode $O_{2(g)} + 2H_2O_{(l)} + 4e^- \rightarrow 4OH^-_{(aq)}$ $E_r° = +0.40$ V
        anode $2[H_{2(g)} + 2OH^-_{(aq)} \rightarrow 2H_2O_{(l)} + 2e^-]$ $E_r° = -0.83$ V

        net $O_{2(g)} + 2H_{2(g)} \rightarrow 2H_2O_{(l)}$ $\Delta E° = +1.23$ V

24. (a) $Pb_{(s)} \mid Pb^{2+}_{(aq)} \parallel Cu^{2+}_{(aq)} \mid Cu_{(s)}$
        anode   cathode $\Delta E° = +0.47$ V

    (b) $Ni_{(s)} \mid Ni^{2+}_{(aq)} \parallel Zn^{2+}_{(aq)} \mid Zn_{(s)}$
        cathode   anode $\Delta E° = +0.50$ V

    (c) $C_{(s)} \mid Fe^{2+}_{(aq)}, Fe^{3+}_{(aq)} \parallel H^+_{(aq)}, H_{2(g)} \mid Pt_{(s)}$
        cathode   anode $\Delta E° = +0.77$ V

25. cathode $Au^{3+}_{(aq)} + 3e^- \rightarrow Au_{(s)}$ $E_r° = +1.50$ V
    anode $In_{(s)} \rightarrow In^{3+}_{(aq)} + 3e^-$ $E_r° = ?$

    net $Au^{3+}_{(aq)} + In_{(s)} \rightarrow Au_{(s)} + In^{3+}_{(aq)}$ $\Delta E° = +1.84$ V

$\Delta E° = E_r°_{\text{cathode}} - E_r°_{\text{anode}}$

1.84 V = 1.50 V $- E_r°$

$E_r° = -0.34$ V for $In^{3+}_{(aq)}/In_{(s)}$

26. cathode $\quad 2[Ag^+_{(aq)} + e^- \rightarrow Ag_{(s)}]$ $\quad\quad E_r° = +0.80$ V
    anode $\quad\quad X_{(s)} \rightarrow X^{2+}_{(aq)} + 2e^-$ $\quad\quad E_r° = ?$
    ───
    net $\quad\quad 2Ag^+_{(aq)} + X_{(s)} \rightarrow 2Ag_{(s)} + X^{2+}_{(aq)}$ $\quad\quad \Delta E° = +1.08$ V

    1.08 V $= 0.80$ V $- E_r°$
    $E_r° = -0.28$ V

    Therefore, the unknown redox pair is likely $Co^{2+}_{(aq)}/Co_{(s)}$.

27. If the standard lithium cell is chosen as the reference half-cell with its reduction potential defined as 0.00 V, then 3.04 V must be added to each reduction potential in the table. Therefore, the reduction potential for the copper half-cell becomes +3.38 V and the reduction potential for the zinc half-cell becomes +2.28 V.
    Note that the difference of +1.10 V remains the same as the difference in values referenced to the standard hydrogen electrode.

Lab Exercise 13B
Creating a Table of Redox Half-Reactions (page 406)

Problem

What is the relative strength, in decreasing order, of four oxidizing agents?

Analysis

cathode $\quad\quad\quad\quad$ **anode**
$C_{(s)} \mid Cr_2O_7^{2-}{}_{(aq)}, H^+_{(aq)} \parallel Pd^{2+}_{(aq)} \mid Pd_{(s)}$ $\quad\quad 0.28$ V $= 1.23$ V $- E_r°$
$E_r° = +1.23$ V $\quad\quad E_r° = +0.95$ V

cathode $\quad\quad\quad\quad$ **anode**
$Pd_{(s)} \mid Pd^{2+}_{(aq)} \parallel Tl^+_{(aq)} \mid Tl_{(s)}$ $\quad\quad 1.29$ V $= 0.95$ V $- E_r°$
$E_r° = 0.95$ V $\quad\quad E_r° = -0.34$ V

cathode $\quad\quad\quad\quad$ **anode**
$Tl_{(s)} \mid Tl^+_{(aq)} \parallel Ti^{2+}_{(aq)} \mid Ti_{(s)}$ $\quad\quad 1.29$ V $= -0.34$ V $- E_r°$
$E_r° = -0.34$ V $\quad\quad E_r° = -1.63$ V

According to the evidence, the relative strength of the four oxidizing agents in decreasing order is:

SOA $\quad Cr_2O_7^{2-}{}_{(aq)} + 14H^+_{(aq)} + 6e^- \rightarrow 2Cr^{3+}_{(aq)} + 7H_2O_{(l)}$ $\quad E_r° = +1.23$ V
$\quad\quad\quad Pd^{2+}_{(aq)} + 2e^- \rightarrow Pd_{(s)}$ $\quad\quad E_r° = +0.95$ V
$\quad\quad\quad Tl^+_{(aq)} + e^- \rightarrow Tl_{(s)}$ $\quad\quad E_r° = -0.34$ V
$\quad\quad\quad Ti^{2+}_{(aq)} + 2e^- \rightarrow Ti_{(s)}$ $\quad\quad E_r° = -1.63$ V

Lab Exercise 13C
Series Cells (Enrichment) (page 406)

Problem

What is the electric potential difference between two cells connected in series?

Prediction

Since the two cells are connected in series, the electric potential difference between the two cells is predicted to equal the sum of the $\Delta E°$ values of each cell.
copper-silver cell $\Delta E° = +0.80\ V - 0.34\ V = +0.46\ V$
copper-zinc cell $\Delta E° = +0.34\ V - (-0.76\ V) = +1.10\ V$
predicted voltmeter reading $= +0.46\ V + 1.10\ V = +1.56\ V$
The voltmeter measures the electric potential difference between the silver half-cell and and the zinc half-cell. Referring to the table of redox half-reactions, the difference in position between the silver and zinc half-reactions equals the sum of the differences between the silver-copper and copper-zinc reactions.
Alternatively, a diagram like Figure 13.20, page 403, could be used to show the reasoning.

INVESTIGATION 13.4

Testing Voltaic Cells (page 407)

Problem

In cells constructed from various combinations of copper, lead, silver, and zinc half-cells, what are the standard cell potentials, and which is the anode and cathode in each case?

Prediction

According to redox concepts and the table of redox half-reactions,

Cathode (+)		Anode (−)	
$Cu_{(s)} \mid Cu^{2+}_{(aq)}$	$\parallel$	$Pb^{2+}_{(aq)} \mid Pb_{(s)}$	$\Delta E° = + 0.47\ V$
$Ag_{(s)} \mid Ag^{+}_{(aq)}$	$\parallel$	$Cu^{2+}_{(aq)} \mid Cu_{(s)}$	$\Delta E° = + 0.46\ V$
$Cu_{(s)} \mid Cu^{2+}_{(aq)}$	$\parallel$	$Zn^{2+}_{(aq)} \mid Zn_{(s)}$	$\Delta E° = + 1.10\ V$
$Ag_{(s)} \mid Ag^{+}_{(aq)}$	$\parallel$	$Pb^{2+}_{(aq)} \mid Pb_{(s)}$	$\Delta E° = + 0.93\ V$
$Pb_{(s)} \mid Pb^{2+}_{(aq)}$	$\parallel$	$Zn^{2+}_{(aq)} \mid Zn_{(s)}$	$\Delta E° = + 0.63\ V$
$Ag_{(s)} \mid Ag^{+}_{(aq)}$	$\parallel$	$Zn^{2+}_{(aq)} \mid Zn_{(s)}$	$\Delta E° = + 1.56\ V$

Experimental Design

Individual metal-metal ion half-cells are constructed. Different combinations are connected with a salt bridge and the electrodes, and the cell potentials are determined. Controlled variables are temperature and electrolyte concentration.

Procedure

1. Clean metal strips with steel wool.
2. Assemble two of the four metal-metal ion half-cells, for example, copper and lead.
3. Connect the copper half-cell with the lead half-cell using the salt bridge.
4. Use the voltmeter and connecting wires to determine the cathode and anode of the cell.
5. With the voltmeter connected to the cell, measure the initial voltmeter reading.
6. Remove and rinse the salt bridge.
7. Repeat steps 1 to 5 for the remaining combinations of half-cells.

Evidence/Analysis

(+) Cathode	(−) Anode	Predicted Potential (V)	Measured Potential (V)	Percent Difference (%)
$Cu_{(s)} \mid Cu^{2+}_{(aq)} \parallel Pb^{2+}_{(aq)} \mid Pb_{(s)}$		+0.47	+0.46	2
$Ag_{(s)} \mid Ag^{+}_{(aq)} \parallel Cu^{2+}_{(aq)} \mid Cu_{(s)}$		+0.46	+0.42	9
$Cu_{(s)} \mid Cu^{2+}_{(aq)} \parallel Zn^{2+}_{(aq)} \mid Zn_{(s)}$		+1.10	+1.06	4
$Ag_{(s)} \mid Ag^{+}_{(aq)} \parallel Pb^{2+}_{(aq)} \mid Pb_{(s)}$		+0.93	+0.88	5
$Pb_{(s)} \mid Pb^{2+}_{(aq)} \parallel Zn^{2+}_{(aq)} \mid Zn_{(s)}$		+0.63	+0.59	6
$Ag_{(s)} \mid Ag^{+}_{(aq)} \parallel Zn^{2+}_{(aq)} \mid Zn_{(s)}$		+1.56	+1.47	6

Evaluation

The design is adequate to answer the problem with no obvious flaws. However, the procedure used does have some inadequacies. The conditions of temperature and concentration were controlled, but were not set at standard values. This should be improved to eliminate some uncertainties in the results. The technological skills were relatively simple and the results agreed with other groups in the class. Overall, I am only moderately certain of the results. Measurement uncertainties in the solution preparation and the voltmeter readings are predictable, but the uncertainty caused by non-standard conditions is not known.

The difference between the measured and predicted values ranged from 0.01 V to 0.09 V. This range corresponds to a percent difference of 2 to 9%. It was noted that all measured values were lower than predicted values. This suggests a systematic error, perhaps due to non-standard (lower) conditions. It seems reasonable that the small differences could be accounted for by the uncertainties discussed. Therefore, the prediction is judged to be verified pending further measurements to be made. The redox concepts and table appear to be acceptable for predicting electrodes and cell potentials.

INVESTIGATION 13.5

A Potassium Iodide Electrolytic Cell (page 408)

Problem

What are the products of the reaction during the operation of an aqueous potassium iodide electrolytic cell?

Evidence

Before the cell was connected, the solution was completely colorless and both litmus and halogen tests gave negative results. After the cell was connected and in operation for a few minutes,
at the negative electrode and nearby solution:
- many colorless gas bubbles were continuously produced.
- red litmus turned blue.
- the trichlorotrifluoroethane layer remained colorless.

at the positive electrode and nearby solution:
- a yellow-brown color of solution and some black precipitate formed.
- litmus did not change in color.
- the trichlorotrifluoroethane layer was purple.

Analysis

According to the evidence collected, a colorless gas and hydroxide ions were likely produced at the negative electrode. The observations strongly suggest the formation of iodine at the positive electrode.

Exercise (page 410)

28. (a) cathode $\quad Ni^{2+}_{(aq)} + 2e^- \rightarrow Ni_{(s)} \qquad E_r° = -0.26$ V
    anode $\qquad 2I^-_{(aq)} \rightarrow I_{2(s)} + 2e^- \qquad E_r° = +0.54$ V

    net $\qquad Ni^{2+}_{(aq)} + 2I^-_{(aq)} \rightarrow Ni_{(s)} + I_{2(s)} \qquad \Delta E° = -0.80$ V
    minimum potential difference = +0.80 V

    (b) cathode $\quad 2[2H_2O_{(l)} + 2e^- \rightarrow H_{2(g)} + 2OH^-_{(aq)}] \qquad E_r° = -0.83$ V
    anode $\qquad 4OH^-_{(aq)} \rightarrow O_{2(g)} + 2H_2O_{(l)} + 4e^- \qquad E_r° = +0.40$ V

    net $\qquad 2H_2O_{(l)} \rightarrow O_{2(g)} + 2H_{2(g)} \qquad \Delta E° = -1.23$ V
    minimum potential difference = +1.23 V

29. (a) $\Delta E° = -0.41$ V $- (+1.07$ V$) = -1.48$ V
    minimum potential difference = +1.48 V

    (b) $\Delta E° = +0.34$ V $- (+0.34$ V$) = 0.00$ V
    minimum potential difference = 0.00 V
    Any applied voltage greater than 0.00 V will cause the electrodeposition of copper. Higher voltages will increase the rate of deposition.

INVESTIGATION 13.6

Demonstration of Electrolysis (page 411)

Problems

What are the products of electrolytic cells containing
- aqueous copper(II) sulfate
- aqueous sodium sulfate
- aqueous sodium chloride

Prediction

According to redox concepts and the redox table of half-reactions,

SOA $\quad$ OA $\quad$ OA
$Cu^{2+}_{(aq)}, \quad SO_4^{2-}_{(aq)}, \quad H_2O_{(l)}$
$\qquad\qquad\qquad$ SRA

cathode $\quad 2[Cu^{2+}_{(aq)} + 2e^- \rightarrow Cu_{(s)}] \qquad E_r° = +0.34$ V
anode $\qquad 2H_2O_{(l)} \rightarrow O_{2(g)} + 4H^+_{(aq)} + 4e^- \qquad E_r° = +1.23$ V

net $\qquad 2Cu^{2+}_{(aq)} + 2H_2O_{(l)} \rightarrow 2Cu_{(s)} + O_{2(g)} + 4H^+_{(aq)} \qquad \Delta E° = -0.89$ V

OA $\quad$ OA $\quad$ SOA
$Na^+_{(aq)}, \quad SO_4^{2-}_{(aq)}, \quad H_2O_{(l)}$
$\qquad\qquad\qquad$ SRA

cathode $\quad 2[2H_2O_{(l)} + 2e^- \rightarrow H_{2(g)} + 2OH^-_{(aq)}] \qquad E_r° = -0.83$ V
anode $\qquad 2H_2O_{(l)} \rightarrow O_{2(g)} + 4H^+_{(aq)} + 4e^- \qquad E_r° = +1.23$ V

net $\qquad 6H_2O_{(l)} \rightarrow 2H_{2(g)} + 4OH^-_{(aq)} + O_{2(g)} + 4H^+_{(aq)} \qquad \Delta E° = -2.06$ V

$$\begin{array}{ll} & \text{OA} \qquad\qquad \text{SOA} \\ \text{Na}^+_{(aq)}, & \text{Cl}^-_{(aq)}, \quad \text{H}_2\text{O}_{(l)} \\ & \qquad\quad \underbrace{}_{\text{RA}} \underbrace{}_{\text{SRA}} \end{array}$$

cathode	$2[2\text{H}_2\text{O}_{(l)} + 2e^- \rightarrow \text{H}_{2(g)} + 2\text{OH}^-_{(aq)}]$	$E_r° = -0.83\text{V}$
anode	$2\text{H}_2\text{O}_{(l)} \rightarrow \text{O}_{2(g)} + 4\text{H}^+_{(aq)} + 4e^-$	$E_r° = +1.23\text{ V}$
net	$6\text{H}_2\text{O}_{(l)} \rightarrow 2\text{H}_{2(g)} + 4\text{OH}^-_{(aq)} + \text{O}_{2(g)} + 4\text{H}^+_{(aq)}$	$\Delta E° = -2.06\text{ V}$

Evidence

Cell	Cathode	Anode
$\text{C}_{(s)}|\text{CuSO}_{4(aq)}|\text{C}_{(s)}$ applied voltage 6 V	• blue litmus turned red before and after the cell operated • red-brown solid formed on $\text{C}_{(s)}$	• blue litmus turned red before and after the cell operated • gas bubbles formed on $\text{C}_{(s)}$
$\text{Pt}_{(s)}|\text{Na}_2\text{SO}_{4(aq)}|\text{Pt}_{(s)}$ applied voltage 6 V	• no change in litmus before the cell operated • red litmus turned blue in final solution • about 31 mL of gas collected • a loud, squeaky sound heard when gas sample was ignited	• no change in litmus before the cell operated • blue litmus turned red in final solution • about 16 mL of gas collected • a glowing match burst into flames in gas sample
$\text{Pt}_{(s)}|\text{NaCl}_{(aq)}|\text{Pt}_{(s)}$ applied voltage 6 V	• no change in litmus before the cell operated • red litmus turned blue in final solution • about 31 mL of gas collected • a loud, squeaky sound heard when gas sample was ignited	• no change in litmus before the cell operated • both red and blue litmus turned white in final solution • a strong odor of bleach detected • yellow-brown color appeared on addition of $\text{NaI}_{(aq)}$ to a sample of the solution • purple color of trichlorotrifluoroethane layer observed

Analysis

According to the evidence collected,

	at the cathode	at the anode
$\text{CuSO}_{4(aq)}$ cell	$\text{Cu}_{(s)}$	gas and $\text{H}^+_{(aq)}$
$\text{Na}_2\text{SO}_{4(aq)}$ cell	$\text{H}_{2(g)}$ and $\text{OH}^-_{(aq)}$	$\text{O}_{2(g)}$ and $\text{H}^+_{(aq)}$
$\text{NaCl}_{(aq)}$ cell	$\text{H}_{2(g)}$ and $\text{OH}^-_{(aq)}$	$\text{Cl}_{2(aq)}$ (after 10 min)

Evaluation

The experimental design is judged to be adequate with no obvious flaws. The problem was answered with reasonable certainty and the controls for the litmus test were adequate. There are two ways to improve the electrolysis of aqueous copper(II) sulfate. One is to use an apparatus so that the gas can be collected and identified. The other is to use pH paper or pH meter instead of litmus paper. These changes would increase the certainty of the results.

The predictions for the cathode and anode products of copper(II) sulfate and sodium sulfate solutions were verified because the predicted products agree with the experimental results. The prediction for the cathode products of sodium chloride was also verified but the prediction for the anode products was falsified by the evidence obtained. It is necessary to determine how unique the result for sodium chloride is. The redox concepts and procedures remain acceptable until further results are obtained. However, it may be necessary to restrict or revise the procedure used to predict the products.

Exercise (page 414)

30. (a) Many ionic compounds have low solubility in water. Water is a stronger oxidizing agent than the cations of active metals, and, therefore, is predicted to react before the active metal cation.
    (b) Three designs to overcome these difficulties are: electrolysis of a molten salt, electrolysis using a solvent other than water, and use of high voltages to "overpower" the reduction of water from an aqueous solution.

31. SOA
    $Sc^{3+}_{(l)}, Cl^-_{(l)}$
    SRA
    cathode $\quad\quad 2[Sc^{3+}_{(l)} + 3e^- \rightarrow Sc_{(s)}]$
    anode $\quad\quad\quad 3[2Cl^-_{(l)} \rightarrow Cl_{2(g)} + 2e^-]$
    net $\quad\quad\quad 2Sc^{3+}_{(l)} + 6Cl^-_{(l)} \rightarrow 2Sc_{(s)} + 3Cl_{2(g)}$

32. affirmative: ecologically sound, economically viable relative to the cost of making a new metal, ethically correct
    negative: ecological problems elsewhere, economical implications for industries and workers in Canada

33. (Extension)
    (a) $Ca(OH)_{2(s)} + MgCl_{2(aq)} \rightarrow Mg(OH)_{2(s)} + CaCl_{2(aq)}$
    (b) $2HCl_{(aq)} + Mg(OH)_{2(s)} \rightarrow MgCl_{2(aq)} + 2H_2O_{(l)}$
    (c) SOA
    $Mg^{2+}_{(l)}, Cl^-_{(l)}$
    SRA
    cathode $\quad\quad Mg^{2+}_{(l)} + 2e^- \rightarrow Mg_{(s)}$
    anode $\quad\quad\quad 2Cl^-_{(l)} \rightarrow Cl_{2(g)} + 2e^-$
    net $\quad\quad\quad Mg^{2+}_{(l)} + 2Cl^-_{(l)} \rightarrow Mg_{(s)} + Cl_{2(g)}$
    (d) advantages:
    - Fewer steps are needed since a solid magnesium salt is already available.
    - Dolomite is readily available inland (not close to any ocean).
    disadvantages:
    - Oceans contain an almost limitless supply of magnesium salts, whereas dolomite is likely present in much more limited and localized quantities.
    - Mining involves higher energy and costs compared with pumping sea water.

INVESTIGATION 13.7

Copper Plating (page 417)

Problem
Which procedure causes a smooth layer of copper metal to adhere to a conducting object?
This is a technological problem-solving exercise that is quite open-ended. Possible variables include voltage, current, current to surface area ratio, con-

centrations, and cleaning and surface preparations. Alternative ways of handling this investigation may be as follows:
- Students do preliminary research and planning out of class. One class period of work is scheduled, followed by further planning out of class, and a second class period at a later time.
- Since the materials are simple and commercially available, students could do this as a home project (perhaps as an alternative to one of the library reports in this chapter).

Exercise (page 418)

34. $Zn_{(s)} \rightarrow Zn^{2+}_{(aq)} + 2e^-$
    m \qquad\qquad 0.500 A
    65.38 g/mol \qquad 10.0 min

$$n_{e^-} = \frac{It}{F} = \frac{0.500 \text{ C/s} \times (10.0 \text{ min} \times 60 \text{ s/min})}{9.65 \times 10^4 \text{ C/mol}} = 3.11 \text{ mmol}$$

$$n_{Zn} = 3.11 \text{ mmol} \times \frac{1}{2} = 1.55 \text{ mmol}$$

$$m_{Zn} = 1.55 \text{ mmol} \times \frac{65.38 \text{ g}}{1 \text{ mol}} = 102 \text{ mg}$$

According to the stoichiometric method and the laws of electrolysis, the mass of zinc oxidized is 102 mg.

35. (a) $Cr^{3+}_{(aq)} + 3e^- \rightarrow Cr_{(s)}$
    \qquad\qquad 54 A \qquad\qquad m
    \qquad 45 min 30 s \quad 52.00 g/mol

$$n_{e^-} = \frac{It}{F} = \frac{54 \text{ C/s} \times (45 \text{ min} \times 60 \text{ s/min} + 30\text{s})}{9.65 \times 10^4 \text{ C/mol}} = 1.5 \text{ mol}$$

$$n_{Cr} = 1.5 \text{ mol} \times \frac{1}{3} = 0.51 \text{ mol}$$

$$m_{Cr} = 0.51 \text{ mol} \times \frac{52.00 \text{ g}}{1 \text{ mol}} = 26 \text{ g}$$

According to the stoichiometric method and the laws of electrolysis, the mass of chromium deposited on the bumper is 26 g.

(b) $Ni^{2+}_{(aq)} + 2e^- \rightarrow Ni_{(s)}$
    \qquad 0.540 A \quad 0.25 g
    \qquad t \qquad\qquad 58.69 g/mol

$$n_{Ni} = 0.25 \text{ g} \times \frac{1 \text{ mol}}{58.69 \text{ g}} = 0.0043 \text{ mol}$$

$$n_{e^-} = 0.0043 \text{ mol} \times \frac{2}{1} = 0.0085 \text{ mol}$$

$$t = \frac{nF}{I} = \frac{0.0085 \text{ mol} \times 9.65 \times 10^4 \text{ C/mol}}{0.540 \text{ C/s}} = 1.5 \text{ ks} \times \frac{1 \text{ min}}{60 \text{ s}} = 25 \text{ min}$$

According to the stoichiometric method and the laws of electrolysis, the required thickness will be plated in 25 min.

(c) $Ag^+_{(aq)} + e^- \rightarrow Ag_{(s)}$
         84 min   10.00 g
         I   107.87 g/mol

$$n_{Ag} = 10.00 \text{ g} \times \frac{1 \text{ mol}}{107.87 \text{ g}} = 0.09270 \text{ mol}$$

$$n_{e^-} = 0.09270 \text{ mol} \times \frac{1}{1} = 0.09270 \text{ mol}$$

$$I = \frac{nF}{t} = \frac{0.09270 \text{ mol} \times 9.65 \times 10^4 \text{ C/mol}}{84 \text{ min} \times 60 \text{ s/min}} = 1.8 \text{ A}$$

According to the stoichiometric method and the laws of electrolysis, the average current used to plate 10.00 g of silver in 84 min is 1.8 A.

36. (a) $Pb_{(s)} \rightarrow Pb^{2+}_{(aq)} + 2e^-$
         m   1.00 A
         207.20 g/mol   120 h

$$n_{e^-} = \frac{It}{F} = \frac{1.00 \text{ C/s} \times (120 \text{ h} \times 3600 \text{ s/h})}{9.65 \times 10^4 \text{ C/mol}} = 4.48 \text{ mol}$$

$$n_{Pb} = 4.48 \text{ mol} \times \frac{1}{2} = 2.24 \text{ mol}$$

$$m_{Pb} = 2.24 \text{ mol} \times \frac{207.20 \text{ g}}{1 \text{ mol}} = 464 \text{ g}$$

According to the stoichiometric method and the laws of electrolysis, the mass of lead oxidized is 464 g.

(b) $Al_{(s)} \rightarrow Al^{3+}_{(aq)} + 3e^-$
    m   1.00 A
    26.98 g/mol   120 h

$$n_{e^-} = \frac{It}{F} = \frac{1.00 \text{ C/s} \times (120 \text{ h} \times 3600 \text{ s/h})}{9.65 \times 10^4 \text{ C/mol}} = 4.48 \text{ mol}$$

$$n_{Al} = 4.48 \text{ mol} \times \frac{1}{3} = 1.49 \text{ mol}$$

$$m_{Al} = 1.49 \text{ mol} \times \frac{26.98 \text{ g}}{1 \text{ mol}} = 40.3 \text{ g}$$

According to the stoichiometric method and the laws of electrolysis, the mass of aluminum oxidized would be 40.3 g.

37. $2Al^{3+}_{(l)} + 6Cl^-_{(l)} \rightarrow 2Al_{(l)} + 3Cl_{2(g)}$
                                  5.40 g   m
                                  26.98 g/mol   70.90 g/mol

$$n_{Al} = 5.40 \text{ g} \times \frac{1 \text{ mol}}{26.98 \text{ g}} = 0.200 \text{ mol}$$

$$n_{Cl_2} = 0.200 \text{ mol} \times \frac{3}{2} = 0.300 \text{ mol}$$

$$m_{Cl_2} = 0.300 \text{ mol} \times \frac{70.90 \text{ g}}{1 \text{ mol}} = 21.3 \text{ g}$$

According to the method of stoichiometry, the mass of chlorine is 21.3 g.

Lab Exercise 13D
Quantitative Electrolysis (page 419)

Problem
What is the mass of tin plated at the cathode of a tin-plating cell by a current of 3.46 A for 6.0 min?

Prediction
According to the stoichiometric method and the laws of electrolysis, the mass of tin plated at the cathode is predicted to be 0.77 g based on the following calculations.

$$Sn^{2+}_{(aq)} + 2e^- \rightarrow Sn_{(s)}$$
$$\phantom{Sn^{2+}_{(aq)} + }3.46\ A m$$
$$\phantom{Sn^{2+}_{(aq)} + }6.0\ min 118.69\ g/mol$$

$$n_{e^-} = \frac{It}{F} = \frac{3.46\ C/s \times (6.0\ min \times 60\ s/min)}{9.65 \times 10^4\ C/mol} = 0.013\ mol$$

$$n_{Sn} = 0.013\ mol \times \frac{1}{2} = 0.0065\ mol$$

$$m_{Sn} = 0.0065\ mol \times \frac{118.69\ g}{1\ mol} = 0.77\ g$$

Analysis
$m_{Sn} = 118.05\ g - 117.34\ g = 0.71\ g$

According to the evidence, the mass of tin plated at the cathode is 0.71 g.

Evaluation
The prediction is accurate to within 7%.

$$\%\ difference = \frac{|0.71\ g - 0.77\ g|}{0.77\ g} \times 100 = 7\%$$

The prediction is judged to be verified because the 7% difference can be accounted for by various measurement uncertainties in the experiment. The method of stoichiometry for electrolytic cells is, therefore, acceptable.

Overview (page 420)
Review

1. The three essential parts of an electric cell are two electrodes and an electrolyte.
2. Technological problem solving involves a systematic trial-and-error approach to develop a product or process. Scientific problem solving usually involves answering questions to test a scientific concept.
3. Batteries could be made rechargeable (that is, they could be secondary cells), or they could be made so that the fuel can be continuously added (that is, they could be fuel cells).
4. Carbon and platinum are two commonly used inert electrodes.
5. Porous boundaries are provided by a porcelain cup and by a salt bridge containing an inert electrolyte.
6. The components of the hydrogen reference half-cell are a 1.00 mol/L hydrogen ion solution and hydrogen gas at 100 kPa bubbling over a platinum electrode, with all components at 25°C.
7. Three technological applications of electrolytic cells are the production of elements, the refining of metals, and the plating of metals onto other objects.

8. Problems might arise because some ionic compounds have a low solubility in water or because water is a stronger oxidizing agent compared with active metal cations. If the compound has a low solubility in water, it could be dissolved in an ionic compound that has a low melting point. If the metal cation is less reactive than water, electrolysis could be carried out using the molten compound.

9. Voltaic cells convert chemical energy into electrical energy, while electrolytic cells convert electrical energy into chemical energy.

10. (a) Voltaic/Electrolytic Cells

	Anode	Cathode
Half-reaction	oxidation	reduction
Agent reacted	reducing agent	oxidizing agent
Anions	move toward	move away
Cations	move away	move toward
Electrons	move away	move toward

(b)

	Voltaic Cell	Electrolytic Cell
Agents in redox table	SOA above SRA	SOA below SRA
Cell potential	positive	negative

Applications

11. (a) $\Delta E° = 0.00\ V - (-0.14\ V) = +0.14\ V$
    (b) $\Delta E° = +1.51\ V - (+0.80\ V) = +0.71\ V$
    (c) $\Delta E° = -0.14\ V - (-0.76\ V) = +0.62\ V$

12. (a) $\Delta E° = -0.26\ V - (+1.23\ V) = -1.49\ V$ $V_{min} = 1.49\ V$
    (b) $\Delta E° = 0.00\ V - (+0.80\ V) = -0.80\ V$ $V_{min} = 0.80\ V$
    (c) $\Delta E° = -0.14\ V - (-0.14\ V) = 0.00\ V$ $V_{min} = 0.00\ V$

13. A standard cell contains two pairs of oxidizing agents and reducing agents. Therefore, an oxidizing agent will always be listed above a reducing agent in a redox table.

14. *If the standard iodine-iodide half-cell were used as the standard reference, all of the values on the current table of redox half-reactions (page 552) would be lowered by 0.54 V.*
    (a) In the iodine standard table, the reduction potential of a standard silver-silver ion half-cell would be $(+0.80\ V - 0.54\ V) = +0.26\ V$.
    $$Ag^+_{(aq)} + e^- \rightarrow Ag_{(s)} \qquad E_r° = +0.26\ V$$
    (b) In the iodine standard table, the reduction potential of a standard zinc-zinc ion half-cell would be $(-0.76\ V - 0.54\ V) = -1.30\ V$. Therefore, the oxidation potential for the standard zinc-zinc ion half-cell would be $+1.30\ V$.
    $$Zn_{(s)} \rightarrow Zn^{2+}_{(aq)} + 2e^- \qquad E_o° = +1.30\ V$$
    (c) $\Delta E° = E_r°_{\text{cathode}} - E_r°_{\text{anode}}$
    $= +0.26\ V - (-1.30\ V) = +1.56\ V$

15. (a) Lead will be the cathode because it is immersed in the strongest oxidizing agent, $Pb^{2+}_{(aq)}$. Cobalt will be the anode because $Co_{(s)}$ is the strongest reducing agent.

(b)

OA   SOA   OA
$Co^{2+}_{(aq)}, Co_{(s)}, Pb^{2+}_{(aq)}, Pb_{(s)}, H_2O_{(l)}$
    SRA   RA   RA

cathode	$Pb^{2+}_{(aq)} + 2e^- \rightarrow Pb_{(s)}$	$E_r° = -0.13\ V$
anode	$Co_{(s)} \rightarrow Co^{2+}_{(aq)} + 2e^-$	$E_r° = -0.28\ V$
net	$Pb^{2+}_{(aq)} + Co_{(s)} \rightarrow Pb_{(s)} + Co^{2+}_{(aq)}$	$\Delta E° = +0.15\ V$

(c)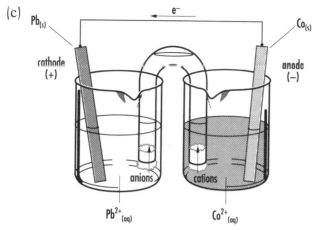

16. cathode $HgO_{(s)} + H_2O_{(l)} + 2e^- \rightarrow Hg_{(l)} + 2OH^-_{(aq)}$ $E_r° = +0.10$ V
    anode $Zn_{(s)} + 2OH^-_{(aq)} \rightarrow ZnO_{(s)} + H_2O_{(l)} + 2e^-$ $E_r° = -1.25$ V
    _____
    net $HgO_{(s)} + Zn_{(s)} \rightarrow Hg_{(l)} + ZnO_{(s)}$ $\Delta E° = +1.35$ V

    Note that the SOA has the more positive reduction potential and will react at the cathode.

17. cathode $2[O_{2(g)} + 2H_2O_{(g)} + 4e^- \rightarrow 4OH^-_{(l)}]$ $E_r° = +0.40$ V
    anode $CH_{4(g)} + 10OH^-_{(l)} \rightarrow CO_3^{2-}_{(l)} + 7H_2O_{(g)} + 8e^-$ $E_r° = +0.17$ V
    _____
    net $CH_{4(g)} + 2O_{2(g)} + 2OH^-_{(l)} \rightarrow CO_3^{2-}_{(l)} + 3H_2O_{(g)}$ $\Delta E° = +0.23$ V

18. Volta's invention of the fuel cell occurred before most of the scientific explanation of electricity and redox theory. Humphry Davy's invention of molten salt electrolysis preceded most of modern atomic theory including redox theory. (Chromium plating is best done from a solution of chromic acid in a process which is not well understood.)

19. $\Delta E° = E_r°_{\text{cathode}} - E_r°_{\text{anode}}$
     $+1.28$ V $= E_r° - (-0.76$ V$)$
     $E_r° = +0.52$ V

20. cathode $Cu^{2+}_{(aq)} + 2e^- \rightarrow Cu_{(s)}$ $E_r° = +0.34$ V
    anode $\rightarrow X^{2+}_{(aq)} + 2e^-$ $E_r° = ?$
    _____
    net $Cu^{2+}_{(aq)} + X_{(s)} \rightarrow Cu_{(s)} + X^{2+}_{(aq)}$ $\Delta E° = +0.48$ V
    $+0.48$ V $= +0.34$ V $- E_r°$
    $E_r° = -0.14$ V

    The redox pair is probably $Sn^{2+}_{(aq)} | Sn_{(s)}$.

21. (a) The entities in a solution of cadmium nitrate are $H_2O_{(l)}$, $Cd^{2+}_{(aq)}$, and $NO_3^-_{(aq)}$. The strongest oxidizing agent, $Cd^{2+}_{(aq)}$, is not strong enough to oxidize the strongest reducing agent, $H_2O_{(l)}$, to cause a spontaneous reaction. Therefore, an external voltage must be applied to inert electrodes to observe evidence of a redox reaction.

    (b) The entities in a solution of iron(III) iodide are $H_2O_{(l)}$, $Fe^{3+}_{(aq)}$, and $I^-_{(aq)}$. The strongest oxidizing agent, $Fe^{3+}_{(aq)}$, is strong enough to oxidize the strongest reducing agent, $I^-_{(aq)}$. Therefore, a redox reaction should be evident without the application of an external voltage.

    (c) The entities in solutions of iron(III) bromide and tin(II) sulfate are $H_2O_{(l)}$, $Fe^{3+}_{(aq)}$, $Br^-_{(aq)}$, $Sn^{2+}_{(aq)}$, and $SO_4^{2-}_{(aq)}$. The strongest oxidizing agent, $Fe^{3+}_{(aq)}$, is strong enough to oxidize the strongest reducing agent, $Sn^{2+}_{(aq)}$. Therefore, a redox reaction should be evident without the application of an external voltage.

(d) The entities in solutions of potassium iodide and zinc nitrate are $H_2O_{(l)}$, $K^+_{(aq)}$, $I^-_{(aq)}$, $Zn^{2+}_{(aq)}$, and $NO_3^-{}_{(aq)}$. The strongest oxidizing agent, $Zn^{2+}_{(aq)}$, is not strong enough to oxidize the strongest reducing agent, $I^-_{(aq)}$, to cause a spontaneous reaction. Therefore, an external voltage must be applied to inert electrodes to observe evidence of a redox reaction.

22. (a) SOA OA OA
$Zn^{2+}_{(aq)}$, $SO_4^{2-}_{(aq)}$, $H_2O_{(l)}$
SRA

cathode	$2[Zn^{2+}_{(aq)} + 2e^- \rightarrow Zn_{(s)}]$	$E_r^\circ = -0.76$ V
anode	$2H_2O_{(l)} \rightarrow O_{2(g)} + 4H^+_{(aq)} + 4e^-$	$E_r^\circ = +1.23$ V
net	$2Zn^{2+}_{(aq)} + 2H_2O_{(l)} \rightarrow 2Zn_{(s)} + O_{2(g)} + 4H^+_{(aq)}$	$\Delta E^\circ = -1.99$ V

(b) $V_{min} = 1.99$ V

23. SOA
$Sr^{2+}_{(l)}$, $O^{2-}_{(l)}$
SRA

cathode	$2[Sr^{2+}_{(l)} + 2e^- \rightarrow Sr_{(l)}]$
anode	$2O^{2-}_{(l)} \rightarrow O_{2(g)} + 4e^-$
net	$2Sr^{2+}_{(l)} + 2O^{2-}_{(l)} \rightarrow 2Sr_{(l)} + O_{2(g)}$

24. $Ni^{2+}_{(aq)} + 2e^- \rightarrow Ni_{(s)}$
    3.50 A   500 mL
    t 0.125 mol/L

$n_{Ni} = 500 \text{ mL} \times \dfrac{0.125 \text{ mol}}{1 \text{ L}} = 62.5$ mmol

$n_{e^-} = 62.5 \text{ mmol} \times \dfrac{2}{1} = 125$ mmol

$t_{e^-} = \dfrac{nF}{I} = \dfrac{125 \text{ mmol} \times 9.65 \times 10^4 \text{ C/mol}}{3.50 \text{ C/s}} = 3.45 \text{ ks} \times \dfrac{1 \text{ min}}{60 \text{ s}} = 57.4$ min

25. $Al_{(s)} \rightarrow Al^{3+}_{(aq)} + 3e^-$
    15 kg I
    26.98 g/mol   1.00 h

$n_{Al} = 15 \text{ kg} \times \dfrac{1 \text{ mol}}{26.98 \text{ g}} = 0.56$ kmol

$n_{e^-} = 0.56 \text{ kmol} \times \dfrac{3}{1} = 1.7$ kmol

$I_{e^-} = \dfrac{nF}{t} = \dfrac{1.7 \text{ kmol} \times 9.65 \times 10^4 \text{ C/mol}}{3600 \text{ s}} = 45$ kA

Extensions

26. *Contact provincial and civic offices to find out about local recycling programs. Scrap metal dealers may also provide useful information.*

27. Turn off the engine of the vehicle containing the booster battery. Connect one end of the red jumper cable to the positive terminal (cathode) of the discharged battery and the other end to the positive terminal of the booster battery. Next connect one end of the black jumper cable to the negative terminal (anode) of the booster battery. The other end of the black jumper cable should be connected to the engine being started (Not to the negative terminal of the discharged battery!). Start the engine of the vehicle containing the booster battery and let

it run for a few minutes. Then attempt to start the engine of the vehicle with the discharged battery.

The final connection should always be made at some distance from both batteries because a spark usually occurs at the final connection. Hydrogen gas in either battery may explode if a spark is produced nearby.

28. A faraday is the charge of one mole of electrons or any other singly charged entities. Since calcium ions have two units of elementary charge in one ion, the molar charge is twice the value of a faraday, i.e., 1.93×10^5 C per mole of Ca^{2+}.

29. Design 1: Set up a cell with a known half-cell. Measure the cell potential and determine the reduction potential of the unknown metal cation. Comparison with reduction potentials from a reference may identify the cation.
Design 2: Test for spontaneous redox reactions using a variety of oxidizing agents. Determine the position of the unknown metal cation in a redox table and consult a reference for identification.
Design 3: Electrolyze the solution with inert electrodes. From the cathode, scrape off some of the plated metal and determine its identity using physical properties such as density, melting point, etc.
Design 4: Compare the solution color and the flame color with the known colors of various transition metal ions.

30. Design 1: Place a measured mass of zinc into a known volume of the solution and allow complete reaction to take place. Extract the unreacted solid zinc and determine the mass remaining.
Design 2: Precipitate the copper(II) ions with excess sodium sulfide solution and filter the mixture to determine the mass of solid.
Design 3: Titrate a known volume of the copper(II) ion solution with a basic sodium sulfite solution. Determine the volume of basic sulfite solution required for the colored copper(II) ion solution to disappear.
Design 4: Electrolyze a copper(II) ion solution using an inert cathode and an inert anode. After all the copper has reacted at the cathode, determine the mass change of the cathode to obtain the mass of copper deposited.

31. There are a variety of half-reactions involving copper entities that may occur.

$Cu^+_{(aq)} + e^- \rightarrow Cu_{(s)}$ $E_r° = +0.52$ V
$Cu^{2+}_{(aq)} + 2e^- \rightarrow Cu_{(s)}$ $E_r° = +0.34$ V
$Cu^{2+}_{(aq)} + e^- \rightarrow Cu^+_{(aq)}$ $E_r° = +0.15$ V

The fact that the cell potential is higher than 0.89 V after a period of time suggests the possibility of some copper(I) ion formation and subsequent oxidation. If only dichromate ion and copper(I) ion react, a maximum voltage of 1.08 V could be obtained. It is likely that a number of reactions may be occurring simultaneously. The green color is difficult to explain since there are many colored species present; for example,

$Cr_2O_7^{2-}{}_{(aq)}$ orange (concentrated); strong yellow (dilute)
$Cr^{3+}_{(aq)}$ bright green
$Cu^{2+}_{(aq)}$ blue
$Cu^+_{(aq)}$ green

With the likelihood of simultaneous reactions and ion migration, several combinations could give a green color. *(Note that the observations in the question were made using a porous cup cell design.)*

32. *An important point here is that batteries are currently the major economically viable technology that provides a portable source of stored energy. (See Table 13.2 and pages 392–393.)*

33. The Nernst equation is used to calculate ΔE values for redox reactions taking place at non-standard conditions. The general form of the equation is

$$\Delta E = \Delta E° - \frac{2.303RT}{nF} \log K$$

where R = universal gas constant
T = absolute temperature in kelvins
n = amount in moles
F = Faraday's constant
K = equilibrium constant for the redox reaction

An example is to calculate the ΔE for a silver-zinc cell operating at 0.0°C rather than at 25.0°C. Note that the equilibrium constant must be known, or at least the concentrations of ions, so that it can be calculated.

Lab Exercise 13E
Testing a Voltaic Cell (page 423)

Problem
What are the components, reactions, and electric potential difference of a standard lead-dichromate cell?

Prediction
According to redox concepts and the table of redox half-reactions, a standard lead-dichromate cell has a cell potential of 1.36 V at SATP and the following reactions and components.

$$\underset{\text{RA}}{\underset{\text{SOA}}{Cr_2O_7^{2-}{}_{(aq)}},\ \underset{\text{OA}}{H^+{}_{(aq)}},\ \underset{\text{RA}}{Cr^{3+}{}_{(aq)}},\ \underset{}{H_2O_{(l)}}\quad \underset{\text{SRA}}{Pb^{2+}{}_{(aq)},\ Pb_{(s)},}\quad C_{(s)}}$$

cathode	$Cr_2O_7^{2-}{}_{(aq)} + 14H^+{}_{(aq)} + 6e^- \rightarrow 2Cr^{3+}{}_{(aq)} + 7H_2O_{(l)}$	$E_r° = +1.23$ V
anode	$3[Pb_{(s)} \rightarrow Pb^{2+}{}_{(aq)} + 2e^-]$	$E_r° = -0.13$ V
net	$Cr_2O_7^{2-}{}_{(aq)} + 14H^+{}_{(aq)} + 3Pb_{(s)} \rightarrow 2Cr^{3+}{}_{(aq)} + 7H_2O_{(l)} + 3Pb^{2+}{}_{(aq)}$	$\Delta E° = +1.36$ V

Materials
lab apron, safety glasses, lead strip, graphite rod, 1.0 mol/L lead(II) nitrate solution, an acidic solution containing 1.0 mol/L potassium dichromate and 1.0 mol/L chromium(III) nitrate, two beakers, salt bridge, voltmeter, connecting wires

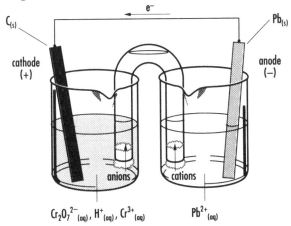

Experimental Design (Diagnostic Tests)

- If a voltmeter is connected to the electrodes (red — $C_{(s)}$, black — $Pb_{(s)}$), and a positive potential of 1.36 V is measured, then carbon is the cathode, lead is the anode, and the half-reactions listed are probably correct.
- If the electrodes of the cell are connected with a wire and an ammeter (red — $C_{(s)}$, black — $Pb_{(s)}$), and a positive current is measured, then the electron flow is from the lead to the carbon electrode.
- If the pH of the dichromate half-cell is measured while the cell is connected with a wire, and the pH increases, then the hydrogen ion concentration is decreasing according to the dichromate half-reaction.
- If the concentration of the lead(II) ions in the lead half-cell is analyzed by precipitation with sodium sulfate, and the concentration is higher than 1.0 mol/L, then lead is undergoing oxidation in this half-cell.
- If a sample of the solution from the dichromate half-cell is analyzed for lead(II) ions by precipitation with sodium sulfate, and a precipitate is observed, then lead(II) cations have moved toward the cathode.

Lab Exercise 13F
Cell Competition (page 423)

Problem
Which of the voltaic cells that can be constructed using only the materials provided has the highest possible electric potential difference?

Prediction
According to redox concepts and the table of redox half-reactions, the following voltaic cell has the highest cell potential.

$C_{(s)}$ | $Cu^{2+}_{(aq)}$, $SO_4^{2-}_{(aq)}$ || $Na^+_{(aq)}$, $Cl^-_{(aq)}$ | $Al_{(s)}$ $\Delta E° = 2.00$ V

SOA   OA   OA
$Cu^{2+}_{(aq)}$, $SO_4^{2-}_{(aq)}$, $H_2O_{(l)}$ $Al_{(s)}$,
                          RA   SRA

cathode	$3[Cu^{2+}_{(aq)} + 2e^- \rightarrow Cu_{(s)}]$	$E_r° = +0.34$ V
anode	$2[Al_{(s)} \rightarrow Al^{3+}_{(aq)} + 3e^-]$	$E_r° = -1.66$ V
net	$3Cu^{2+}_{(aq)} + 2Al_{(s)} \rightarrow 3Cu_{(s)} + 2Al^{3+}_{(aq)}$	$\Delta E° = +2.00$ V

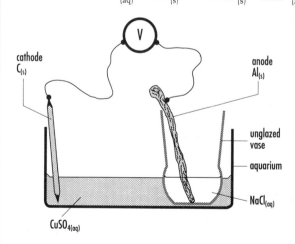

UNIT VII
CHEMICAL SYSTEMS AND EQUILIBRIUM

14 Equilibrium in Chemical Systems

INVESTIGATION 14.1

Extent of a Chemical Reaction (page 427)

Problem

What are the limiting and excess reagents in the chemical reaction of selected quantities of aqueous sodium sulfate and aqueous calcium chloride?

Prediction

According to the method of stoichiometry, which assumes that reactions are quantitative, if equal amounts of reactants are used in a reaction, both reactants are limiting reagents; if different amounts of reactants are used, the reactant in lesser amount is the limiting reagent and the reactant in the greater amount is the excess reagent. The particular reaction of this investigation is assumed to be quantitative. The mole ratio is 1:1, and the limiting and excess reagents are as shown.

$CaCl_{2(aq)}$ + $Na_2SO_{4(aq)}$ → $CaSO_{4(s)}$ + $2NaCl_{(aq)}$
10 mL   5 mL
1.0 mol/L   1.0 mol/L
excess   limiting
reagent   reagent

Procedure

1. Measure 10 mL (or 5 mL) of $CaCl_{2(aq)}$ in a graduated cylinder.
2. Pour the solution into a clean 50 mL or 100 mL beaker.
3. Measure 5 mL (or 10 mL) of $Na_2SO_{4(aq)}$ in a clean graduated cylinder.
4. Slowly add this quantity of $Na_2SO_{4(aq)}$ to the $CaCl_{2(aq)}$ while stirring.
5. Filter the precipitate from the mixture.
6. Collect about 5 mL of the filtrate into a small clean test tube.
7. Test the filtrate by adding a few drops of $Na_2CO_{3(aq)}$.
8. Collect another 5 mL of the filtrate in step 5 into a test tube and test with $Ba(NO_3)_{2(aq)}$.
9. Repeat the experiment again if time and supplies permit.

Evidence

- A white precipitate formed by mixing solutions of $CaCl_{2(aq)}$ and $Na_2SO_{4(aq)}$.
- White precipitates formed in both diagnostic tests of the filtrate.
- The same evidence was obtained when the experiment was repeated.

Analysis

According to the evidence obtained, neither reactant appears to be limiting. Both reactants appear to be in excess. Unexpected positive tests were obtained for the presence of both calcium and sulfate ions in the final (filtrate) solution.

> *Evaluation*
>
> The experimental design combining precipitation with diagnostic tests for excess ions is adequate since clear evidence was obtained to answer the question. Because the experiment required little time, it was repeated. To make the testing complete, at least one diagnostic test for one ion from each reactant is performed. Testing for sodium and chloride ions is not appropriate since these ions are expected to be in the filtrate regardless of the quantitative nature of the reaction.
>
> The procedure appears to be adequate to answer the question. The results obtained from both trials agreed with the rest of the class. Initially we were uncertain of the need for varying the quantity of reactants, performing two diagnostic tests to the filtrates, and repeating the experiment. However, these steps now appear necessary. The technological skills required to perform the test were adequate as no practice was needed to obtain reliable results.
>
> Overall, I am very confident of the experimental results, even though the results did not agree with my prediction. The only source of uncertainty may be the purity of the original solutions. An impurity could have caused the positive diagnostic tests — although a logical explanation of the unexpected results has been created (in the Synthesis).
>
> The prediction is falsified since the evidence indicates the presence of both reactants after the reaction was completed. The assumption of a quantitative reaction is judged to be unacceptable for this reaction. The method of stoichiometry needs to be restricted or revised. I am very confident about this judgment.
>
> *Synthesis*
> See page 429.

Exercise (page 428)

1. (a) $Na_2SO_{4(aq)} + CaCl_{2(aq)} \rightarrow CaSO_{4(s)} + 2NaCl_{(aq)}$
   (b) When equal amounts (25 mmol) of $Na_2SO_{4(aq)}$ and $CaCl_{2(aq)}$ were mixed, a white precipitate of $CaSO_{4(s)}$ was produced. Diagnostic tests indicate the presence of both $SO_4^{2-}{}_{(aq)}$ and $Ca^{2+}{}_{(aq)}$ in the final mixture.
   (c) Yes. A reverse reaction can be accounted for by collision-reaction theory.

2. Yes, other groups in the class collected similar evidence.

Exercise (page 429)

3. It has constant macroscopic properties.

4. At equilibrium, the forward and reverse processes are occurring, although the reaction appears to have stopped.

5. The assumptions that all reactions are spontaneous, fast, and quantitative have been found invalid in some cases.

Exercise (page 431)

6. kinetic molecular theory, collision-reaction theory, reversibility, and dynamic equilibrium

7. (a) $3H_{2(g)} + N_{2(g)} \overset{11\%}{\rightleftharpoons} 2NH_{3(g)}$

   (b) $C_{(s)} + H_2O_{(g)} \overset{>50\%}{\rightleftharpoons} CO_{(g)} + H_{2(g)}$

   (c) $2Ag^+{}_{(aq)} + Cu_{(s)} \overset{>99\%}{\rightleftharpoons} 2Ag_{(s)} + Cu^{2+}{}_{(aq)}$

(d) $2SO_{2(g)} + O_{2(g)} \underset{}{\overset{65\%}{\rightleftharpoons}} 2SO_{3(g)}$

8. $Ni_{(s)} + 2Cr^{3+}_{(aq)} \underset{}{\overset{<50\%}{\rightleftharpoons}} Ni^{2+}_{(aq)} + 2Cr^{2+}_{(aq)}$

Lab Exercise 14A
The Synthesis of an Equilibrium Law (page 432)

Problem

What mathematical formula, using equilibrium concentrations of reactants and products, gives a constant for the iron(III)-thiocyanate reaction system?

Analysis

Let $[Fe^{3+}_{(aq)}] = a$, $[SCN^-_{(aq)}] = b$, and $[FeSCN^{2+}_{(aq)}] = c$

Trial	$a + b + c$ ($\times 10^{-2}$ mol/L)	$c - (a + b)$ ($\times 10^{-2}$ mol/L)	$a \times b \times c$ ($\times 10^{-9}$ mol³/L³)	$c/(a \times b)$ (mol/L)$^{-1}$
1	4.01	−3.83	2.89	294
2	1.58	−1.42	2.34	293
3	0.73	−0.60	1.51	288
4	0.30	−0.23	0.41	307
5	0.27	−0.21	0.35	296

According to the evidence and mathematical trial-and-error analysis, the mathematical formula that gives a constant for the equilibrium is shown below.

$$\frac{[FeSCN^{2+}_{(aq)}]}{[Fe^{3+}_{(aq)}][SCN^-_{(aq)}]}$$ *(The reciprocal also represents a constant.)*

Exercise (page 434)

9. The equilibrium law does not explain why systems reach equilibrium. The equilibrium law compares the equilibrium concentration of products with reactants and provides a quantitative index of the state of equilibrium.

10. (a) $K = \dfrac{[HCl_{(g)}]^2}{[H_{2(g)}][Cl_{2(g)}]}$, from $H_{2(g)} + Cl_{2(g)} \rightleftharpoons 2HCl_{(g)}$

(b) $K = \dfrac{[NH_{3(g)}]^2}{[H_{2(g)}]^3[N_{2(g)}]}$, from $N_{2(g)} + 3H_{2(g)} \rightleftharpoons 2NH_{3(g)}$

(c) $K = \dfrac{[H_2O_{(g)}]^2}{[H_{2(g)}]^2[O_{2(g)}]}$, from $2H_{2(g)} + O_{2(g)} \rightleftharpoons 2H_2O_{(g)}$

(d) $K = \dfrac{[Ni(NH_3)_6^{2+}{}_{(aq)}]}{[Ni^{2+}_{(aq)}][NH_{3(aq)}]^6}$, from $Ni^{2+}_{(aq)} + 6NH_{3(aq)} \rightleftharpoons Ni(NH_3)_6^{2+}{}_{(aq)}$

11. $K = \dfrac{[HI_{(g)}]^2}{[H_{2(g)}][I_{2(g)}]}$, from $H_{2(g)} + I_{2(g)} \rightleftharpoons 2HI_{(g)}$

$K_1 = \dfrac{(1.56 \text{ mmol/L})^2}{(0.22 \text{ mmol/L})(0.22 \text{ mmol/L})} = 50$

$$K_2 = \frac{(2.10 \text{ mmol/L})^2}{(0.30 \text{ mmol/L})(0.30 \text{ mmol/L})} = 49$$

$$K_3 = \frac{(2.50 \text{ mmol/L})^2}{(0.35 \text{ mmol/L})(0.35 \text{ mmol/L})} = 51$$

Lab Exercise 14B
Determining an Equilibrium Constant (page 434)

Problem

What is the value of the equilibrium constant for the decomposition of phosphorus pentachloride gas to phosphorus trichloride gas and chlorine gas?

Analysis

$PCl_{5(g)} \rightleftharpoons PCl_{3(g)} + Cl_{2(g)} \quad t = 200°C$

$$K = \frac{[PCl_{3(g)}][Cl_{2(g)}]}{[PCl_{5(g)}]}$$

$$= \frac{(0.014 \text{ mol/L})^2}{4.3 \times 10^{-4} \text{ mol/L}}$$

$$= 0.46 \text{ mol/L}$$

According to the evidence provided, the value of the equilibrium constant is 0.46 mol/L.

INVESTIGATION 14.2

Demonstration of Equilibrium Shifts (page 439)

Problem

How does a change in temperature affect the nitrogen dioxide-dinitrogen tetraoxide equilibrium system? How does a change in pressure affect the carbon dioxide-carbonic acid equilibrium system?

Prediction

According to Le Châtelier's principle, an increase in temperature of the nitrogen oxide equilibrium system will shift the equilibrium to produce more nitrogen dioxide while a decrease in temperature will shift the equilibrium to produce more dinitrogen tetraoxide. Reducing the pressure on the carbon dioxide equilibrium system will shift the equilibrium toward the carbon dioxide side of the equilibrium equation.

Evidence

- When the temperature of the nitrogen oxide equilibrium was increased, the color of the mixture darkened. When the temperature was decreased, the color of the chemical system lightened.
- When the pressure on the carbon dioxide equilibrium system was decreased, the methyl red indicator in the system turned yellow. When the pressure was increased, the indicator turned red.

Analysis

Based upon the evidence gathered in this experiment, an increase in temperature of the nitrogen oxide equilibrium system (as represented by the equilibrium equation) causes the equilibrium to shift to the right to increase the concentration of nitrogen dioxide. This answer is based on the interpretation that a darkening of the brown color means an increase in the amount of nitrogen dioxide present. A decrease in pressure of the carbon dioxide equilibrium system (as represented by the equilibrium equation) shifts the equilibrium to the left to increase the concentration (partial pressure) of the carbon dioxide. This answer is based on the interpretation that a change in the color of methyl red from red to yellow means an increase in the pH and a decrease in the hydrogen ion concentration.

Evaluation

The experimental design based on observing color changes is judged to be adequate because sufficient evidence was gathered to answer the problem with a high degree of certainty. Alternative interpretations of the evidence are not apparent. The procedure and skills were adequate because the steps were simple and easy to complete. The experimental results seem very certain.

The prediction is judged to be verified because the evidence agrees with the predicted results of the tests. Le Châtelier's principle is judged to be acceptable because the prediction was verified. I feel very confident about the results of this experiment and the judgments made thereon.

INVESTIGATION 14.3

Testing Le Châtelier's Principle (page 440)

Problem

How does changing the temperature affect this chemical equilibrium system? How does changing the concentration affect this chemical equilibrium system?

Prediction

According to Le Châtelier's principle, increasing the temperature of the cobalt(II) ion equilibrium should shift the equilibrium to the left of the equation shown, and decreasing the temperature should shift the equilibrium to the right. Increasing the chloride ion concentration should shift the equilibrium to the left and decreasing the chloride ion concentration should shift the equilibrium to the right. In each case, the reasoning is that the equilibrium will shift to counteract (undo) the change to the chemical system.

Procedure

1. Obtain about 50 mL of the cobalt(II) ion equilibrium mixture in a 100 mL beaker.
2. Fill two small test tubes about half full with the cobalt(II) ion equilibrium mixture.
3. Place some hot tap water in a 400 mL beaker and hold one of the test tubes of the mixture in the hot water for several minutes. Record any change in color of the mixture.
4. Place some ice and water in a 400 mL beaker and hold the other test tube of the mixture in the ice water for several minutes. Record any color change of the mixture.

5. Rinse the test tubes into the waste container, or obtain two other clean test tubes, and half fill each of them with the equilibrium mixture.
6. Add a few grains of solid sodium chloride to one of the test tubes, and then stopper and shake the tube. Record any observations.
7. Add a few grains of solid silver nitrate to the other test tube. Stopper and shake the tube, and record any changes in the equilibrium mixture.
8. Dispose of the waste solutions in the appropriate container.

Evidence

- The cobalt(II) ion equilibrium mixture in the test tube became more blue and less pink when placed in the hot water.
- The equilibrium mixture became more pink and less blue when placed in cold water.
- The equilibrium mixture turned to blue when solid sodium chloride was added.
- A white precipitate was formed and the equilibrium mixture turned to pink when solid silver nitrate was added.

Analysis

Based upon the evidence gathered in this experiment, increasing the temperature shifts the cobalt(II) ion equilibrium to the left while decreasing the temperature shifts the equilibrium to the right. Increasing the chloride ion concentration shifts the equilibrium to the left while decreasing the chloride ion concentration shifts the equilibrium to the right. In each case, the interpretation is based on the color change that occurred.

Evaluation

The experimental design involving observation of color changes is judged to be adequate. The design provided a definite answer to the problem. It was important that the same equilibrium mixture and the same volume be used in each case. The changes in temperature were sufficient to cause a color change and adding solid sodium chloride and silver nitrate produced the predicted effect. I am very confident about the experimental results obtained using this experimental design. The procedure and skills are also judged to be adequate because the experiment was easy to perform. A white background and light would help in making the observations.

The prediction was verified because the experimental results agree with the predicted results. The results were quite definite. The authority used to make the prediction, Le Châtelier's principle, is judged to be acceptable because the prediction was verified. The results lend further support to Le Châtelier's principle and add confidence to its future use.

Exercise (page 440)

12. concentration, temperature, and pressure
13. (a) shifts to the right, producing more $H_2O_{(g)}$
    (b) shifts to the left, reducing $[H^+_{(aq)}]$ and $[OH^-_{(aq)}]$, although the final $[OH^-_{(aq)}]$ is higher than the original equilibrium concentration
    (c) shifts to the right, producing more $CaO_{(s)}$ and $CO_{2(g)}$
    (d) shifts to the right, increasing $[H^+_{(aq)}]$ and $[CH_3COO^-_{(aq)}]$
14. (a) shifts to the right, producing more $NO_{(g)}$ and $H_2O_{(g)}$
    (b) shifts to the left, consuming $NO_{(g)}$ and $H_2O_{(g)}$
    (c) shifts to the right, producing more $NO_{(g)}$ and $H_2O_{(g)}$

(d) shifts to the left, consuming $NO_{(g)}$ and $H_2O_{(g)}$

15. decrease temperature, increase pressure, increase $[N_{2(g)}]$ and increase $[H_{2(g)}]$

16. (a) It indicates that the equilibrium favors the reactants.
    (b) As more water is added to the system the equilibrium shifts to the right, increasing the concentration of $Cu(H_2O)_4^{2+}{}_{(aq)}$.
    (c) Place a test tube of the solution containing approximately equal concentrations of $CuCl_4^{2-}{}_{(aq)}$ and $Cu(H_2O)_4^{2+}{}_{(aq)}$ (green color) into an ice-water bath. If the solution turns blue (equilibrium shifts to the right), then the reaction is exothermic. If the solution turns darker green or even yellow (equilibrium shifts to the left), then the reaction is endothermic.

Exercise (page 445)

17. $[OH^-_{(aq)}] = \dfrac{1.0 \times 10^{-14} \ (mol/L)^2}{4.40 \times 10^{-3} \ mol/L} = 2.3 \times 10^{-12} \ mol/L$

18. $[H^+_{(aq)}] = \dfrac{1.0 \times 10^{-14} \ (mol/L)^2}{0.299 \times 10^{-3} \ mol/L} = 3.3 \times 10^{-11} \ mol/L$

19. $HCl_{(g)} \rightarrow H^+_{(aq)} + Cl^-_{(aq)}$

    $n_{HCl} = 0.37 \ g \times \dfrac{1 \ mol}{36.46 \ g} = 1.0 \times 10^{-2} \ mol = 10 \ mmol$

    $n_{H^+} = 10 \ mmol \times \dfrac{1}{1} = 10 \ mmol$

    $[H^+_{(aq)}] = \dfrac{10 \ mmol}{250 \ mL} = 0.040 \ mol/L$

    $[OH^-_{(aq)}] = \dfrac{1.0 \times 10^{-14} \ (mol/L)^2}{0.040 \ mol/L} = 2.5 \times 10^{-13} \ mol/L$

20. $Ca(OH)_{2(s)} \rightarrow Ca^{2+}_{(aq)} + 2OH^-_{(aq)}$

    $[OH^-_{(aq)}] = 6.9 \ mmol/L \times \dfrac{2}{1} = 14 \ mmol/L$

    $[H^+_{(aq)}] = \dfrac{1.0 \times 10^{-14} \ (mol/L)^2}{1.4 \times 10^{-2} \ mol/L} = 7.2 \times 10^{-13} \ mol/L$

21. $KOH_{(s)} \rightarrow K^+_{(aq)} + OH^-_{(aq)}$

    $n_{KOH} = 20.0 \ g \times \dfrac{1 \ mol}{56.11 \ g} = 0.356 \ mol$

    $n_{OH^-} = 0.356 \ mol \times \dfrac{1}{1} = 0.356 \ mol \quad \text{or} \quad n_{OH^-} = n_{KOH} = 0.356 \ mol$

    $[OH^-_{(aq)}] = \dfrac{0.356 \ mol}{0.500 \ L} = 0.713 \ mol/L$

    $[H^+_{(aq)}] = \dfrac{1.0 \times 10^{-14} \ (mol/L)^2}{0.713 \ mol/L} = 1.4 \times 10^{-14} \ mol/L$

22. (Enrichment)

    $H_2O_{(l)} \rightleftharpoons H^+_{(aq)} + OH^-_{(aq)}$

    $p = \dfrac{n_{ionized}}{n_{total}} \times 100$

    $= \dfrac{1.0 \times 10^{-7} \ mol}{1000 \ g / 18.02 \ g/mol} \times 100 = \dfrac{1.0 \times 10^{-5} \ mol}{55.49 \ mol} = 1.8 \times 10^{-7}\%$

> # Lab Exercise 14C
> ## The Chromate-Dichromate Equilibrium (page 445)
>
> *Problem*
>
> How does changing the hydrogen ion concentration affect the chromate-dichromate equilibrium?
>
> *Prediction*
>
> According to Le Châtelier's principle and the equilibrium equation, as the hydrogen ion concentration increases, the equilibrium will shift to the right, increasing the dichromate ion concentration. As the hydrogen ion concentration decreases, the equilibrium will shift to the left, increasing the chromate ion concentration. In both cases, the equilibrium shifts to oppose the change: to the right to try to use up the added hydrogen ions, and to the left to try to replace the hydrogen ions that are removed.
>
> *Experimental Design*
>
> Equal samples of a chromate-dichromate equilibrium mixture are placed in three separate test tubes. One test tube acts as a control. The added acid or base (to remove hydrogen ions) is the manipulated variable, and the color of the mixture is the responding variable. Controlled variables are temperature and volume.
>
> If a few drops of a strong acid are added to one test tube, and the color becomes more orange, then the dichromate ion concentration has increased. If a few drops of a strong base are added to another test tube, and the color becomes more yellow, then the chromate ion concentration has increased.

Exercise (page 447)

23. (a)

Food	$[H^+_{(aq)}]$ (mol/L)	$[OH^-_{(aq)}]$ (mol/L)	pH	pOH
oranges	5.5×10^{-3}	1.8×10^{-12}	2.26	11.74
asparagus	4×10^{-9}	3×10^{-6}	8.4	5.6
olives	5.0×10^{-4}	2.0×10^{-11}	3.30	10.70
blackberries	4.0×10^{-4}	2.5×10^{-11}	3.40	10.60

(b) oranges

24. $n_{NaOH} = 26 \text{ g} \times \dfrac{1 \text{ mol}}{40.0 \text{ g}} = 0.65 \text{ mol}$

$n_{OH^-} = n_{NaOH}$

$[OH^-_{(aq)}] = \dfrac{0.65 \text{ mol}}{0.150 \text{ L}} = 4.3 \text{ mol/L}$

pOH = $-\log[OH^-_{(aq)}] = -\log(4.3) = -0.64$
pH = $14.00 - \text{pOH} = 14.00 - (-0.64) = 14.64$
An alternative solution uses K_w.

25. pOH = $14.00 - 11.5 = 2.5$
$[OH^-_{(aq)}] = 10^{-2.5}$ mol/L = 3×10^{-3} mol/L
$[KOH_{(aq)}] = [OH^-_{(aq)}] = 3 \times 10^{-3}$ mol/L
$n_{KOH} = 3 \times 10^{-3}$ mol/L $\times 0.500$ L = 0.002 mol
$m_{KOH} = 0.002$ mol $\times 56.11$ g/mol = 0.09 g
An alternative solution uses K_w.

INVESTIGATION 14.4

pH of Common Substances (page 448)

Problem
What generalizations can be made about the pH of foods and cleaning agents?

Evidence

Substance	pH
vinegar	2.4
antacid	10.6
soft drink	3.8
tap water	7.6
household ammonia	12.2
oven cleaner	13.9
shampoo	6.3
milk	6.5
orange juice	4.1
liquid soap	8.2

Analysis
Based on the evidence gathered in this investigation, the pH of foods is generally less than 7 (acidic) and the pH of cleaning agents is generally greater than 7 (basic). These generalizations are based upon a limited number of foods and cleaners tested but the trend seems obvious.

Lab Exercise 14D
Strengths of Acids (page 449)

Problem
What is the order of several common acids in terms of decreasing acidity?

Experimental Design
Equally concentrated solutions of several common acids are prepared and the pH of each solution measured.
manipulated variable: the acid
responding variable: the pH of the solution
controlled variables: temperature, concentration, and volume

Analysis
According to the evidence collected and the definition of pH, the order of decreasing acidity is hydrochloric and nitric acids (equal acidity), hydrofluoric acid, methanoic acid, ethanoic acid, and hydrocyanic acid.

Lab Exercise 14E
Qualitative Analysis (page 453)

Problem
Which of the unknown solutions provided is $HBr_{(aq)}$, $CH_3COOH_{(aq)}$, $NaCl_{(aq)}$, $C_{12}H_{22}O_{11(aq)}$, $Ba(OH)_{2(aq)}$, and $KOH_{(aq)}$?

Analysis
According to the evidence provided,
solution 1 is $Ba(OH)_{2(aq)}$,
solution 2 is $CH_3COOH_{(aq)}$,
solution 3 is $C_{12}H_{22}O_{11(aq)}$,
solution 4 is $HBr_{(aq)}$,
solution 5 is $NaCl_{(aq)}$, and
solution 6 is $KOH_{(aq)}$.

Exercise (page 454)

26. Other properties of acids and bases could be used to identify the solutions, for example, reaction with zinc, neutralization of acids/bases.

27. Acid strength depends on the extent to which an acid ionizes in water. The differing hydrogen ion concentration of acids of equal concentration is evidence for differing acid strength.

28. (a) $[H^+_{(aq)}] = \dfrac{7.8}{100} \times 0.10 \text{ mol/L}$
$= 7.8 \times 10^{-3} \text{ mol/L} = 7.8 \text{ mmol/L}$

(b) $[H^+_{(aq)}] = \dfrac{100}{100} \times 2.3 \text{ mmol/L} = 2.3 \text{ mmol/L}$

(c) $[H^+_{(aq)}] = \dfrac{0.0078}{100} \times 0.10 \text{ mol/L}$
$= 7.8 \times 10^{-6} \text{ mol/L} = 7.8 \text{ μmol/L}$

(d) The solution in (a) is most acidic.

29. $p = \dfrac{1.16 \times 10^{-3} \text{ mol/L}}{0.100 \text{ mol/L}} \times 100 = 1.16\%$

30. $[H^+_{(aq)}] = 10^{-pH} = 10^{-2.43} \text{ mol/L} = 3.7 \times 10^{-3} \text{ mol/L}$
$p = \dfrac{3.7 \times 10^{-3}}{0.100} \times 100 = 3.7\%$

31. $K_a = \dfrac{[H^+_{(aq)}][F^-_{(aq)}]}{[HF_{(aq)}]}$
$6.6 \times 10^{-4} \text{ mol/L} = \dfrac{[H^+_{(aq)}]^2}{2.0 \text{ mol/L}}$
$[H^+_{(aq)}] = 3.6 \times 10^{-2} \text{ mol/L}$
$[F^-_{(aq)}] = [H^+_{(aq)}] = 3.6 \times 10^{-2} \text{ mol}$

32. $K_a = \dfrac{[H^+_{(aq)}][H_2PO_4^-{}_{(aq)}]}{[H_3PO_{4(aq)}]}$ *Note that only the first ionization is considered.*
$7.1 \times 10^{-3} \text{ mol/L} = \dfrac{[H^+_{(aq)}]^2}{1.0 \text{ mol/L}}$
$[H^+_{(aq)}] = 8.4 \times 10^{-2} \text{ mol/L}$
$pH = -\log[H^+_{(aq)}] = -\log(8.4 \times 10^{-2}) = 1.07$

33. $[H^+_{(aq)}] = 10^{-2.40} \text{ mol/L} = 4.0 \times 10^{-3} \text{ mol/L}$
$K_a = \dfrac{(4.0 \times 10^{-3} \text{ mol/L})^2}{0.20 \text{ mol/L}} = 7.9 \times 10^{-5} \text{ mol/L}$

Overview (page 456)

Review

1. Chemical equilibrium is a state of a closed system in which all macroscopic properties are constant.
2. Chemical equilibrium is explained as a balance between forward and reverse processes occurring at the same rate.
3. The relative amounts of reactants and products in a chemical reaction at equilibrium can be described by a percent reaction, or an equilibrium constant.
4. (a) When a soft drink bottle has just been opened it is in a non-equilibrium state. Carbon dioxide gas escapes from the solution as the rate of decomposition of carbonic acid into carbon dioxide exceeds the rate of its formation of carbonic acid from carbon dioxide gas and water.
   (b) When the bottle is sealed and at a constant temperature, it is in an equilibrium state. Carbon dioxide gas and water are in equilibrium with carbonic acid.
5. When a chemical system at equilibrium is disturbed by a change in some property of the system, the system adjusts in a way that opposes the change.
6. Temperature, volume, and concentration are commonly manipulated when an equilibrium system is shifted.
7. Decreasing the volume increases the pressure, and increasing the volume decreases the pressure.
8. (a) If a solution is neutral, the hydrogen ion concentration equals the hydroxide ion concentration.
   (b) If a solution is acidic, the hydrogen ion concentration is greater than the hydroxide ion concentration.
   (c) If a solution is basic, the hydroxide ion concentration is greater than the hydrogen ion concentration.
9. Both increasing the concentration of the reactant (by adding more) and decreasing the concentration of the product (by removing some) will increase the yield of the product.
10. A catalyst does not affect the state of equilibrium. It decreases the time required to reach equilibrium.
11. Conductivity and pH tests can distinguish a weak acid from a strong acid if the temperature and the initial solute concentrations are the same.
12. According to Arrhenius's theory, all bases increase the hydroxide ion concentration of solutions.
13. $K = \dfrac{[C]^c[D]^d}{[A]^a[B]^b}$

    $K_w = [H^+_{(aq)}][OH^-_{(aq)}]$

    $[H^+_{(aq)}] = \dfrac{p}{100} \times [HA_{(aq)}]$

    $pH = -\log[H^+_{(aq)}]$

    $pOH = -\log[OH^-_{(aq)}]$

    $[H^+_{(aq)}] = 10^{-pH}$

    $[OH^-_{(aq)}] = 10^{-pOH}$

    $pH + pOH = 14.00$

    $K_a = \dfrac{[H^+_{(aq)}][A^-_{(aq)}]}{[HA_{(aq)}]}$

Applications

14. (a) $N_{2(g)} + 3H_{2(g)} \underset{<10\%}{\rightleftharpoons} 2NH_{3(g)}$ $K = \dfrac{[NH_{3(g)}]^2}{[N_{2(g)}][H_{2(g)}]^3}$

    (b) $2H_{2(g)} + O_{2(g)} \rightarrow 2H_2O_{(g)}$ $K = \dfrac{[H_2O_{(g)}]^2}{[H_{2(g)}]^2[O_{2(g)}]}$

    (c) $CO_{(g)} + H_2O_{(g)} \underset{67\%}{\rightleftharpoons} CO_{2(g)} + H_{2(g)}$ $K = \dfrac{[CO_{2(g)}][H_{2(g)}]}{[CO_{(g)}][H_2O_{(g)}]}$

15. (a) $1.15 \text{ L/mol} = \dfrac{[N_2O_{4(g)}]}{[NO_{2(g)}]^2}$ $t = 55°C$

    (b) $[N_2O_{4(g)}] = 1.15 \text{ L/mol} \times [NO_{2(g)}]^2$
    $= 1.15 \text{ L/mol} \times (0.050 \text{ mol/L})^2$
    $= 0.0029 \text{ mol/L}$
    $= 2.9 \text{ mmol/L}$

    (c) According to Le Châtelier's principle, if the concentration of nitrogen dioxide is increased, then the equilibrium would shift to the right. The system shifts in such a way as to reduce the concentration of nitrogen dioxide (it reacts to produce dinitrogen tetraoxide).

16. At equilibrium the concentration of the product is much greater than the concentration of the reactants. The large equilibrium constant suggests a quantitative reaction equilibrium.

17. (a) $HCN_{(aq)} \underset{0.0078\%}{\rightleftharpoons} H^+_{(aq)} + CN^-_{(aq)}$

    (b) $[H^+_{(aq)}] = \dfrac{p}{100} \times [HA_{(aq)}]$
    $= \dfrac{0.0078}{100} \times 0.10 \text{ mol/L}$
    $= 7.8 \times 10^{-6} \text{ mol/L}$

    Using K_a, the $[H^+_{(aq)}]$ is 7.9×10^{-6} mol/L.
    pH $= -\log(7.8 \times 10^{-6})$
    $= 5.11$
    Using K_a, the pH is 5.10.

18. (a) The equilibrium system will shift to the left, partially counteracting the applied heat by undergoing an endothermic reaction.
    (b) The equilibrium system will shift to the left, partially counteracting the decreased total pressure by producing more gas phase molecules.
    (c) The equilibrium system will shift to the right, partially counteracting the increased concentration of oxygen by reacting it with $HCl_{(g)}$.
    (d) Catalysts have no effect on the position of an equilibrium system.

19. (a) high concentration of $C_2H_{6(g)}$, low concentration of $C_2H_{4(g)}$ and $H_{2(g)}$, high temperature, low pressure
    (b) high concentration of $CO_{(g)}$ and $H_{2(g)}$, low concentration of $CH_3OH_{(g)}$, low temperature, high pressure

20. $[OH^-_{(aq)}] = \dfrac{1.0 \times 10^{-14} \text{ (mol/L)}^2}{1.3 \times 10^{-3} \text{ mol/L}}$
    $= 7.7 \times 10^{-12} \text{ mol/L}$

21. $[H^+_{(aq)}] = \dfrac{1.0 \times 10^{-14} \ (mol/L)^2}{2.5 \times 10^{-7} \ mol/L}$

    $\phantom{[H^+_{(aq)}]} = 4.0 \times 10^{-8} \ mol/L$

    $pH = -\log(4.0 \times 10^{-8})$

    $ = 7.40$

22. $[H^+_{(aq)}] = 10^{-5.6} \ mol/L$

    $\phantom{[H^+_{(aq)}]} = 3 \times 10^{-6} \ mol/L$

23. by a factor of 1000 (10^3)

24. $n_{NaOH} = 8.50 \ g \times \dfrac{1 \ mol}{40.00 \ g}$

    $\phantom{n_{NaOH}} = 0.213 \ mol$

    $[NaOH_{(aq)}] = 0.213 \ mol \ / \ 0.500 \ L$

    $\phantom{[NaOH_{(aq)}]} = 0.425 \ mol/L$

    $[OH^-_{(aq)}] = [NaOH_{(aq)}] = 0.425 \ mol/L$

    $pOH = -\log(0.425)$

    $ = 0.372$

25. $[H^+_{(aq)}] = 10^{-1.57} \ mol/L$

    $\phantom{[H^+_{(aq)}]} = 0.027 \ mol/L$

    Since $HCl_{(aq)}$ is a strong acid, $[HCl_{(aq)}] = [H^+_{(aq)}] = 0.027 \ mol/L$

    $n_{HCl} = 0.250 \ L \times 0.027 \ mol/L$

    $\phantom{n_{HCl}} = 0.0067 \ mol$

    $m_{HCl} = 0.0067 \ mol \times \dfrac{36.46 \ g}{1 \ mol} = 0.25 \ g$

26. $[H^+_{(aq)}] = \dfrac{0.054}{100} \times 0.10 \ mol/L$

    $\phantom{[H^+_{(aq)}]} = 5.4 \times 10^{-5} \ mol/L$

    $pH = -\log(5.4 \times 10^{-5})$

    $ = 4.27$

    Using K_a, the pH is the same.

27. $n_{HCl} = 30.5 \ kg \times \dfrac{1 \ mol}{36.46 \ g} = 837 \ mol$

    $[HCl_{(aq)}] = 837 \ mol \ / \ 806 \ L$

    $\phantom{[HCl_{(aq)}]} = 1.04 \ mol/L$

    $[H^+_{(aq)}] = [HCl_{(aq)}] = 1.04 \ mol/L$

    $pH = -\log(1.04)$

    $ = -0.016$

    $pOH = 14.00 - (-0.016)$

    $ = 14.02$

    It is necessary to assume that the percent reaction of hydrogen chloride is the same as a 0.10 mol/L solution and that the volume of the solution is equal to the volume of water used.

28. $n_{CH_3COOH} = 60.0 \ kg \times \dfrac{1 \ mol}{60.06 \ g}$

    $\phantom{n_{CH_3COOH}} = 0.999 \ kmol$

    $[CH_3COOH_{(aq)}] = 0.999 \ kmol \ / \ 1.25 \ kL$

    $\phantom{[CH_3COOH_{(aq)}]} = 0.799 \ mol/L$

Since the concentration of the acid is not equal to 0.10 mol/L, the percent dissociation given in Appendix F, page 553, does not apply. However, an approximate answer can be obtained using this percent dissociation.

$$[H^+_{(aq)}] = \frac{1.3}{100} \times 0.799 \text{ mol/L}$$
$$= 0.010 \text{ mol/L}$$
$$pH = -\log(0.10)$$
$$= 1.98$$
$$pOH = 14.00 - 1.98$$
$$= 12.02$$

A more reliable answer can be obtained using the K_a of acetic acid, which is reported to be 1.8×10^{-5} mol/L. The reaction of acetic acid with water can be represented as:

$$CH_3COOH_{(aq)} \rightleftharpoons H^+_{(aq)} + CH_3COO^-_{(aq)}$$
$$K_a = \frac{[H^+_{(aq)}][CH_3COO^-_{(aq)}]}{[CH_3COOH_{(aq)}]}$$

Because $[H^+_{(aq)}] = [CH_3COO^-_{(aq)}]$, and very little of the acetic acid is expected to react with water, this expression can be reduced to

$$1.8 \times 10^{-5} \text{ mol/L} = \frac{[H^+_{(aq)}]^2}{0.799 \text{ mol/L}}$$
$$[H^+_{(aq)}] = 3.8 \times 10^{-3} \text{ mol/L}$$
$$pH = -\log(3.8 \times 10^{-3})$$
$$= 2.42$$
$$pOH = 14.00 - 2.42$$
$$= 11.58$$

29. pOH = 14.00 − 10.35 = 3.65
   $[OH^-_{(aq)}] = 10^{-3.65}$ mol/L = 2.2×10^{-4} mol/L
   n_{OH^-} = 2.00 L × 2.2×10^{-4} mol/L = 4.5×10^{-4} mol
   $n_{NaOH} = n_{OH^-} = 4.5 \times 10^{-4}$ mol
   $m_{NaOH} = 4.5 \times 10^{-4}$ mol × 40.00 g/mol = 0.018 g = 18 mg
   An alternative solution uses K_w.

30.
Diagnostic Test	Strong Acid	Weak Acid	Neutral Molecular	Neutral Ionic
Conductivity	high	low	none	high
Blue Litmus	turns red	turns red	no change	no change
Red Litmus	no change	no change	no change	no change

31.
$$pH = -\log[H^+_{(aq)}] \qquad K_W = [H^+_{(aq)}][OH^-_{(aq)}]$$
$$pH \rightleftharpoons [H^+_{(aq)}] \rightleftharpoons [OH^-_{(aq)}]$$
$$[H^+_{(aq)}] = 10^{-pH}$$

$$[H^+_{(aq)}] = \frac{p}{100} \times [HA_{(aq)}] \qquad [HA_{(aq)}] = \frac{100}{p} \times [H^+_{(aq)}]$$

$$[HA_{(aq)}]$$

32. $$K_a = \frac{[H^+_{(aq)}][ASA^-_{(aq)}]}{[HASA_{(aq)}]}$$

$$3.27 \times 10^{-4} \text{ mol/L} = \frac{[H^+_{(aq)}]^2}{0.018 \text{ mol/L}}$$

$$[H^+_{(aq)}] = 2.4 \times 10^{-3} \text{ mol/L}$$

$$\text{pH} = -\log(2.4 \times 10^{-3})$$
$$= 2.62$$

As the temperature changes to 37°C, more acetylsalicylic acid might dissolve, increasing the concentration of hydrogen ions and decreasing the pH.

33. $$K_a = \frac{[H^+_{(aq)}][H_2BO_3^-{}_{(aq)}]}{[H_3BO_{3(aq)}]}$$

$$5.8 \times 10^{-10} \text{ mol/L} = \frac{[H^+_{(aq)}]^2}{0.50 \text{ mol/L}}$$

$$[H^+_{(aq)}] = 1.7 \times 10^{-5} \text{ mol/L}$$

$$\text{pH} = -\log(1.7 \times 10^{-5})$$
$$= 4.77$$

A pH of 4.42 obtained by using the percent ionization in the table of acids and bases is not valid because the solution concentration is 0.50 mol/L, not 0.1 mol/L.

34. Concentration of saturated solution of salicylic acid:

$$n_{HSA} = 1 \text{ g} \times \frac{1 \text{ mol}}{138.13 \text{ g}} = 7 \times 10^{-3} \text{ mol}$$

$$[HSA_{(aq)}] = \frac{7 \times 10^{-3} \text{ mol}}{0.460 \text{ L}} = 0.02 \text{ mol/L}$$

$$[H^+_{(aq)}] = 10^{-\text{pH}}$$
$$= 10^{-2.4} \text{ mol/L}$$
$$= 4 \times 10^{-3} \text{ mol/L}$$

$$K_a = \frac{[H^+_{(aq)}][SA^-_{(aq)}]}{[HSA_{(aq)}]}$$

$$= \frac{(4 \times 10^{-3} \text{ mol/L})^2}{0.02 \text{ mol/L}}$$

$$= 1 \times 10^{-3} \text{ mol/L}$$

Extensions

35. The wolfram (tungsten) and iodine reaction to produce wolfram(II) iodide is exothermic. A high temperature of the system forces the equilibrium to shift to the left, depositing wolfram on the filament. This deposition reverses the tendency of the wolfram to be lost gradually from the filament. The presence of the halogen (iodine) helps to establish an equilibrium which at high temperature, restores the filament.

36. Reducing the pressure of a solution-gas equilibrium (blood and air in the body) reduces the quantity of gas (oxygen) that is dissolved in the solution (blood). At high altitudes, the human body can compensate the loss of oxygen by increasing the number of red blood cells in the blood. This adaptation to increase the red blood cell count usually takes about 3–4 weeks, but it can be speeded up by transfusion.

INVESTIGATION 14.5

Studying a Chemical Equilibrium System (page 459)

Problem

Is the iron(III) thiocyanate equilibrium endothermic or exothermic?

Prediction

Based upon the theory that forming bonds releases energy and that breaking bonds requires energy, the equilibrium system as written is exothermic.
$$Fe^{3+}_{(aq)} + SCN^-_{(aq)} \rightleftharpoons FeSCN^{2+}_{(aq)} + energy$$

Experimental Design

Aqueous solutions containing iron(III) ions and thiocyanate ions are mixed and then heated and cooled while observing any color change.

Materials

lab apron
safety glasses
iron(III) nitrate solution
potassium thiocyanate solution
100 mL beaker
distilled water
ice water
hot water
(2) medium test tubes with rack

Procedure

1. Mix the iron(III) nitrate solution with the potassium thiocyanate solution to obtain an equilibrium mixture that is light red. Dilute the mixture if necessary
2. Half fill each of the two test tubes with the equilibrium mixture.
3. Place one test tube of the mixture into hot water and the other into ice water.
4. Observe and record any color changes.
5. All solutions are non-toxic and can be poured down the sink.

Evidence

- The equilibrium mixture placed in hot water decreased in color intensity.
- The equilibrium mixture placed in ice water increased in color intensity, that is, the solution became a darker red than it was.

Analysis

According to the evidence gathered in this investigation and Le Châtelier's principle, the reaction as written above is exothermic. The decrease in color intensity when the temperature is increased indicates that the equilibrium system shifts to the left to counteract the added energy. This would only happen if energy is a product of the reaction. An interpretation of the evidence from the cold water bath provides the same answer.

Evaluation

The experimental design using hot and cold water to shift the equilibrium system was adequate because it provides an answer to the problem. A third test tube of the equilibrium mixture should have been used as a control — this would have increased the certainty of the results. The procedure and skills are judged to be adequate because they were simple and easy to complete. Using a light or a white background could have helped in observing color changes.

The prediction based on bond energy theory is judged to be verified because the experimental results agree with the predicted results. Because the prediction was verified, the bond energy concept used to make the prediction is judged to be acceptable. These judgments are made with a high degree of confidence.

15 Acid-Base Chemistry

INVESTIGATION 15.1

Testing Arrhenius's Acid-Base Definitions (page 462)

Problem
Which of the substances tested may be classified as acid, base, or neutral?

Experimental Design
Equally concentrated samples of the solutions are tested for electrical conductivity and with litmus to determine if they exhibit properties of acids and bases. The controlled variables are concentration and temperature. The manipulated variable is the substance tested and the responding variables are conductivity and the litmus color. The control used in this experiment is pure water.

Prediction
According to Arrhenius's definitions, of the substances to be tested, the acids are $HCl_{(aq)}$ and $CH_3COOH_{(aq)}$, the bases are $NaOH_{(aq)}$ and $Ca(OH)_{2(aq)}$, and all others are neutral. The reasoning behind this prediction is grounded in the theory that acids are hydrogen compounds that ionize to produce $H^+_{(aq)}$ ions and that bases are ionic hydroxides that dissociate to produce $OH^-_{(aq)}$ ions. The other substances tested are classified as neutral compounds that do not yield hydrogen or hydroxide ions when they dissociate in water.

$HCl_{(g)} \rightarrow H^+_{(aq)} + Cl^-_{(aq)}$
$CH_3COOH_{(l)} \rightarrow H^+_{(aq)} + CH_3COO^-_{(aq)}$
$NaOH_{(s)} \rightarrow Na^+_{(aq)} + OH^-_{(aq)}$
$Ca(OH)_{2(s)} \rightarrow Ca^{2+}_{(aq)} + 2OH^-_{(aq)}$
$Na_2CO_{3(s)} \rightarrow 2Na^+_{(aq)} + CO_3^{2-}_{(aq)}$
$NaHCO_{3(s)} \rightarrow Na^+_{(aq)} + HCO_3^-_{(aq)}$
$NaHSO_{4(s)} \rightarrow Na^+_{(aq)} + HSO_4^-_{(aq)}$
$SO_{2(g/aq)} \rightarrow SO_{2(aq)}$
$MgO_{(s)} \rightarrow MgO_{(aq)}$
$NH_{3(g)} \rightarrow NH_{3(aq)}$

Procedure

1. Test and record the conductivity of pure water.
2. Test pure water with both red and blue litmus paper.
3. Repeat steps 1 and 2 for all the solutions provided.

Evidence / Analysis

DIAGNOSTIC TEST RESULTS AND ANALYSIS				
Substance	Conductivity	Red Litmus	Blue Litmus	Analysis
$H_2O_{(l)}$	very slight	no change	no change	neutral
$HCl_{(aq)}$	high	no change	pink	acid
$Na_2CO_{3(aq)}$	high	blue	no change	base
$NaHCO_{3(aq)}$	high	blue	no change	base
$NaHSO_{4(aq)}$	high	no change	pink	acid
$NaOH_{(aq)}$	high	blue	no change	base

Ca(OH)$_{2(aq)}$	low	blue	no change	base
SO$_{2(aq)}$	low	no change	pink	acid
MgO$_{(aq)}$	low	blue	no change	base
NH$_{3(aq)}$	low	blue	no change	base
CH$_3$COOH$_{(aq)}$	low	no change	pink	acid

Based upon the evidence gathered in this investigation, the substances tested are acids and bases as listed below. None of the substances, other the water control, tested as being neutral.

Acids	**Bases**	**Neutral**
HCl$_{(aq)}$	Na$_2$CO$_{3(aq)}$	H$_2$O$_{(l)}$ (the control)
NaHSO$_{4(aq)}$	NaHCO$_{3(aq)}$	
SO$_{2(aq)}$	NaOH$_{(aq)}$	
CH$_3$COOH$_{(aq)}$	Ca(OH)$_{2(aq)}$	
	MgO$_{(s)}$	
	NH$_{3(aq)}$	

Evaluation

The experimental design of testing the solutions for electrical conductivity and with litmus paper appears adequate since the question could be answered with high confidence. Conductivity is not a useful diagnostic test for classifying acids, bases, and neutral substances and is thus an unnecessary part of the design. Using pH paper to test the pH of the solutions would be a more efficient, but expensive, method than using litmus paper. The procedure and technological skills appear to be adequate since no difficulties were encountered and the evidence agrees with that obtained by the rest of the class. Some of the changes in color of the litmus (for example, for magnesium oxide) were slow — from my experience, a minimum time of, say, 5 min should be added as a controlled variable. Except for this concern of minimum time requirement, I am very confident in the experimental results obtained.

Of the ten results, six did not agree with the prediction and only four agreed with the prediction. None of the solutions tested were neutral. Arrhenius's definitions are unacceptable since the prediction was falsified. Restricting the definitions is not an option as there are too many exceptions that are common chemicals. The evidence warrants a revision or replacement of Arrhenius's definitions. I am very confident that the results of this experiment and the subsequent evaluation of the Arrhenius concept of acid and base will be replicated by others.

Synthesis
See pages 462–467.

Exercise (page 464)

1. (a) HCO$_3^-{}_{(aq)}$ + H$_2$O$_{(l)}$ → H$_2$CO$_{3(aq)}$ + OH$^-{}_{(aq)}$
   (b) HSO$_4^-{}_{(aq)}$ + H$_2$O$_{(l)}$ → SO$_4^{2-}{}_{(aq)}$ + H$_3$O$^+{}_{(aq)}$
   (c) CO$_3^{2-}{}_{(aq)}$ + H$_2$O$_{(l)}$ → HCO$_3^-{}_{(aq)}$ + OH$^-{}_{(aq)}$
   (d) O$^{2-}{}_{(aq)}$ + H$_2$O$_{(l)}$ → OH$^-{}_{(aq)}$ + OH$^-{}_{(aq)}$

2. (a) $SO_{3(g)} + H_2O_{(l)} \rightarrow H_2SO_{4(aq)}$
   $H_2SO_{4(aq)} + H_2O_{(l)} \rightarrow HSO_4^-{}_{(aq)} + H_3O^+{}_{(aq)}$
   (b) $3NO_{2(g)} + H_2O_{(l)} \rightarrow 2HNO_{3(aq)} + NO_{(g)}$ (from page 96)
   $HNO_{3(aq)} + H_2O_{(l)} \rightarrow NO_3^-{}_{(aq)} + H_3O^+{}_{(aq)}$
   (c) Nitrogen oxides are produced in high temperature combustion reactions, for example, automobile engines, where the temperature is high enough for nitrogen and oxygen in the atmosphere to react with each other. Sulfur oxides are produced by smelting sulfide ores and burning fossil fuels that contain traces of sulfur. Legislation to limit emissions of nitrogen and sulfur oxides could force industries to implement technology to reduce the emissions. (See p. 107.)
   (d) These emissions might be controlled by improved technology, for example, catalytic converters and $SO_{2(g)}$ scrubbers.
3. The predictions have improved somewhat, but there are still many situations where it is impossible to predict whether an acidic, a basic, or a neutral solution will be formed.
4. (Enrichment) In an aqueous solution, the concentration of $H_2O_{(l)}$ molecules is much higher than the concentration of $H_3O^+{}_{(aq)}$ ions. It is more likely that a proton would bond to a $H_2O_{(l)}$ molecule than to a $H_3O^+{}_{(aq)}$ ion, which repels the positively charged proton.

Exercise (page 467)

5. (a) According to the Arrhenius concept, acids are substances that ionize in aqueous solution to produce hydrogen ions.
   (b) According to the revised Arrhenius concept, acids are substances that react with water to form hydronium ions.
   (c) According to the Brønsted-Lowry concept, acids are substances that donate protons.
6. The Arrhenius's definition is restricted to aqueous solutions, while the Brønsted-Lowry definition is more general.
7. (a) $HF_{(aq)} + SO_3^{2-}{}_{(aq)} \rightleftharpoons F^-{}_{(aq)} + HSO_3^-{}_{(aq)}$
       acid   base
   (b) $CO_3^{2-}{}_{(aq)} + CH_3COOH_{(aq)} \rightleftharpoons CH_3COO^-{}_{(aq)} + HCO_3^-{}_{(aq)}$
       base   acid
   (c) $H_3PO_{4(aq)} + OCl^-{}_{(aq)} \rightleftharpoons H_2PO_4^-{}_{(aq)} + HOCl_{(aq)}$
       acid   base
   (d) $HCO_3^-{}_{(aq)} + HSO_4^-{}_{(aq)} \rightleftharpoons SO_4^{2-}{}_{(aq)} + H_2CO_{3(aq)}$
       base   acid
8. (a) $HCO_3^-{}_{(aq)} + OH^-{}_{(aq)} \rightarrow CO_3^{2-}{}_{(aq)} + H_2O_{(l)}$
       acid   base
   (b) $HCO_3^-{}_{(aq)} + H_3O^+{}_{(aq)} \rightarrow H_2CO_{3(aq)} + H_2O_{(l)}$
       base   acid
9. The Brønsted-Lowry definitions remove the restriction of aqueous solutions for acid-base reactions.
10. The Brønsted-Lowry concept does not explain *why* a proton is accepted or donated.

Exercise (page 470)

11. (a) $HCO_3^-{}_{(aq)}/CO_3^{2-}{}_{(aq)}$ and $HS^-{}_{(aq)}/S^{2-}{}_{(aq)}$
    (b) $H_2CO_{3(aq)}/HCO_3^-{}_{(aq)}$ and $H_2O_{(l)}/OH^-{}_{(aq)}$
    (c) $HSO_4^-{}_{(aq)}/SO_4^{2-}{}_{(aq)}$ and $H_2PO_4^-{}_{(aq)}/HPO_4^{2-}{}_{(aq)}$
    (d) $H_2O_{(l)}/OH^-{}_{(aq)}$ and $H_3O^+{}_{(aq)}/H_2O_{(l)}$
12. $HCO_3^-{}_{(aq)}/CO_3^{2-}{}_{(aq)}$ and $H_2CO_{3(aq)}/HCO_3^-{}_{(aq)}$

Exercise (page 472)

13. (a) red
    (b) yellow
    (c) blue
    (d) orange
14. (a) If methyl red is added to a solution, and the color of the solution turns red, then the pH is less than 4.8.
    (b) If alizarin yellow is added to a solution, and the color of the solution turns red, then the pH is greater than 12.0.
    (c) If bromocresol green is added to a solution, and the color of the solution turns blue, then the pH is greater than 5.4.
    (d) If bromothymol blue is added to a solution, and the color of the solution turns green, then the pH is between 6.0 and 7.6.
15. (a) between 5.4 and 6.0
    (b) $[H_3O^+_{(aq)}]$ between $10^{-5.4}$ mol/L and $10^{-6.0}$ mol/L, or
    $[H_3O^+_{(aq)}]$ between 1×10^{-6} mol/L and 4×10^{-6} mol/L

Lab Exercise 15A
Using Indicators to Determine pH (page 472)

Problem
What is the approximate pH of three unknown solutions?

Analysis
According to the evidence listed below and the table of acid-base indicators, the pH of solution A is between 4.4 and 4.8.

Indicator	Indicator Color	pH
methyl violet	blue	1.6 or higher
methyl orange	yellow	4.4 or higher
methyl red	red	4.8 or lower
phenolphthalein	colorless	8.2 or lower

According to the evidence listed below and the table of acid-base indicators, the pH of solution B is between 6.0 and 6.6.

Indicator	Indicator Color	pH
indigo carmine	blue	11.4 or lower
phenol red	yellow	6.6 or lower
bromocresol green	blue	5.4 or higher
methyl red	yellow	6.0 or higher

According to the evidence listed below and the table of acid-base indicators, the pH of solution C is between 3.2 and 3.8.

Indicator	Indicator Color	pH
phenolphthalein	colorless	8.2 or lower
thymol blue	yellow	between 2.8 and 8.0
bromocresol green	yellow	3.8 or lower
methyl orange	orange	between 3.2 and 4.4

Lab Exercise 15B
Designing an Indicator Experiment (page 473)

Problem

Which of three unknown solutions labelled X, Y, and Z have pH values of 3.5, 5.8, and 7.8?

Experimental Design

If separate samples of X, Y, and Z are tested with bromothymol blue, and the indicator turns a solution blue, then that solution has a pH of 7.8. If separate samples of X, Y, and Z are tested with bromocresol green, and the indicator turns a solution yellow, then that solution has a pH of 3.5. The third solution has a pH of 5.8.
Alternative: If separate samples of X, Y, and Z are tested with chlorophenol red, the solution that is yellow has a pH of 3.5; the solution that is orange has a pH of 5.8; the solution that is red has a pH of 7.8.

Lab Exercise 15C
Position of Acid-Base Equilibria (page 473)

Problem

How do the positions of the reactant acid and base in the acid-base table relate to the position of equilibrium?

Analysis

Reactions 1, 3, and 5 all have <50% reaction. According to the acid-base table, the reactant acid is listed below the reactant base for each of these equilibria. Reactions 2, 4, and 6 all have >50% reaction. According to the acid-base table, the reactant acid is listed above the reactant base for each of these equilibria.

Exercise (page 476)

16. SA   A
$HF_{(aq)}$, $Na^+_{(aq)}$, $SO_4^{2-}_{(aq)}$, $H_2O_{(l)}$
             SB   B

$HF_{(aq)} + SO_4^{2-}_{(aq)} \overset{<50\%}{\rightleftharpoons} F^-_{(aq)} + HSO_4^-_{(aq)}$

17. SA   A
$H_3O^+_{(aq)}$, $ClO_4^-_{(aq)}$, $Na^+_{(aq)}$, $OH^-_{(aq)}$, $H_2O_{(l)}$
           B   SB   B

$H_3O^+_{(aq)} + OH^-_{(aq)} \rightarrow 2H_2O_{(l)}$

18.   SA   A
$Na^+_{(aq)}$, $ClO^-_{(aq)}$, $CH_3COOH_{(aq)}$, $H_2O_{(l)}$
      SB   B

$ClO^-_{(aq)} + CH_3COOH_{(aq)} \overset{>50\%}{\rightleftharpoons} HClO_{(aq)} + CH_3COO^-_{(aq)}$

19. SA   A   A
$HCOOH_{(aq)}$, $Na^+_{(aq)}$, $HS^-_{(aq)}$, $H_2O_{(l)}$
                  SB   B

$HCOOH_{(aq)} + HS^-_{(aq)} \overset{>50\%}{\rightleftharpoons} HCOO^-_{(aq)} + H_2S_{(aq)}$

20.
$$\underset{B}{\overset{SA}{NH_4^+{}_{(aq)}}},\ Cl^-{}_{(aq)},\ \underset{SB}{Na^+{}_{(aq)}},\ \underset{B}{\overset{A}{NO_2^-{}_{(aq)}}},\ H_2O_{(l)}$$

$$NH_4^+{}_{(aq)} + NO_2^-{}_{(aq)} \underset{<50\%}{\rightleftharpoons} NH_{3(aq)} + HNO_{2(aq)}$$

21.
$$\underset{B}{\overset{SA}{H_3O^+{}_{(aq)}}},\ NO_3^-{}_{(aq)},\ \underset{SB}{Na^+{}_{(aq)}},\ \underset{B}{\overset{A}{CH_3COO^-{}_{(aq)}}},\ H_2O_{(l)}$$

$$H_3O^+{}_{(aq)} + CH_3COO^-{}_{(aq)} \rightarrow H_2O_{(l)} + CH_3COOH_{(aq)}$$

22.
$$Na^+{}_{(aq)},\ \underset{B}{\overset{SA}{HSO_4^-{}_{(aq)}}},\ \underset{SB}{Na^+{}_{(aq)}},\ \underset{B}{\overset{A}{OH^-{}_{(aq)}}},\ H_2O_{(l)}$$

$$HSO_4^-{}_{(aq)} + OH^-{}_{(aq)} \underset{>50\%}{\rightleftharpoons} SO_4^{2-}{}_{(aq)} + H_2O_{(l)}$$

23.
$$\underset{B}{\overset{A}{NH_4^+{}_{(aq)}}},\ NO_3^-{}_{(aq)},\ \underset{B}{\overset{SA}{H_3O^+{}_{(aq)}}},\ \underset{SB}{\overset{A}{Cl^-{}_{(aq)}}},\ H_2O_{(l)}$$

$$H_3O^+{}_{(aq)} + H_2O_{(l)} \rightleftharpoons H_2O_{(l)} + H_3O^+{}_{(aq)} \text{ (no reaction)}$$

Therefore, $NH_4NO_{3(aq)}$ cannot be used to neutralize $HCl_{(aq)}$.

Lab Exercise 15D
Testing the Five-Step Method (page 476)

Problem
What are the products and position of the equilibrium for sodium hydrogen carbonate with stomach acid, vinegar, household ammonia, and lye, respectively?

Prediction
According to the five-step method for predicting acid-base reactions, the following reactions are expected.

1. Reaction with $HCl_{(aq)}$

$$\underset{SB}{\overset{A}{Na^+{}_{(aq)}}},\ \underset{B}{HCO_3^-{}_{(aq)}},\ \underset{B}{\overset{SA}{H_3O^+{}_{(aq)}}},\ \overset{A}{Cl^-{}_{(aq)}},\ H_2O_{(l)}$$

$$H_3O^+{}_{(aq)} + HCO_3^-{}_{(aq)} \underset{>50\%}{\rightleftharpoons} H_2O_{(l)} + H_2CO_{3(aq)}$$

2. Reaction with $CH_3COOH_{(aq)}$

$$\underset{SB}{\overset{A}{Na^+{}_{(aq)}}},\ \underset{B}{HCO_3^-{}_{(aq)}},\ \underset{}{\overset{SA}{CH_3COOH_{(aq)}}},\ \underset{B}{\overset{A}{H_2O_{(l)}}}$$

$$CH_3COOH_{(aq)} + HCO_3^-{}_{(aq)} \underset{>50\%}{\rightleftharpoons} CH_3COO^-{}_{(aq)} + H_2CO_{3(aq)}$$

3. Reaction with $NH_{3(aq)}$

$$\underset{B}{\overset{SA}{Na^+{}_{(aq)}}},\ HCO_3^-{}_{(aq)},\ \underset{SB}{\overset{}{NH_{3(aq)}}},\ \underset{B}{\overset{A}{H_2O_{(l)}}}$$

$$HCO_3^-{}_{(aq)} + NH_{3(aq)} \underset{<50\%}{\rightleftharpoons} CO_3^{2-}{}_{(aq)} + NH_4^+{}_{(aq)}$$

4. Reaction with NaOH$_{(aq)}$

$$Na^+_{(aq)}, \underset{B}{HCO_3^-_{(aq)}}, \underset{SB}{\overset{SA}{OH^-_{(aq)}}}, \underset{B}{\overset{A}{H_2O_{(l)}}}$$

$$HCO_3^-_{(aq)} + OH^-_{(aq)} \overset{>50\%}{\rightleftharpoons} CO_3^{2-}_{(aq)} + H_2O_{(l)}$$

Analysis

1. The evidence indicates that baking soda reacts appreciably with hydrochloric acid, producing gas bubbles (assumed to be carbon dioxide from the decomposition of carbonic acid) and increasing the pH of the solution. No evidence was obtained for the formation of water.

2. The evidence indicates that baking soda reacts appreciably with vinegar, producing gas bubbles (assumed to be carbon dioxide from the decomposition of carbonic acid), increasing the pH of the solution, and causing the disappearance of the odor of vinegar. No evidence was obtained for the production of acetate ions.

3. The decrease in pH is an indirect indication of a possible reaction of baking soda with ammonia. The characteristic ammonia smell remaining after the reaction indicates that the reaction was not complete, agreeing with the less than 50% position of equilibrium predicted. No evidence for any predicted products was obtained.

4. The decrease in pH is evidence that the baking soda reacts with lye. No evidence is given for the position of the equilibrium or the presence of the predicted products.

Evaluation

The experimental design is judged to be inadequate because the problem cannot be completely answered. More evidence of better quality with specific diagnostic tests for the predicted products is required. I am not very confident with this design and only moderately certain of the results.

The prediction for the reaction of baking soda and hydrochloric acid, and baking soda and acetic acid appears to be verified since the evidence in each case does suggest a significant reaction with clear indication of one major product. The prediction for the other two reactions is inconclusive because the evidence is too limited. Based upon this judgment, the five-step method remains acceptable. No evidence was obtained that definitely falsified the prediction. Further tests are needed in order to test the validity of this five-step method.

INVESTIGATION 15.2

Testing Brønsted-Lowry Predictions (page 477)

Problem

What reactions occur when the following substances are mixed?
1. ammonium chloride and sodium hydroxide solutions
2. hydrochloric acid and sodium acetate solutions
3. sodium benzoate and sodium hydrogen sulfate solutions
4. hydrochloric acid and aqueous ammonium chloride
5. solid sodium chloride added to water
6. solid aluminum sulfate added to water

7. solid sodium phosphate added to water
8. solid sodium hydrogen sulfate added to water
9. solid sodium hydrogen carbonate added to hydrochloric acid
10. solid sodium hydrogen carbonate added to sodium hydroxide solution
11. solid sodium hydrogen carbonate added to sodium hydrogen sulfate solution

Prediction

According to the Brønsted-Lowry concept of acids and bases and the five-step procedure, the following reactions and diagnostic test results are predicted.

1.   SA   A
   $NH_4^+{}_{(aq)}$, $Cl^-{}_{(aq)}$, $Na^+{}_{(aq)}$, $OH^-{}_{(aq)}$, $H_2O_{(l)}$
                B   SB   B

   $NH_4^+{}_{(aq)} + OH^-{}_{(aq)} \overset{>50\%}{\rightleftharpoons} NH_{3(aq)} + H_2O_{(l)}$
   (Ammonia odor is predicted.)

2.   SA   A
   $H_3O^+{}_{(aq)}$, $Cl^-{}_{(aq)}$, $Na^+{}_{(aq)}$, $CH_3COO^-{}_{(aq)}$, $H_2O_{(l)}$
                B   SB   B

   $H_3O^+{}_{(aq)} + CH_3COO^-{}_{(aq)} \overset{>50\%}{\rightleftharpoons} H_2O_{(l)} + CH_3COOH_{(aq)}$
   (Vinegar odor is predicted.)

3.   SA   A
   $Na^+{}_{(aq)}$, $C_6H_5COO^-{}_{(aq)}$, $HSO_4^-{}_{(aq)}$, $H_2O_{(l)}$
        SB   B   B

   $HSO_4^-{}_{(aq)} + C_6H_5COO^-{}_{(aq)} \overset{>50\%}{\rightleftharpoons} SO_4^{2-}{}_{(aq)} + C_6H_5COOH_{(s)}$
   (Precipitate is predicted.)

4.   SA   A   A
   $H_3O^+{}_{(aq)}$, $Cl^-{}_{(aq)}$, $NH_4^+{}_{(aq)}$, $H_2O_{(l)}$
                B   B

   $H_3O^+{}_{(aq)} + H_2O_{(l)} \rightleftharpoons H_2O_{(l)} + H_3O^+{}_{(aq)}$
   (No change in odor is predicted.)

5.   SA
   $Na^+{}_{(aq)}$, $Cl^-{}_{(aq)}$, $H_2O_{(l)}$
         B   SB

   $H_2O_{(l)} + H_2O_{(l)} \overset{>50\%}{\rightleftharpoons} H_3O^+{}_{(aq)} + OH^-{}_{(aq)}$
   (No litmus change is predicted.)

6.   SA
   $Al^{3+}{}_{(aq)}$, $SO_4^{2-}{}_{(aq)}$, $H_2O_{(l)}$
         SB   B

   $H_2O_{(l)} + SO_4^{2-}{}_{(aq)} \overset{<50\%}{\rightleftharpoons} OH^-{}_{(aq)} + HSO_4^-{}_{(aq)}$
   (Red litmus should turn blue.)

7.   SA
   $Na^+{}_{(aq)}$, $PO_4^{3-}{}_{(aq)}$, $H_2O_{(l)}$
         SB   B

   $H_2O_{(l)} + PO_4^{3-}{}_{(aq)} \overset{<50\%}{\rightleftharpoons} OH^-{}_{(aq)} + HPO_4^{2-}{}_{(aq)}$
   (Red litmus should turn blue.)

8.   SA   A
   $Na^+{}_{(aq)}$, $HSO_4^-{}_{(aq)}$, $H_2O_{(l)}$
         B   SB

$$HSO_4^-{}_{(aq)} + H_2O_{(l)} \overset{>50\%}{\rightleftharpoons} SO_4^{2-}{}_{(aq)} + H_3O^+{}_{(aq)}$$
(Blue litmus should turn red.)

9. $\underset{}{\text{A}} \quad \underset{\text{SB}}{\text{SA}} \quad \underset{\text{B}}{\text{A}} \quad \underset{\text{B}}{} \quad$
$Na^+{}_{(aq)},\ HCO_3^-{}_{(aq)},\ H_3O^+{}_{(aq)},\ Cl^-{}_{(aq)},\ H_2O_{(l)}$

$$H_3O^+{}_{(aq)} + HCO_3^-{}_{(aq)} \overset{>50\%}{\rightleftharpoons} H_2O_{(l)} + H_2CO_{3(aq)}$$
(pH of $HCl_{(aq)}$ should increase.)

10. $\underset{\text{B}}{\text{SA}} \quad \underset{\text{SB}}{\text{A}} \quad \underset{\text{B}}{}$
$Na^+{}_{(aq)},\ HCO_3^-{}_{(aq)},\ OH^-{}_{(aq)},\ H_2O_{(l)}$

$$HCO_3^-{}_{(aq)} + OH^-{}_{(aq)} \overset{>50\%}{\rightleftharpoons} CO_3^{2-}{}_{(aq)} + H_2O_{(l)}$$
(pH of $NaOH_{(aq)}$ should decrease.)

11. $\underset{\text{SB}}{\text{A}} \quad \underset{\text{B}}{\text{SA}} \quad \underset{\text{B}}{\text{A}}$
$Na^+{}_{(aq)},\ HCO_3^-{}_{(aq)},\ HSO_4^-{}_{(aq)},\ H_2O_{(l)}$

$$HSO_4^-{}_{(aq)} + HCO_3^-{}_{(aq)} \overset{>50\%}{\rightleftharpoons} SO_4^{2-}{}_{(aq)} + H_2CO_{3(aq)}$$
(pH of $NaHSO_{4(aq)}$ should increase.)

Materials

lab apron
safety glasses
(2) 18 × 150 mm test tubes
50 mL beaker
wash bottle of distilled water
laboratory scoop
vial of red litmus paper
vial of blue litmus paper
pH paper roll
pH meter

$NH_4Cl_{(aq)}$, $NaOH_{(aq)}$, $HCl_{(aq)}$, $NaCH_3COO_{(aq)}$, $NaHSO_{4(aq)}$, $NaC_6H_5COO_{(aq)}$
$NaCl_{(s)}$, $Al_2(SO_4)_{3(s)}$, $Na_3PO_{4(s)}$, $NaHSO_{4(s)}$, $NaHCO_{3(s)}$

Procedure

Problems 1–4

1. Rinse test tubes with distilled water.
2. Add each solution to a separate test tube to a depth of one-third of the test tube.
3. Pour one solution into the other and note evidence of a reaction.
4. Dispose of the final mixture in the sink.

Caution: Do not stopper a test tube if a gas is being produced.

Problems 5–11

5. Rinse beaker with distilled water.
6. Add about 10 mL of the liquid to the beaker.
7. Test the liquid with litmus or measure the pH.
8. Add a small quantity of solid and swirl to dissolve. Observe any changes.
9. Test the final mixture with litmus or measure the pH.
10. Repeat steps 8 and 9 if no change is observed.
11. Dispose of the litmus or pH paper in a waste container and dispose of the liquid in the sink.

Evidence/Analysis

SUMMARY OF OBSERVATIONS AND ANALYSIS		
Reactants	Observations	Analysis*
1. $NH_4Cl_{(aq)} + NaOH_{(aq)}$	ammonia odor	yes
2. $HCl_{(aq)} + NaCH_3COO_{(aq)}$	vinegar odor	yes

3. $NaHSO_{4(aq)} + NaC_6H_5COO_{(aq)}$	white precipitate	yes	
4. $HCl_{(aq)} + NH_4Cl_{(aq)}$	no change	yes	
5. $NaCl_{(s)} + H_2O_{(l)}$	no change in litmus	yes	
6. $Al_2(SO_4)_{3(s)} + H_2O_{(l)}$	blue litmus turned red	no	
7. $Na_3PO_{4(s)} + H_2O_{(l)}$	red litmus turned blue	yes	
8. $NaHSO_{4(s)} + H_2O_{(l)}$	blue litmus turned red	yes	
9. $NaHCO_{3(s)} + HCl_{(aq)}$	pH increased, gas bubbles	yes	
10. $NaHCO_{3(s)} + NaOH_{(aq)}$	pH decreased	yes	
11. $NaHCO_{3(s)} + NaHSO_{4(aq)}$	pH increased, gas bubbles	yes	

* consistent with predicted reaction

Evaluation

The experimental design is barely adequate since the evidence obtained for the products was limited and no direct evidence of the extent of reaction was obtained. The evidence is not sufficient to confirm the prediction and answer the problem. The procedure adequately covered a very limited experimental design. More detailed diagnostic tests should be conducted to verify predicted products and to test the position of the acid-base equilibrium. The technological skills were adequate for the procedure since the required skills were minimal. I am only moderately certain of the interpretations made from the evidence gathered. The major source of uncertainty is the very restricted experimental design employed.

Although the evidence obtained was very limited, the experimental results of ten out of eleven reactions agree with the prediction. The gas bubbles formed in reactions 9 and 11 indicated the presence of the predicted carbonic acid product. Overall, the prediction is judged to be inconclusive since the evidence for reaction 6 clearly disagrees with the prediction. However, the five-step procedure based on Brønsted-Lowry definitions and the empirical generalization on the position of acid-base equilibria is judged to be acceptable since the prediction was mostly verified. The Brønsted-Lowry definition should be revised or restricted to deal with the result of reaction 6. Again, my level of confidence in this judgment is low.

Lab Exercise 15E
An Acid-Base Table (page 477)

Problem
What is the order of acid strength for the first four members of the carboxylic acid family?

Analysis
According to the evidence provided, the order of strength is shown in the following table of carboxylic acids and their conjugate bases.

	Acid	Conjugate Base
SA	$HCOOH_{(aq)}$	$HCOO^-_{(aq)}$
	$CH_3COOH_{(aq)}$	$CH_3COO^-_{(aq)}$
	$C_3H_7COOH_{(aq)}$	$C_3H_7COO^-_{(aq)}$
	$C_2H_5COOH_{(aq)}$	$C_2H_5COO^-_{(aq)}$

INVESTIGATION 15.3

Demonstration of pH curves (page 479)

Problem
What are the shapes of the pH curves for the continuous addition of hydrochloric acid to a sample of a sodium hydroxide solution and to a sample of a carbonate solution?

Evidence

50.00 mL OF 0.10 mol/L $NaOH_{(aq)}$ TITRATED WITH 0.10 mol/L $HCl_{(aq)}$			
Volume of $HCl_{(aq)}$ (mL)	pH	Volume of $HCl_{(aq)}$ (mL)	pH
0.00	12.92	49.60	7.25
1.00	12.90	49.80	6.24
10.00	12.86	50.00	5.46
20.20	12.73	50.20	3.76
30.20	12.51	50.40	3.26
40.30	12.15	50.60	3.04
44.00	11.92	50.80	2.88
46.00	11.28	51.80	2.47
47.20	11.28	52.60	2.30
48.20	10.83	54.00	2.11
48.80	10.29	56.00	1.92
49.00	10.00	60.00	1.73
49.24	9.38	70.00	1.48
49.44	8.37	80.00	1.33

50.00 mL OF 0.10 mol/L $Na_2CO_{3(aq)}$ TITRATED WITH 0.10 mol/L $HCl_{(aq)}$			
Volume of $HCl_{(aq)}$ (mL)	pH	Volume of $HCl_{(aq)}$ (mL)	pH
0.00	11.20	66.00	6.70
2.00	11.00	71.00	6.50
4.00	10.80	82.00	6.10
13.20	10.30	90.00	5.80
20.00	10.10	94.00	5.60
30.00	9.80	96.00	5.40
40.00	9.50	98.00	5.10
44.00	9.20	98.80	4.90
48.00	8.90	100.00	4.20
49.20	8.70	100.60	3.60
50.00	8.60	101.20	3.30
50.40	8.50	102.40	3.00
51.00	8.30	104.00	2.70
52.00	7.90	106.00	2.60
54.80	7.50	112.00	2.30
58.00	7.10	120.00	2.10
62.00	6.90	130.00	2.00

Analysis

Based upon the evidence gathered by this experimental design, the shapes of the two pH curves are shown below.

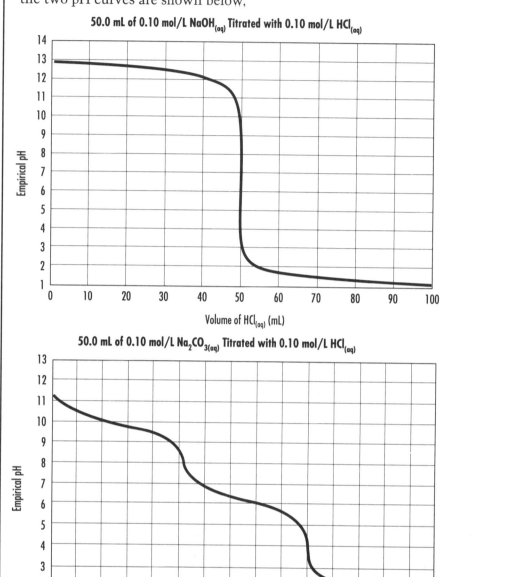

Exercise (page 483)

24. Buffering action is displayed by a nearly horizontal portion of a pH curve, indicating that the pH changes very little when small amounts of acid or base are added.

25. Quantitative reactions are displayed by a nearly vertical portion of a pH curve, indicating an abrupt change in pH around the equivalence point.

26. A pH curve is used to determine the pH of the solution at the equivalence point of a titration, so that a suitable indicator that will change color at that endpoint can be chosen.

27. (a) The endpoint is at pH 9 and the equivalence point is at 26 mL.
    (b) thymol blue or phenolphthalein

(c) $CH_3COOH_{(aq)} + OH^-_{(aq)} \rightarrow CH_3COO^-_{(aq)} + H_2O_{(l)}$

28. (a) The third reaction is not quantitative.
    (b) $PO_4^{3-}{}_{(aq)} + H_3O^+{}_{(aq)} \rightarrow HPO_4^{2-}{}_{(aq)} + H_2O_{(l)}$
    $HPO_4^{2-}{}_{(aq)} + H_3O^+{}_{(aq)} \rightarrow H_2PO_4^-{}_{(aq)} + H_2O_{(l)}$
    $$H_2PO_4^-{}_{(aq)} + H_3O^+{}_{(aq)} \overset{>50\%}{\rightleftharpoons} H_3PO_{4(aq)} + H_2O_{(l)}$$

29. **Oxalic Acid Reacted with Sodium Hydroxide**

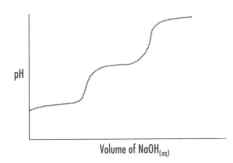

INVESTIGATION 15.4

Buffers (page 486)

Problem

What effect does the addition of small amounts of a strong acid or strong base have on a dihydrogen phosphate ion-hydrogen phosphate ion buffer?

Prediction

According to the empirical definition of a buffer, the addition of small amounts of a strong acid or strong base will produce only a slight change in the pH of the mixture. The reasoning is that the added hydronium ion or hydroxide ion will be quantitatively removed by reaction with a component of the buffer as shown below.

$H_3O^+{}_{(aq)} + HPO_4^{2-}{}_{(aq)} \rightarrow H_2O_{(l)} + H_2PO_4^-{}_{(aq)}$
$OH^-{}_{(aq)} + H_2PO_4^-{}_{(aq)} \rightarrow H_2O_{(l)} + HPO_4^{2-}{}_{(aq)}$

These reactions are quantitative according to the evidence in Figures 15.23 (page 484), and 15.21 (page 483), respectively.

Experimental Design

A small volume of a strong acid or strong base is added, one drop at at time, to a fixed volume of the buffer and the same volume of distilled water. The volume of the acid or base is the manipulated variable and the resulting pH is the responding variable. The controlled variables are temperature, volume and concentration of the buffer, and concentration of the added acid or base. The same procedure carried out with distilled water acts as the control.

Materials

lab apron
safety glasses
$Na_2HPO_4 \cdot 2H_2O_{(s)}$
$NaH_2PO_4 \cdot 2H_2O_{(s)}$
1.0 mol/L $HCl_{(aq)}$
1.0 mol/L $NaOH_{(aq)}$
distilled water
(2) medicine droppers

pH meter
pH 7 calibration buffer
laboratory scoop
centigram balance
(5) 100 mL beakers
100 mL volumetric flask and stopper
50 mL graduated cylinder
stirring rod

Procedure
1. Measure 1.56 g of $NaH_2PO_4 \cdot 2H_2O_{(s)}$ and 1.78 g of $Na_2HPO_4 \cdot 2H_2O_{(s)}$ into a clean, dry 100 mL beaker.
2. Dissolve the solid mixture to make a 100 mL solution in the volumetric flask.
3. Using a 50 mL graduated cylinder, measure 40 mL of the buffer solution into each of two clean, dry, labelled 100 mL beakers.
4. Measure 40 mL of distilled water into a clean, dry 100 mL beaker.
5. Obtain about 5 mL of 1.0 mol/L $HCl_{(aq)}$ in a clean, dry beaker.
6. Calibrate the pH meter with the pH 7 calibration solution.
7. Measure and record the initial pH of the buffer and distilled water.
8. Add two drops of $HCl_{(aq)}$ to the buffer solution, stir, and measure the pH.
9. Continue adding $HCl_{(aq)}$, two drops at a time, until 10 drops have been added, measuring the pH after each addition.
10. Repeat steps 8 and 9 using distilled water in place of the buffer.
11. Repeat steps 4 to 10 using 1.0 mol/L $NaOH_{(aq)}$ in place of the $HCl_{(aq)}$.
12. Dispose of all solutions in the sink.

Evidence

Volume of $HCl_{(aq)}$ (drops)	0	2	4	6	8	10
pH of buffer solution	6.7	6.7	6.7	6.6	6.6	6.6
pH of distilled water	7.8	2.7	2.4	2.2	2.1	2.0

Volume of $NaOH_{(aq)}$ (drops)	0	2	4	6	8	10
pH of buffer solution	6.7	6.7	6.7	6.7	6.8	6.8
pH of distilled water	7.8	10.9	11.3	11.5	11.7	11.9

Analysis

According to the evidence collected, the addition of small amounts of a strong acid or strong base produced a very small change of pH in the buffer solution, a 0.1 pH unit change compared with a 4 to 5.8 change in distilled water.

Evaluation

The experimental design is judged to be adequate because the problem was clearly answered with no obvious flaws. The use of distilled water as a control demonstrated the pH change of a non-buffer solution. The design was simple and efficient and I am quite confident with this design. The procedure is judged to be adequate because sufficient evidence was collected to answer the problem. A possible improvement would be to add the acid or base in larger increments to a total volume of 2–3 mL. This would provide a more rigorous test of the buffer. The technological skills required were minimal and adequate. Overall, I am very certain of the results I have obtained.

The prediction is judged to be verified because the slight change in pH predicted clearly agrees with the very small change in pH observed. The empirical definition of a buffer appears acceptable based on the test conducted in this experiment.

Exercise (page 486)

30. A buffer is a mixture of a conjugate weak acid/base pair that maintains nearly constant pH when diluted or when small amounts of a strong acid or base are added.

31. $H_2PO_4^-{}_{(aq)} - HPO_4^{2-}{}_{(aq)}$ and $H_2CO_{3(aq)} - HCO_3^-{}_{(aq)}$

32. (a)
      A   A   A   SA
$H_2CO_{3(aq)}$, $HCO_3^-{}_{(aq)}$, $H_2O_{(l)}$, $H_3O^+{}_{(aq)}$, $Cl^-{}_{(aq)}$
    SB   B   B

$H_3O^+{}_{(aq)} + HCO_3^-{}_{(aq)} \rightarrow H_2O_{(l)} + H_2CO_{3(aq)}$

(b) SA   A   A
$H_2CO_{3(aq)}$, $HCO_3^-{}_{(aq)}$, $H_2O_{(l)}$, $Na^+{}_{(aq)}$, $OH^-{}_{(aq)}$
     B   B   SB

$H_2CO_{3(aq)} + OH^-{}_{(aq)} \rightarrow HCO_3^-{}_{(aq)} + H_2O_{(l)}$

33. A large amount of strong acid would completely neutralize the conjugate base in the buffer and the pH of the solution would drop. A large amount of a strong base would completely neutralize the conjugate acid in the buffer and the pH of the solution would rise.

34. $CH_3COOH_{(aq)} + H_2O_{(l)} \rightleftharpoons CH_3COO^-{}_{(aq)} + H_3O^+{}_{(aq)}$

   (a) Adding $HCl_{(aq)}$ introduces $H_3O^+{}_{(aq)}$ ions which cause the equilibrium to shift to the left, producing more $CH_3COOH_{(aq)}$.

   (b) Adding $NaOH_{(aq)}$ introduces $OH^-{}_{(aq)}$ ions which react with $H_3O^+{}_{(aq)}$ and cause the equilibrium to shift to the right, producing more $CH_3COO^-{}_{(aq)}$.

Exercise (page 492)

35. Additional perspectives that can be adopted on acid rain are esthetic, emotional, ethical, legal, militaristic, mystical, and social.

36. *Answers will vary. One example is given below.*

Perspective	For Statements	Against Statements
scientific	• Empirical work indicates that the main causes of acid deposition in North America are sulfur dioxide, SO_2, and nitorgen oxides, NO_x. • In the atmosphere, SO_2 reacts with water in the atmosphere to produce sulfurous acid, $H_2SO_{3(aq)}$.	• A 1988 U.S. federal task force maintained that damage to human health, crops, and forests by acid deposition is yet to be proven.
technological	• High-tech instruments provide evidence that bands of smoke from tall stacks sometimes travelled hundreds of kilometres.	• Taller smoke stacks were introduced as an inexpensive technological fix for local air pollution problems.
ecological	• Sulfuric and sulfurous acids cause considerable damage when they fall to Earth in the form of acid deposition. • Some environmental groups claimed that 14 000 Canadian lakes have been damaged by acid deposition. • Acid deposition is also suspected as one of the causes of forest decline, particularly in forests at high altitudes and colder altitudes.	• Some researchers report that acid deposition painted on seedlings in soil with inadequate nutrients actually has a beneficial effect on growth.
economic	• Acid deposition is endangering fishing, tourism, agriculture, and forestry. • It is estimated that acid deposition causes about one billion dollars worth of damage in Canada annally.	• Regulations could severely hinder economic growth.
political	• The pro-development and pro-environment lobby groups both tell the voters that they want to protect the environment.	• The fact that acid deposition is not subject to legislation regarding international boundaries makes it a contentious political issue between Canada and the United States.

legal	• Both the Canadian and the American governments have passed legislation to reduce the discharge of sulfur and nitrogen oxides.	• Passing laws will not solve the problem but only creates difficulties for companies.
social	• Failure to reduce acid rain will destroy the social fabric of fishing and logging villages and towns.	• Shutting down acid rain-producing industries will severely affect the social fabric of the industry towns and cities.

37. *Answers will vary. Students should assign a value (1 to 10) to each statement, based upon their own kowledge and values. They should tally the total for and against scores and use these scores to support their opinion.*

38. (a) Scientific research on catalysis assisted the development of catalytic converter technology.
    (b) Sensitive detection devices have helped scientific research in the disbursements of sulfur oxide emissions.
    (c) The technology that produces sulfur oxides (smelting and power generating) has a harmful impact on human health.
    (d) Society provides the resources, through government funding, that enable scientific research on the causes and effects of acid deposition.
    (e) Society affects technology by purchasing its products, such as cars with catalytic converters.

39. Scientific research can increase our understanding of the natural phenomena associated with the STS issue.

40. Scientific communication that accurately reflects the uncertainty of scientific knowledge can be a drawback outside the scientific community. Courts rely heavily on "proof" and uncertainty works against this approach.

41. "the evidence suggests," "according to the current theory," "testing the hypothesis that ..."

Exercise (page 494)

42. $Na_2SO_{3(aq)} + HCl_{(aq)} \rightarrow NaHSO_{3(aq)} + NaCl_{(aq)}$
    10.00 mL   14.2 mL
    C 0.225 mol/L
    n_{HCl} = 0.225 mol/L × 14.2 mL = 3.20 mmol
    $n_{Na_2SO_3}$ = 3.20 mmol × 1/1 = 3.20 mmol
    $C_{Na_2SO_3}$ = 3.20 mmol/10.00 mL = 0.320 mol/L

43. $HOOCCOOH_{(aq)} + 2KOH_{(aq)} \rightarrow K_2OOCCOO_{(aq)} + 2H_2O_{(l)}$
    25.0 mL   15.6 mL
    C 0.485 mol/L
    n_{KOH} = 0.485 mol/L × 15.6 ml = 7.58 mmol
    $n_{HOOCCOOH}$ = 7.58 mmol × 1/2 = 3.79 mmol
    $C_{HOOCCOOH}$ = 3.79 mmol/25.0 mL = 0.152 mol/L

44. $NaHSO_{4(aq)} + NaOH_{(aq)} \rightarrow Na_2SO_{4(aq)} + H_2O_{(l)}$
    25.0 mL   10.2 mL
    C 0.500 mol/L
    n_{NaOH} = 0.500 mol/L × 10.2 mL = 5.10 mmol
    n_{NaHSO_4} = 5.10 mmol × 1/1 = 5.10 mmol
    C_{NaHSO_4} = 5.10 mmol/25.0 mL = 0.204 mol/L

45. Using values from Figure 15.27,
    $2NaOH_{(aq)} + H_3PO_{4(aq)} \rightarrow 2H_2O_{(l)} + Na_2HPO_{4(aq)}$
    11.95 mL   10.0 mL
    0.123 mol/L C

$n_{NaOH} = 11.95 \text{ mL} \times 0.123 \text{ mol/L} = 1.47 \text{ mmol}$
$n_{H_3PO_4} = 1.47 \text{ mmol} \times 1/2 = 0.735 \text{ mmol}$
$C_{H_3PO_4} = 0.735 \text{ mmol}/10.0 \text{ mL} = 0.0735 \text{ mol/L}$

46. $Na_2S_{(aq)} + 2HCl_{(aq)} \rightarrow H_2S_{(aq)} + 2NaCl_{(aq)}$
    10.00 mL 16.8 mL
    C 0.150 mol/L
    $n_{HCl} = 0.150 \text{ mol/L} \times 16.8 \text{ mL} = 2.52 \text{ mmol}$
    $n_{Na_2S} = 2.52 \text{ mmol} \times 1/2 = 1.26 \text{ mmol}$
    $C_{Na_2S} = 1.26 \text{ mmol}/10.00 \text{ mL} = 0.126 \text{ mol/L}$

47. $H_2SO_{4(aq)} + 2NaOH_{(aq)} \rightarrow Na_2SO_{4(aq)} + 2H_2O_{(l)}$
    10.0 mL   11.48 mL
    C 0.484 mol/L
    $n_{NaOH} = 0.484 \text{ mol/L} \times 11.48 \text{ mL} = 5.56 \text{ mmol}$
    $n_{H_2SO_4} = 5.56 \times 1/2 = 2.78 \text{ mmol}$
    $C_{H_2SO_4} = 2.78 \text{ mmol}/10.0 \text{ mL} = 0.278 \text{ mol/L}$

Lab Exercise 15F
Mass Percent of Sodium Phosphate (page 495)

Problem
What is the mass-by-volume percent of sodium phosphate in a cleaning solution?

Analysis
Standardization of $HCl_{(aq)}$:

$n_{Na_2CO_3} = 1.36 \text{ g} \times \dfrac{1 \text{ mol}}{105.99 \text{ g}} = 12.8 \text{ mmol}$

$C_{Na_2CO_3} = \dfrac{12.8 \text{ mmol}}{100.0 \text{ mL}} = 0.128 \text{ mol/L}$

$Na_2CO_{3(aq)} + 2HCl_{(aq)} \rightarrow H_2CO_{3(aq)} + 2NaCl_{(aq)}$
10.00 mL   12.7 mL
0.128 mol/L C

$n_{Na_2CO_3} = 10.00 \text{ mL} \times \dfrac{0.128 \text{ mol}}{1 \text{ L}} = 1.28 \text{ mmol}$

$n_{HCl} = 1.28 \text{ mmol} \times \dfrac{2}{1} = 2.57 \text{ mmol}$

$C_{HCl} = \dfrac{2.57 \text{ mmol}}{12.7 \text{ mL}} = 0.202 \text{ mol/L}$

Chemical analysis of the sodium phosphate solution:
$Na_3PO_{4(aq)} + 2HCl_{(aq)} \rightarrow NaH_2PO_{4(aq)} + 2NaCl_{(aq)}$
10.00 mL   18.4 mL
c (W/V)   0.202 mol/L

$n_{HCl} = 18.4 \text{ mL} \times \dfrac{0.202 \text{ mol}}{1 \text{ L}} = 3.72 \text{ mmol}$

$n_{Na_3PO_4} = 3.72 \text{ mmol} \times \dfrac{1}{2} = 1.86 \text{ mmol}$

$m_{Na_3PO_4} = 1.86 \text{ mmol} \times \dfrac{163.94 \text{ g}}{1 \text{ mol}} = 305 \text{ mg}$

$$c_{Na_3PO_4} = \frac{0.305 \text{ g}}{10.00 \text{ mL}} \times 100 = 3.05\%$$

According to the evidence provided and the stoichiometric method, the mass-by-volume percent of sodium phosphate in a cleaning solution is 3.05%.

Lab Exercise 15G
Identifying an Unknown Acid (page 495)

Problem

What is the molar mass of an unknown acid?

Analysis

Since a second endpoint was measured, the unknown acid must be at least diprotic. Suppose it has the formula H_2X.

$$H_2X_{(aq)} + 2NaOH_{(aq)} \rightarrow Na_2X_{(aq)} + 2H_2O_{(l)}$$
$$\phantom{H_2X_{(aq)} + 2}0.217 \text{ g} \quad 16.1 \text{ mL}$$
$$M 0.182 \text{ mol/L}$$

$$n_{NaOH} = 16.1 \text{ mL} \times \frac{0.182 \text{ mol}}{1 \text{ L}} = 2.93 \text{ mmol}$$

$$n_{H_2X} = 2.93 \text{ mmol} \times \frac{1}{2} = 1.47 \text{ mmol}$$

$$M_{H_2X} = \frac{217 \text{ mg}}{1.47 \text{ mmol}} = 148 \text{ g/mol}$$

According to the evidence presented, the molar mass of the unknown acid was found to be 148 g/mol.

Possible Experimental Designs

1. Determine the melting point of the solid acid and use a reference to identify the acid.
2. Do a titration with a pH meter to determine the pH curve. Determine the first and second ionization constants and then use a reference to identify the acid.
3. Based on the hypothesis that the unknown acid is an organic (carboxylic) acid, a sample of the acid is oxidized with a strong oxidizing agent to produce carbon dioxide gas. If a sample of the gas is bubbled into limewater, and the limewater turns cloudy, then carbon dioxide is likely present.
4. If Design 3 is successful then a sample of the unknown acid is analyzed using a combustion analyzer to determine the molecular formula. The most likely acid is identified from this molecular formula.
5. A sample of the unknown acid is analyzed using a mass spectrometer. A possible structure is determined from the mass fragments observed.

INVESTIGATION 15.5

Ammonia Analysis (page 496)

Problem

What is the molar concentration of the household ammonia sample provided?

Experimental Design

Hydrochloric acid, approximately 0.25 mol/L, was standardized by titrating a known sodium carbonate standard solution using the methyl orange endpoint (second reaction step). A sample of household ammonia was diluted by a factor of ten and then titrated with the standardized hydrochloric acid using methyl red as an indicator. The evidence obtained was used to evaluate the concentration of the household ammonia predicted from literature sources.

Materials

lab apron
safety glasses
approx. 0.25 mol/L hydrochloric acid
household ammonia
vial of $Na_2CO_{3(s)}$
(1) 250 mL beaker
(2) 250 mL Erlenmeyer flasks
(3) 100 mL beakers
wash bottle of distilled water
medicine dropper

meniscus finder
10 mL pipet and bulb
50 mL buret and stand
100 mL volumetric flask
stopper for flask
funnel
stirring rod
laboratory scoop
centigram balance

Procedure

1. Measure 1.52 g of $Na_2CO_{3(s)}$.
2. Prepare a 100 mL standard solution of $Na_2CO_{3(aq)}$ using the accepted procedure.
3. Pipet 10.00 mL of $Na_2CO_{3(aq)}$ into an Erlenmeyer flask and add 2 drops of methyl orange indicator.
4. Titrate $Na_2CO_{3(aq)}$ with $HCl_{(aq)}$.
5. Repeat Steps 3 to 4 until three consistent readings are obtained.
6. Dilute a 10.00 mL sample of household ammonia to 100.0 mL in a volumetric flask.
7. Pipet 10.00 mL of diluted ammonia into an Erlenmeyer flask and add 2 drops of methyl red indicator.
8. Titrate $NH_{3(aq)}$ with $HCl_{(aq)}$.
9. Repeat steps 7 to 8 until three consistent readings are obtained.
10. Dispose of all chemicals in the sink, flushing with cold water.

Evidence

mass of $Na_2CO_{3(s)}$ = 1.52 g
volume of $Na_2CO_{3(aq)}$ standard solution = 100.0 mL

TITRATION OF 10.00 mL OF $Na_2CO_{3(aq)}$ WITH $HCl_{(aq)}$				
Trial	1	2	3	4
Final buret reading (mL)	13.4	26.2	39.0	13.3
Initial buret reading (mL)	0.2	13.4	26.2	0.4
Comment on endpoint	overshot	good	good	good

initial volume of household ammonia = 10.00 mL
final volume of diluted ammonia = 100.0 mL

TITRATION OF 10.00 mL OF DILUTE $NH_{3(aq)}$ WITH $HCl_{(aq)}$			
Trial	1	2	3
Final buret reading (mL)	11.8	23.2	34.5
Initial buret reading (mL)	0.4	11.8	23.2
Comment on endpoint	good	good	good

Analysis

$$2\,HCl_{(aq)} + Na_2CO_{3(aq)} \rightarrow H_2CO_{3(aq)} + 2NaCl_{(aq)}$$
12.8 mL   1.52 g/100 mL
C 10.00 mL

$m_{Na_2CO_3} = 0.152$ g in 10.00 mL

$n_{Na_2CO_3} = 0.152\,g \times \dfrac{1\,mol}{105.99\,g} = 1.43$ mmol

$n_{HCl} = 1.43\,mmol \times \dfrac{2}{1} = 2.87$ mmol

$C_{HCl} = \dfrac{2.87\,mmol}{12.8\,mL} = 0.223$ mol/L

$$HCl_{(aq)} + NH_3 \rightarrow NH_4^{+}{}_{(aq)} + Cl^{-}{}_{(aq)}$$
11.4 mL   10.00 mL
0.223 mol/L C

$n_{HCl} = 11.4\,mL \times \dfrac{0.223\,mol}{1\,L} = 2.54$ mmol

$n_{NH_3} = 2.54\,mmol \times \dfrac{1}{1} = 2.54$ mmol

$C_{NH_3} = \dfrac{2.54\,mmol}{10.00\,mL} = 0.254$ mol/L (diluted solution)

$C_{household\,NH_3} \times 10.00\,mL = 0.254\,mol/L \times 100.0\,mL$

$C_{household\,NH_3} = 2.54$ mol/L

According to the evidence collected and the stoichiometry, the molar concentration of the household ammonia provided is 2.54 mol/L.

Evaluation

The titration experimental design is judged to be adequate because the problem was answered with no obvious flaws. From past experience this is known to be an efficient and reliable design for chemical analysis. Therefore, I have a great deal of confidence in this design. The procedure used is adequate. Sufficient evidence was collected. Since the smell of ammonia was noticed, an improvement would be to keep ammonia containers closed and pipet the sample just before titrating. The technological skills were also adequate as shown by the reproducibility of the trials.

Overall, I am quite certain of the results obtained in this experiment. Uncertainties include those arising from measurements with the balance and volumetric glassware, from the estimate of the first distinct color change, and perhaps from the loss of a small quantity of ammonia by evaporation. An acceptable percent difference would be 5%.

It is not possible to calculate the accuracy of my result since a wide range of concentrations was predicted. However, the result obtained clearly falls within the range predicted. Therefore, the prediction is judged to be verified. A 5% difference that was expected would still keep the experimental result within the predicted range. Since the prediction is verified, the literature values are considered to be acceptable.

Overview (page 497)

Review

1. According to the oxygen concept, all acids contain oxygen. This definition is too restricted and has too many exceptions, notably $HCl_{(aq)}$. According to the hydrogen concept, all acids are compounds of hydrogen. This definition is limited because it does not explain why only certain hydrogen compounds are acids. According to Arrhenius's concept, acids are substances that ionize in aqueous solution to produce hydrogen ions. This definition is limited to aqueous solutions and cannot explain or predict the properties of many common substances. According to the Brønsted-Lowry concept, acids are substances that donate protons to bases in a chemical reaction. The main limitation of the Brønsted-Lowry concept is the restriction to protons and the inability to explain and predict the acid nature of ions of multi-valent metals.

2. The theory is restricted, revised, or replaced.

3. Test the explanations and predictions made using the theory.

4. According to the evidence, the hydrogen ion exists as a hydrated proton whose simplest representation is the hydronium ion, $H_3O^+_{(aq)}$.

5. According to Arrhenius's concept, a base is a substance that dissociates in aqueous solution to produce hydroxide ions. According to the revised Arrhenius concept, a base is a substance that reacts with water to produce hydroxide ions. According to the Brønsted-Lowry concept, a base is a proton acceptor that removes protons from an acid.

6. (a) The pH of the nitric acid solution is lower than the pH of the nitrous acid solution.
   (b) Water, acting as a base, can quantitatively remove the proton from the HNO_3 molecule. The proton in HNO_2 is more strongly bonded and not as easily given up to the water.
   $$H_2O_{(l)} + HNO_{2(aq)} \overset{8.1\%}{\rightleftharpoons} H_3O^+_{(aq)} + NO_2^-_{(aq)}$$

7. The position of equilibrium is determined by the result of the competition for protons. Of the forward and reverse reactions, the reaction involving the stronger acid and the stronger base is favored.

8. If the acid is listed above the base in the table of acids and bases, then the products will be favored. If the acid is listed below the base, then the reactants will be favored.

9. $H_2SO_{3(aq)} - HSO_3^-_{(aq)}$ and $HSO_3^-_{(aq)} - SO_3^{2-}_{(aq)}$

10. (a) yellow
    (b) red
    (c) green
    (d) colorless
    (e) yellow

11. (a) Buffering action means a relatively constant pH when small amounts of a strong acid or base are added.
    (b) Buffering action is most noticeable at a volume of titrant that is one-half the first equivalence point or half-way between successive equivalence points for polyprotic acids.
    (c) Quantitative reactions are represented by nearly vertical portions of a pH curve.
    (d) The pH endpoint is the mid-point of the sharp change in pH in an acid-base titration. The equivalence point is the quantity of titrant at the endpoint of the titration.

(e) The mid-point of the pH range of a suitable indicator should equal the pH endpoint and the indicator should complete its color change while the pH is changing abruptly.

(f) Non-quantitative reactions do not have a distinct endpoint because the pH changes gradually in the region where the equivalence point is expected.

12. Buffers are used in making cheese, yogurt, and sour cream, in preserving food, and in the production of antibiotics.

Applications

13. (a) $HBr_{(aq)} + H_2O_{(l)} \rightarrow Br^-_{(aq)} + H_3O^+_{(aq)}$ acidic
   (b) $NO_2^-{}_{(aq)} + H_2O_{(l)} \rightleftharpoons HNO_2(aq) + OH^-_{(aq)}$ basic
   (c) $NH_{3(aq)} + H_2O_{(l)} \rightleftharpoons NH_4^+{}_{(aq)} + OH^-_{(aq)}$ basic
   (d) This is impossible to predict, since the hydrogen sulfate ion may react with water in one of two ways.
   $NaHSO_{4(aq)} \rightarrow Na^+_{(aq)} + HSO_4^-{}_{(aq)}$
   $HSO_4^-{}_{(aq)} + H_2O_{(aq)} \rightleftharpoons SO_4^{2-}{}_{(aq)} + H_3O^+_{(aq)}$ acidic
   $HSO_4^-{}_{(aq)} + H_2O_{(aq)} \rightleftharpoons H_2SO_{4(aq)} + OH^-_{(aq)}$ basic

14. Each substance is tested with litmus paper. The color change is the responding variable and concentration is a controlled variable.

15. Solutions of equal concentration of several bases are prepared and the pH is measured for each solution. Solutions of equal concentration of several bases are prepared and the conductivity is precisely measured for each solution.

16.

compound
- ionic
  - neutral: NaCl, KBr
  - acid: NaHSO$_4$, NaOOCCOOH
  - base: NaOH, Na$_2$CO$_3$
- molecular
  - neutral: H$_2$O, CH$_3$OH
  - acid: HCl, CH$_3$COOH
  - base: NH$_3$

17. (a) A   B   >50%   B   A
   $HCOOH_{(aq)} + CN^-_{(aq)} \rightleftharpoons HCOO^-_{(aq)} + HCN_{(aq)}$
   $HCOOH_{(aq)} - HCOO^-_{(aq)}$ and $HCN_{(aq)} - CN^-_{(aq)}$
   (b) B   A   <50%   A   B
   $HPO_4^{2-}{}_{(aq)} + HCO_3^-{}_{(aq)} \rightleftharpoons H_2PO_4^-{}_{(aq)} + CO_3^{2-}{}_{(aq)}$
   $HCO_3^-{}_{(aq)} - CO_3^{2-}{}_{(aq)}$ and $H_2PO_4^-{}_{(aq)} - HPO_4^{2-}{}_{(aq)}$

18. A possible pH is 5.1. The hydronium ion concentration for this solution is 8×10^{-6} mol/L.

19. (a) $HF_{(aq)} + SO_4^{2-}{}_{(aq)} \overset{<50\%}{\rightleftharpoons} F^-_{(aq)} + HSO_4^-{}_{(aq)}$
   If the pH of the sulfate solution is measured before and after the addition of the $HF_{(aq)}$ and the pH decreases, then the $HF_{(aq)}$ probably reacted with the $SO_4^{2-}{}_{(aq)}$.
   (b) $HSO_4^-{}_{(aq)} + HS^-_{(aq)} \overset{>50\%}{\rightleftharpoons} SO_4^{2-}{}_{(aq)} + H_2S_{(aq)}$
   If the mixture is carefully smelled, and a "rotten egg" odor is noticed, then $H_2S_{(aq)}$ is likely to be present.
   (c) $HCOOOH_{(aq)} + OH^-_{(aq)} \rightarrow HCOO^-_{(aq)} + H_2O_{(l)}$
   If the pH is measured during the titration, and a sharp change in pH is observed, then methanoic acid has reacted quantitatively with the sodium hydroxide.

(d) $H_3O^+_{(aq)} + PO_4^{3-}_{(aq)} \overset{>50\%}{\rightleftharpoons} H_2O_{(l)} + HPO_4^{2-}_{(aq)}$

If the pH of the solution is measured before and after the addition of a strong acid, and the pH remains relatively constant, then the strong acid has reacted.

(e) $HPh_{(aq)} + OH^-_{(aq)} \overset{>50\%}{\rightleftharpoons} Ph^-_{(aq)} + H_2O_{(l)}$

If the color is observed and it changes immediately to red, then the indicator has reacted with the strong base.

20. The solution will be basic.
$CH_3O^-_{(aq)} + H_2O_{(l)} \rightleftharpoons CH_3OH_{(aq)} + OH^-_{(aq)}$

21. $OH^-_{(aq)} + HCO_3^-_{(aq)} \rightleftharpoons H_2O_{(l)} + CO_3^{2-}_{(aq)}$
$H_3O^+_{(aq)} + HCO_3^-(aq) \rightleftharpoons H_2O_{(l)} + H_2CO_{3(aq)}$

22. *Note that there are many correct solutions to this problem — be creative.*
If the solutions are tested with a pH meter, and the pH values are ordered from smallest to largest, then the solutions are sulfuric acid, hydrochloric acid, acetic acid, ethanediol, ammonia, sodium hydroxide, and barium hydroxide, respectively.

Diagnostic Tests on the Unlabelled Solutions

Litmus	Conductivity	Acid/Base Titration	Analysis
red	low	one volume	$CH_3COOH_{(aq)}$
blue	very high	two volumes	$Ba(OH)_{2(aq)}$
blue	low	one volume	$NH_{3(aq)}$
no change	none	not applicable	$C_2H_4(OH)_{2(aq)}$
red	higher	two volumes	$H_2SO_{4(aq)}$
red	high	one volume	$HCl_{(aq)}$
blue	high	one volume	$NaOH_{(aq)}$

23. (a) one

(b) $H_3O^+_{(aq)} + SO_3^{2-}_{(aq)} \rightarrow H_2O_{(l)} + HSO_3^-_{(aq)}$

(c) The endpoint occurs at a pH of 4.0, when 23 mL of $HCl_{(aq)}$ have been added.

(d) congo red or methyl orange

(e) A buffering region on the graph occurs where about 12 mL of $HCl_{(aq)}$ have been added. The entities present are $SO_3^{2-}_{(aq)}$, $HSO_3^-_{(aq)}$, $Cl^-_{(aq)}$, $H_2O_{(l)}$.

24. $2HOCl_{(aq)} + Ba(OH)_{2(aq)} \rightarrow 2H_2O_{(l)} + Ba(OCl)_{2(aq)}$
10.0 mL   0.350 mol/L
C 12.6 mL
$n_{Ba(OH)_2} = 12.6 \text{ mL} \times 0.350 \text{ mol/L} = 4.41 \text{ mmol}$
$n_{HOCl} = 4.41 \text{ mmol} \times 2/1 = 8.82 \text{ mmol}$
$C_{HOCl} = 8.82 \text{ mmol}/10.0 \text{ mL} = 0.882 \text{ mol/L}$

25. $H_3PO_{4(aq)} + 2NaOH_{(aq)} \rightarrow Na_2HPO_{4(aq)} + 2H_2O_{(l)}$
25.0 mL   17.9 mL
C 1.50 mol/L
$n_{NaOH} = 17.9 \times 1.50 \text{ mol/L} = 26.9 \text{ mmol}$
$n_{H_3PO_4} = 26.9 \text{ mmol} \times 1/2 = 13.4 \text{ mmol}$
$C_{H_3PO_4} = 13.4 \text{ mmol}/25.0 \text{ mL} = 0.537 \text{ mol/L}$

26. (a)

Acid	Base	Conjugate Acid/Base Pair
C_4H_4NH	$(C_6H_5)_3C^-$	$C_4H_4NH/C_4H_4N^-$, $(C_6H_5)_3CH/(C_6H_5)_3C^-$
CH_3COOH	HS^-	CH_3COOH/CH_3COO^-, H_2S/HS^-
$(C_6H_5)_3CH$	O^{2-}	$(C_6H_5)_3CH/(C_6H_5)_3C^-$, OH^-/O^{2-}
H_2S	$C_4H_4N^-$	H_2S/HS^-, $C_4H_4NH/C_4H_4N^-$

(b)

Acid	Conjugate Base
CH_3COOH	CH_3COO^-
H_2S	HS^-
C_4H_4NH	$C_4H_4N^-$
$(C_6H_5)_3CH$	$(C_6H_5)_3C^-$
OH^-	O^{2-}

27. (a) There is no third pH endpoint and therefore no third equivalence point in this titration (Figure 15.21, page 483).
    (b) Pure acetic acid is a liquid at SATP and will be driven off by the boiling.
    (c) Removing $OH^-_{(aq)}$ ions by precipitation will cause the equilibrium to shift to the right, producing more $OH^-_{(aq)}$ ions.
    $$NH_{3(aq)} + H_2O_{(l)} \rightleftharpoons NH_4^+_{(aq)} + OH^-_{(aq)}$$
    (d) Hydrochloric acid is not a primary standard.
    (e) Both reactants and products form basic solutions and litmus cannot be used to distinguish among basic solutions.

28. Design 1: Prepare a primary standard of $Na_2CO_{3(aq)}$ and use it to titrate the $HCl_{(aq)}$. Calculate the concentration of the $HCl_{(aq)}$ from the reaction equation.

    Design 2: Use a pH meter to measure the pH of the $HCl_{(aq)}$. Calculate the concentration of the $HCl_{(aq)}$ from the pH.

    Design 3: Use indicators to estimate the pH of the $HCl_{(aq)}$. Calculate the concentration of the $HCl_{(aq)}$ from the pH.

    Design 4: Place an excess of $Zn_{(s)}$ in a measured volume of the $HCl_{(aq)}$, and collect the gas produced at a measured temperature and pressure. Calculate the concentration of the $HCl_{(aq)}$ from the amount of gas using the ideal gas law and the reaction equation.

    Design 5: Place a measured mass of $CaCO_{3(s)}$ in a measured volume of the $HCl_{(aq)}$. After the reaction stops, dry the $CaCO_{3(s)}$ and measure its mass. Calculate the concentration of the $HCl_{(aq)}$ from the reaction equation.

    Design 6: Measure the density of the $HCl_{(aq)}$ and determine the concentration from a graph of density and concentration.

Extensions

29. (a) Acetic acid:
    $$CH_3COOH_{(aq)} + H_2O_{(l)} \rightleftharpoons CH_3COO^-_{(aq)} + H_3O^+_{(aq)}$$
    $$p = \frac{[H_3O^+_{(aq)}]}{[CH_3COOH_{(aq)}]} \times 100 = \frac{10^{-2.89} \text{ mol/L}}{0.10 \text{ mol/L}} \times 100$$
    $$= \frac{1.3 \times 10^{-3} \text{ mol/L}}{0.10 \text{ mol/L}} \times 100 = 1.3\%$$
    $$K_a = \frac{[H_3O^+_{(aq)}]^2}{[CH_3COOH_{(aq)}]} = \frac{(10^{-2.89} \text{ mol/L})^2}{(0.10 - 0.0013) \text{ mol/L}}$$
    $$= \frac{(1.3 \times 10^{-3} \text{ mol/L})^2}{0.0987 \text{ mol/L}} = 1.7 \times 10^{-5} \text{ mol/L}$$

    Chloroacetic acid:
    $$CH_2ClCOOH_{(aq)} + H_2O_{(l)} \rightleftharpoons CH_2ClCOO^-_{(aq)} + H_3O^+_{(aq)}$$
    $$p = \frac{[H_3O^+_{(aq)}]}{[CH_2ClCOOH_{(aq)}]} \times 100 = \frac{10^{-1.94} \text{ mol/L}}{0.10 \text{ mol/L}} \times 100$$
    $$= \frac{1.1 \times 10^{-2} \text{ mol/L}}{0.10 \text{ mol/L}} \times 100 = 11\%$$

$$K_a = \frac{[H_3O^+_{(aq)}]^2}{[CH_2ClCOOH_{(aq)}]} = \frac{(10^{-1.94}\,\text{mol/L})^2}{(0.10 - 0.011)\,\text{mol/L}}$$

$$= \frac{(1.1 \times 10^{-2}\,\text{mol/L})^2}{0.089\,\text{mol/L}} = 1.5 \times 10^{-3}\,\text{mol/L}$$

Dichloroacetic acid:

$$CHCl_2COOH_{(aq)} + H_2O_{(l)} \rightleftharpoons CHCl_2COO^-_{(aq)} + H_3O^+_{(aq)}$$

$$p = \frac{[H_3O^+_{(aq)}]}{[CHCl_2COOH_{(aq)}]} \times 100 = \frac{10^{-1.30}\,\text{mol/L}}{0.10\,\text{mol/L}} \times 100$$

$$= \frac{5.0 \times 10^{-2}\,\text{mol/L}}{0.10\,\text{mol/L}} \times 100 = 50\%$$

$$K_a = \frac{[H_3O^+_{(aq)}]^2}{[CHCl_2COOH_{(aq)}]} = \frac{(10^{-1.30}\,\text{mol/L})^2}{(0.10 - 0.050)\,\text{mol/L}}$$

$$= \frac{(5.0 \times 10^{-2}\,\text{mol/L})^2}{0.050\,\text{mol/L}} = 5.0 \times 10^{-2}\,\text{mol/L}$$

Trichloroacetic acid:

$$CCl_3COOH_{(aq)} + H_2O_{(l)} \rightleftharpoons CCl_3COO^-_{(aq)} + H_3O^+_{(aq)}$$

$$p = \frac{[H_3O^+_{(aq)}]}{[CCl_3COOH_{(aq)}]} \times 100 = \frac{10^{-1.14}\,\text{mol/L}}{0.10\,\text{mol/L}} \times 100$$

$$= \frac{7.2 \times 10^{-2}\,\text{mol/L}}{0.10\,\text{mol/L}} \times 100 = 72\%$$

$$K_a = \frac{[H_3O^+_{(aq)}]^2}{[CCl_3COOH_{(aq)}]} = \frac{(10^{-1.14}\,\text{mol/L})^2}{(0.10 - 0.072)\,\text{mol/L}}$$

$$= \frac{(7.2 \times 10^{-2}\,\text{mol/L})^2}{0.028\,\text{mol/L}} = 1.9 \times 10^{-1}\,\text{mol/L}$$

(b) Because chlorine has a high electronegativity, as the number of chlorine atoms in the molecule increases, the electrons in the molecule are more attracted to them. As a result, the O—H bond in the COOH group becomes more polar and therefore the O—H bond is weaker. This in turn makes the acid stronger because it is then easier to remove a hydrogen ion from the acid molecule.

(c) The relative amounts of the acids in the mixture can be determined by the amounts of base added to get to each endpoint. The more acid present, the more base is required to get to the endoint of that acid.

Reaction of Chloroacetic Acids with a Strong Base

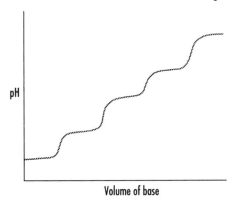

30. $H_2SO_{4(aq)} + Ba(OH)_{2(aq)} \rightarrow 2H_2O_{(l)} + BaSO_{4(s)}$

Place an accurately measured volume of $Ba(OH)_{2(aq)}$ into a beaker and immerse the leads from a conductivity meter in the solution. Measure and record the initial conductivity reading. Titrate with standardized $H_2SO_{4(aq)}$ and record the conductivity readings. A sudden drop in the conductivity serves as the endpoint of the titration. This drop in conductivity corresponds to the removal of most of the ions through the production of water and insoluble barium sulfate. When the $H_2SO_{4(aq)}$ is added in excess, the conductivity increases due to the ions present in this acid.

31. Acidic:

$HCl_{(aq)} + H_2O_{(l)} \rightarrow Cl^-_{(aq)} + H_3O^+_{(aq)}$

$NaHSO_{4(aq)} + H_2O_{(l)} \rightleftharpoons Na^+_{(aq)} + SO_4^{2-}_{(aq)} + H_3O^+_{(aq)}$

$Al(NO_3)_{3(aq)} + 7H_2O_{(l)} \rightleftharpoons 3NO_3^-_{(aq)} + [Al(H_2O)_5OH]^{2+}_{(aq)} + H_3O^+_{(aq)}$

Basic:

$KOH_{(s)} \rightarrow K^+_{(aq)} + OH^-_{(aq)}$

$NH_{3(aq)} + H_2O_{(l)} \rightleftharpoons NH_4^+_{(aq)} + OH^-_{(aq)}$

$Na_2CO_{3(s)} + H_2O_{(l)} \rightleftharpoons 2Na^+_{(aq)} + HCO_3^-_{(aq)} + OH^-_{(aq)}$

$CaO_{(s)} + H_2O_{(l)} \rightarrow Ca^{2+}_{(aq)} + 2OH^-_{(aq)}$

32. *The concept map could be based on a model of scientific inquiry (see Appendix B).*

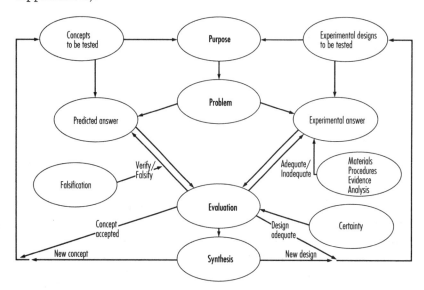

33. In general, the role of science is to improve the understanding of the natural phenomena associated with the STS issue.

34. $3NO_{2(g)} + H_2O_{(l)} \rightarrow 2HNO_{3(aq)} + NO_{(g)}$

$SO_{2(g)} + H_2O_{(l)} \rightarrow H_2SO_{3(aq)}$

$SO_{3(g)} + H_2O_{(l)} \rightarrow H_2SO_{4(aq)}$

These acids can be deposited as rain, snow, hail, or fog. Acid deposition affects virtually everything it comes into contact with — soil, water, plants, and structural materials. The pH of lakes is lowered so that aquatic life is depleted and the growth of trees can be retarded. Toxic metals, such as cadmium, mercury, lead, arsenic, aluminum, and chromium are leached out of rocks and soil and find their way into the water table. The general causes of acid deposition are well understood — the burning of coal with high sulfur content and high temperature combustion lead to the oxides responsible for acid deposition.

Lab Exercise 15H
Interpretation of Results (page 501)

Prediction
According to the single replacement generalization, the products of the reaction between solid aluminum and aqueous copper(II) sulfate are solid copper and aqueous aluminum sulfate.

$$2Al_{(s)} + 3CuSO_{4(aq)} \longrightarrow 3Cu_{(s)} + Al_2(SO_4)_{3(aq)}$$

Analysis
According to the evidence gathered in this investigation, the products of the reaction between solid aluminum and aqueous copper(II) sulfate are solid copper, gaseous hydrogen, and other substances not detected by diagnostic tests.

Evaluation
Although the experimental design was revised in progress to include diagnostic tests on the gas that was unexpectedly produced, the design should have included tests to determine the identity of the solution components and the orange-brown solid. The prediction was falsified. Although copper appeared to be produced as predicted, the gas produced was not predicted. The redox concepts used to make the prediction are judged to be unacceptable because the prediction was falsified.

Synthesis
An explanation that would be consistent with the evidence is that redox and acid-base reactions both occur. The aluminum and copper(II) sulfate single replacement redox reaction may be accompanied by another spontaneous redox reaction between aluminum and hydronium ions. The hydronium ions may be produced by the hydrolysis of aqueous copper(II) ions (i.e., by the acid-base reaction of water with copper(II) ions).

---

40. (a) H_2S

    (b) The electronegativity of oxygen is 3.5 and that of sulfur is 2.5. The lower electronegativity of sulfur could account for less polar bonds between the hydrogen and sulfur which would allow H_2S to donate protons less readily than H_2O. Therefore, electronegativities and Brønsted–Lowry concepts are not sufficient to predict or explain this evidence.

    (c) The shape of the molecule and the relative sizes of H, O, and S atoms are possible factors.

    (d) Need to consult a table of acid strength to check prediction.

41. (a) According to the oxygen concept, all acids contain oxygen. This definition is too restricted and has too many exceptions, notably $HCl_{(aq)}$. According to the hydrogen concept, all acids are compounds of hydrogen. This definition is limited because it does not explain why only certain hydrogen compounds are acids. According to Arrhenius's concept, acids are substances that ionize in aqueous solution to produce hydrogen ions. This definition is limited to aqueous solutions and cannot explain or predict the properties of many common substances. According to the Brønsted-Lowry concept, acids are substances that donate protons to bases in a chemical reaction. The main limitation of the Brønsted-Lowry concept is the restriction to protons and the inability to explain and predict the acid nature of ions of multi-valent metals.

    (b) Scientists usually use the simplest theory that will explain the observation.

35. *Answers will vary.*

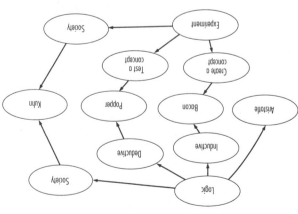

36. (a) $H_2SO_{4(l)} + 2KOH_{(s)} \longrightarrow K_2SO_{4(s)} + 2H_2O_{(l)} + 324.2 \text{ kJ}$

The water produced will dissolve some of the potassium sulfate.

(b) According to Le Châtelier's principle, low temperature would increase the likelihood of a quantitative reaction.

(c) $n_{K_2SO_4} = 100 \text{ kJ} \times \dfrac{1 \text{ mol}}{324.2 \text{ kJ}} = 0.308 \text{ mol}$

$m_{K_2SO_4} = 0.308 \text{ mol} \times 174.26 \text{ g/mol} = 53.8 \text{ g}$

37. (a) Chemical species can be ranked in order of their ability to gain/lose electrons/protons. These rankings can be used to predict the spontaneity/extent of electron/proton transfer reactions. Electron transfers change the oxidation state of atoms, while proton transfers do not.

(b) Both electron and proton transfer reactions often release heat,

$Zn_{(s)} + Br_{2(l)} \longrightarrow ZnBr_{2(s)} + \text{heat}$

$2NaOH_{(s)} + H_2SO_{4(aq)} \longrightarrow Na_2SO_{4(aq)} + 2H_2O_{(l)} + \text{heat}$

Electron transfer reactions can release or absorb electrical energy;

$Zn_{(s)} + Cu^{2+}_{(aq)} \longrightarrow Zn^{2+}_{(aq)} + Cu_{(s)} \quad \Delta E° = +1.10 \text{ V}$

$2H_2O_{(l)} \longrightarrow 2H_{2(g)} + O_{2(g)} \quad \Delta E° = -2.06 \text{ V}$

(c) Nitric acid and basic sodium sulfite solution could produce either an electron transfer reaction or a proton transfer reaction.

$2NO_3^-{}_{(aq)} + 4H^+_{(aq)} + SO_3^{2-}{}_{(aq)} \longrightarrow 2NO_{2(g)} + 3H_2O_{(l)} + SO_4^{2-}{}_{(aq)}$

$H_3O^+_{(aq)} + SO_3^{2-}{}_{(aq)} \longrightarrow H_2O_{(l)} + HSO_3^-{}_{(aq)}$

If a brown gas ($NO_{2(g)}$) is observed, then the electron transfer reaction predominates.

38. (a) redox (b) acid-base
    (c) precipitation (d) acid-base
    (e) redox (f) acid-base

39. (a) NH_4^+
    (b) $NH_2^-{}_{(l)}$
    (c) $NH_3^-{}_{(l)} + NH_{3(l)} \rightleftharpoons NH_2^-{}_{(l)} + NH_4^+{}_{(l)}$

(d) Reaction of NH_2^- with NH_4^+

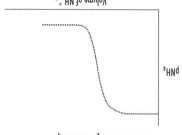